Organic Chemistry Experiment

有机化学实验

（英汉双语版）

主　编　柏一慧

副主编　朱钢国　吕　新　张　岩　黄　鑫

科学出版社

北　京

内 容 简 介

本书根据目前我国高等学校双语教学的需要编写而成。全书共 4 章，包括有机化学实验的一般知识和基本操作、有机化合物制备实验和有机化合物性质实验。其中基本操作实验 10 个，制备实验 30 个，性质实验 7 个。实验还编有注释和思考题，重要的实验配有实验装置图和实验操作流程图。

本书可作为高等学校化学、化工及相关专业本科生的有机化学实验教材，也可作为相关专业学生攻读硕士学位或申请海外高校的实验参考书。

图书在版编目(CIP)数据

有机化学实验=Organic Chemistry Experiment：英、汉/柏一慧主编. —北京：科学出版社，2019.8
 ISBN 978-7-03-061184-0

Ⅰ. ①有··· Ⅱ. ①柏··· Ⅲ. ①有机化学–化学实验–英、汉 Ⅳ. ①O62-33

中国版本图书馆 CIP 数据核字(2019)第 089838 号

责任编辑：丁　里/责任校对：何艳萍
责任印制：赵　博/封面设计：迷底书装

科学出版社 出版
北京东黄城根北街 16 号
邮政编码：100717
http://www.sciencep.com

中煤（北京）印务有限公司印刷
科学出版社发行　各地新华书店经销

*

2019 年 8 月第 一 版　　开本：787×1092　1/16
2024 年 12 月第八次印刷　印张：18
字数：472 000

定价：69.00 元
(如有印装质量问题，我社负责调换)

《有机化学实验》（英汉双语版）编写委员会

主　编　柏一慧

副主编　朱钢国　吕　新　张　岩　黄　鑫

编　委（按姓名汉语拼音排序）

　　　　柏一慧　程存归　黄　鑫　孔黎春

　　　　李士坤　李新生　罗　芳　吕　新

　　　　毛　会　彭　勃　肖孝辉　谢建武

　　　　张　磊　张　岩　朱钢国

前　言

　　党的二十大报告指出："教育、科技、人才是全面建设社会主义现代化国家的基础性、战略性支撑。"教育国际交流是我国教育事业的重要组成部分。随着我国"一带一路"倡议的提出，高校扩大了来华留学教育的招生规模；另外，国内各高校本科生毕业后申请海外留学深造的需要也日益增加，因而相应高校都纷纷推出了全英文授课的教学项目。在此背景下，适合我国高校教学体系的中英文双语教材的出版也提上了日程。目前国内有机化学实验的双语教材不多，难以满足教学需要；而国外的有机化学实验指导书与国内的实验教学内容匹配度不高，因此编者编写了这本双语教材。

　　本书采用英、汉两种语言，英文为先，中英文并重；参照国内各高校成熟的有机化学实验教材内容，结合浙江师范大学有机化学学科多年来的教学实践，同时参考国外重点高校有机化学实验指导书的英文内容编写而成。本书主要面向化学专业的有机化学实验中英文教学，以及国际化专业的发展需要，使学生掌握有机化学实验所涉及的规则、流程、方法、仪器、工艺、报告等基本内容，培养从事有机化学相关工作的基本能力和素养，并具有适应国际化发展的有机化学专业英语知识背景。

　　本书共4章，包括：有机化学实验的一般知识、有机化学实验基本操作、有机化合物制备实验、有机化合物性质实验。内容上从基础到应用，从简单到综合；制备实验所涉及有机化合物按国内大多数高校使用的有机化学教材中的编排顺序编写，尽可能使实验教学进度与理论课程进度相匹配，做到理论指导实践，实验促进理论理解。

　　本书重点在基础、实用、成熟的有机合成实验，尤其注重基本操作，特别强调实验关键步骤的把握及实验安全和环保部分的处理。重要实验项目中增加了详细、明了的实验操作流程图和实验装置图。同时根据浙江师范大学有机化学学科多年的教学积累，对现有教材的实验内容进行了修订，并增加了教学改革的部分内容，提高了实验的成功率和环保性。编者将继续完善本书配套的网络资源，包括电子课件、网站建设、在线答疑，以及视频、动画、实验操作实图等资源的建设。

　　本书由浙江师范大学有机化学学科组织编写，其中柏一慧担任主编并负责编写第1章和第2章；吕新负责编写第3章实验11～23；张岩负责编写第3章实验24～39；黄鑫负责编写第3章实验40和第4章；朱钢国负责全书内容的审定和实验的设计；彭勃负责英文部分的校对工作；浙江师范大学有机化学学科的其他教师均对本书做出直接贡献。本书同时得到了浙江师范大学"十二五"省级实验教学示范中心重点建设项目(软件建设项目)和浙江师范大学化学与生命科学学院教学建设项目的资助，在此一并表示真挚的感谢。

　　因时间有限，经验不足，书中不足之处在所难免，敬请专家和读者批评指正。

编　者

2023年6月

目 录

前言

Chapter 1　General Knowledge of Organic Chemistry Experiments ··············· 1
  1.1　Rules for Organic Chemistry Laboratory ··············· 1
  1.2　Safety Knowledge in the Organic Laboratory ··············· 2
  1.3　Commonly Used Equipment and Apparatus of Organic Chemistry Experiments ··········· 6
  1.4　Cleaning and Drying of Glassware ··············· 12
  1.5　Heating, Cooling and Drying of Reagents in Organic Chemistry Experiments ··········· 13
  1.6　Basic Requirements for Pre-lab Work, Lab Record and Lab Report ··············· 20

Chapter 2　Basic Operations of Organic Chemistry Experiments ··············· 25
  Exp.1　Determination of Melting Point ··············· 25
  Exp.2　Simple Distillation and Determination of Boiling Point ··············· 29
  Exp.3　Recrystallization ··············· 32
  Exp.4　Extraction ··············· 38
  Exp.5　Vacuum Distillation ··············· 44
  Exp.6　Steam Distillation ··············· 49
  Exp.7　Fractional Distillation ··············· 54
  Exp.8　Column Chromatography ··············· 57
  Exp.9　Thin Layer Chromatography ··············· 60
  Exp.10　Paper Chromatography ··············· 64

Chapter 3　Preparation Experiments of Organic Compounds ··············· 68
  Exp.11　Preparation of Cyclohexene ··············· 68
  Exp.12　Preparation of 1-Bromobutane ··············· 71
  Exp.13　Preparation of Bromobenzene ··············· 75
  Exp.14　Preparation of 2-Methyl-2-Butanol ··············· 77
  Exp.15　Preparation of Diphenylmethanol ··············· 80
  Exp.16　Preparation of 1, 1′-Binaphthol (a Reference-based Experiment) ··············· 82
  Exp.17　Preparation of Di-n-butyl Ether ··············· 83
  Exp.18　Preparation of Cyclohexanone ··············· 86
  Exp.19　Preparation of Benzalacetone and Dibenzalacetone ··············· 87
  Exp.20　Preparation of Benzoin ··············· 90
  Exp.21　Preparation of Benzoic Acid ··············· 92
  Exp.22　Preparation of Adipic Acid ··············· 94
  Exp.23　Preparation of Cinnamic Acid ··············· 97
  Exp.24　Preparation of Coumarin-3-carboxylic Acid ··············· 99

Exp.25　Preparation of Furfuryl Alcohol and Furoic Acid ·················102
Exp.26　Preparation of Ethyl Acetate ·················104
Exp.27　Synthesis of Dibutyl Phthalate ·················107
Exp.28　Synthesis of Ethyl Acetoacetate ·················109
Exp.29　Preparation of Acetylsalicylic Acid (Aspirin) ·················110
Exp.30　Preparation of Acetanilide ·················112
Exp.31　Preparation of *p*-Acetamidobenzenesulfonyl Chloride ·················115
Exp.32　Preparation of *p*-Aminobenzenesulfonamide ·················118
Exp.33　Green Synthesis of *N*-bromosuccinimide ·················120
Exp.34　Synthesis of Methyl Orange ·················121
Exp.35　The Preparation and Beckmann Rearrangement of Benzophenone Oxime ·················124
Exp.36　Preparation of Aniline ·················126
Exp.37　Preparation and Resolution of Racemic α-Phenylethylamine ·················129
Exp.38　Optical Resolution of *Trans*-1, 2-diaminocyclohexane ·················132
Exp.39　Preparation of 2, 3; 5, 6-Di-*O*-isopropylidene-α-D-mannofuranose ·················134
Exp.40　Isolation of Caffeine from Tea Leaves ·················136

Chapter 4　Reactivity Experiments of Organic Compounds ·················140
Exp.41　The Preparation and Reactivities of Ethylene and Ethyne ·················140
Exp.42　The Reactivities of Aromatic Compounds ·················143
Exp.43　The Reactivities of Some Alkyl Halides ·················145
Exp.44　The Reactivities of Alcohols and Phenols ·················148
Exp.45　The Reactivities of Aldehydes and Ketones ·················149
Exp.46　The Reactivities of Carboxylic Acids and Their Derivatives ·················151
Exp.47　The Reactivities of Amines ·················152

第1章　有机化学实验的一般知识 ·················155
1.1　有机化学实验室规则 ·················155
1.2　有机化学实验室的安全知识 ·················155
1.3　有机化学实验常用仪器和装置 ·················158
1.4　玻璃仪器的洗涤和干燥 ·················163
1.5　有机化学实验中试剂的加热、冷却和干燥 ·················165
1.6　实验预习、实验记录和实验报告的基本要求 ·················169

第2章　有机化学实验基本操作 ·················174
实验1　熔点测定 ·················174
实验2　简单蒸馏及沸点测定 ·················177
实验3　重结晶 ·················180
实验4　萃取 ·················184
实验5　减压蒸馏 ·················188
实验6　水蒸气蒸馏 ·················192
实验7　分馏 ·················196

实验 8	柱色谱	198
实验 9	薄层色谱	201
实验 10	纸色谱	204

第 3 章 有机化合物制备实验 ··· 207

实验 11	环己烯的制备	207
实验 12	1-溴丁烷的制备	210
实验 13	溴苯的制备	212
实验 14	2-甲基-2-丁醇的制备	215
实验 15	二苯甲醇的制备	217
实验 16	1,1′-联二萘酚的合成（文献实验）	218
实验 17	正丁醚的制备	219
实验 18	环己酮的制备	221
实验 19	苄叉丙酮和二苄叉丙酮的制备	222
实验 20	安息香的制备	224
实验 21	苯甲酸的制备	226
实验 22	己二酸的制备	228
实验 23	肉桂酸的制备	229
实验 24	香豆素-3-羧酸的制备	232
实验 25	呋喃甲醇和呋喃甲酸的制备	234
实验 26	乙酸乙酯的制备	236
实验 27	邻苯二甲酸二丁酯的制备	238
实验 28	乙酰乙酸乙酯的制备	240
实验 29	乙酰水杨酸（阿司匹林）的制备	241
实验 30	乙酰苯胺的制备	242
实验 31	对乙酰氨基苯磺酰氯的制备	245
实验 32	对氨基苯磺酰胺的制备	247
实验 33	N-溴代丁二酰亚胺的绿色合成	248
实验 34	甲基橙的制备	250
实验 35	二苯甲酮肟的制备及贝克曼重排	252
实验 36	苯胺的制备	253
实验 37	外消旋 α-苯乙胺的制备和拆分	255
实验 38	反-1,2-二氨基环己烷的手性拆分	258
实验 39	2,3;5,6-二-O-异丙叉基-α-D-呋喃甘露糖的制备	259
实验 40	从茶叶中提取咖啡因	261

第 4 章 有机化合物性质实验 ··· 264

实验 41	乙烯、乙炔的制备和性质	264
实验 42	芳烃的性质	266
实验 43	卤代烃的性质	268
实验 44	醇和酚的性质	270

实验 45 醛和酮的性质…………………………………………………………271
实验 46 羧酸及其衍生物的性质……………………………………………272
实验 47 胺的性质……………………………………………………………273

参考文献………………………………………………………………………276

Chapter 1 General Knowledge of Organic Chemistry Experiments

The laboratory course is crucial for learning organic chemistry. This lab course aims at training and developing students' expertise in scientific method; creating and reinforcing their good work habits in the lab; training their scientific attitude of seeking truth from facts, their abilities for rigorous and detailed observation; guiding students to put forward scientific exploration problems, stimulating students' scientific interest and innovative consciousness.

The basic tasks of this course are:

(1) To train students to master the basic operating skills of experiments in organic chemistry and enable students to synthesize, isolate, characterize and identify organic compounds properly.

(2) To guide students to understand the general procedures for conducting a research in organic chemistry; develop students' skills in accessing journal articles, extracting information from them, evaluating data, developing and solving problems, and writing scientific reports correctly.

Organic chemistry laboratories contain certain inherent dangers and hazards. When working in a laboratory, students must learn how to work safely with these hazards in order to prevent injury to themselves and others around them. The following guidelines are here to help. Students should understand and follow these guidelines and act according to the principles behind them to help everybody to be as safe as possible. Students are required to be sure that they are familiar with the safety precautions, hazard warnings and procedures before working in the lab.

1.1 Rules for Organic Chemistry Laboratory

The following rules must be strictly followed by the students to ensure that all the organic chemistry experiments are performed normally, effectively and safely.

(1) Prepare in advance and make a preview report prior to an experiment. Anyone who gets into the lab without a preview report will not be allowed to start the experiment.

(2) Know the location and proper use of emergency equipment (water, electricity, safety showers, eye baths, fire extinguishers, and first aid kits). Follow the instructions during an emergency and notify the instructor immediately.

(3) Wear a lab coat (and do it up) in all chemistry labs. Wear clothes that provide maximum protection and cover most of the skin. Wear long pants (no shorts, capris, skirts, or dresses). Footwear must completely cover the foot and heel (no sandals, or open-toed footwear).

(4) No food or drink is allowed in the lab.

(5) No noise or disruptive behavior in the lab. Do not leave experiments without permission.

(6) Only the authorized scheduled experiments can be performed in the lab. You are not allowed to alter the procedure of the lab experiments. Carefully follow all instructions. Any changes to the experiments must be made with the approval of the instructor. Conduct the operation seriously, observe the experimental phenomenon carefully, and record the data truthfully during the experiment.

(7) Keep your own work area clean as well as the shared areas. Wastes should not be discarded into the sewer system; they should be stored in appropriate containers in a timely manner and according to the instructions provided in your lab manual. Return all materials to their place after using them.

(8) After an experiment, you should properly clean all the areas/equipment you worked on. Before leaving the laboratory, notify the instructor and ensure that: water and electrical devices at your lab station are turned off; your desktop and work area is thoroughly clean; all equipment in community drawer is complete and well organized; wash your hands, and any other area of skin that has contacted lab equipment or lab benches.

1.2　Safety Knowledge in the Organic Laboratory

Many of the chemicals used in an organic chemistry laboratory are flammable, explosive or toxic, and may cause accidents, such as fire, explosion or poisoning, if not properly handled. In addition, most parts of the equipment used in an organic chemistry lab are commonly made of glass, cuts and scratches may be caused if one is not careful enough. Therefore, students should pay special attention to safety instructions when performing an organic chemistry experiment.

Most laboratory accidents occur when researchers lack understanding and attention to materials and equipment or do not manage them with strict adherence to safety procedure and precautions. However, with careful preparation beforehand, close attention and careful use of chemicals and equipment, accidents can be avoided. Therefore, all students must read and understand the safety instructions before coming to the laboratory, and strictly follow them when conducting an experiment.

1.2.1　Laboratory Safety Rules

(1) Before starting the experiment, the students should carefully check up each item of the working equipment and apparatus to make sure that the equipment is intact and the apparatus is set up properly and fixedly. And permission from the instructor is required to start the experiment.

(2) Experiments must not be left unattended while in progress. The operator should inspect the reaction and check for leaks routinely.

(3) When conducting experiments that are potentially dangerous, the operator should wear necessary safety equipment, such as safety goggles, facemasks or gloves, etc.

(4) Keep ignition sources away from flammable or combustible dangerous substances.

(5) Do not taste chemicals. Do not smoke, eat, or drink in the laboratory. Wash your hands carefully after the experiment.

(6) Know the location, operation and use of the emergency equipment, such as fire extinguishers, sandbox, safety showers, eye baths, and first aid kit, etc.

(7) Reagents used or generated during the experiment should not be discarded arbitrarily. The waste liquid or solid should be disposed into designated containers for unified treatment.

1.2.2 Accident Prevention

1. Fire Prevention

There are many reasons why a fire can start, such as a low-boiling solvent is heated in an open container, or an incorrect heating method is applied, etc. In order to prevent fire hazards in the experiments, the following basic principles should be followed at all times:

(1) **DO NOT** store volatile or flammable chemicals in an open container. When using and handling of flammable liquids, keep away from naked flames, sparks, fire and heat; and maintain adequate ventilation. Water baths can be used to heat low-boiling liquids.

(2) **DO NOT** place volatile or flammable wastes into a waste container or regular trash. When dealing with a large amount of flammable liquid, use bulk waste containers for specific recycling.

(3) If you are caught in a fire, remain calm and take correct measures as soon as possible to control the fire. First, shut off the power immediately and remove the flammable materials. Next, find out the classes of fires and how best to tackle them safely by choosing the most appropriate type of fire protection equipment. Organic laboratories do not use water (extinguishers), fires with chemicals and electrical equipment should be extinguished with other agents. If the fire is small, cover the fire with a damp cloth; when it is large, use a fire extinguisher.

The most common types of extinguishers in laboratories are carbon dioxide extinguishers, dry powder extinguishers and foam extinguishers.

Standard dry powder extinguishers are used commonly in organic laboratories, and special dry powder extinguishers can be used for flammable metals. However they are not recommended for use in enclosed spaces. This is because the powder can be easily inhaled, and also the residue is very difficult to clean up. When using a dry powder extinguisher, pull the pin to discharge the extinguisher; aim at the base of the fire (if you aim at the flames, the extinguishing agent will fly right through); squeeze the top handle to release the pressurized extinguishing agent in the extinguisher; sweep from side to side until the fire is completely out.

Carbon dioxide extinguisher is also a common fire extinguisher in organic labs, which is predominantly used for electrical fire risks and labs with electric equipment.

Foam extinguishers are ideal where cooling a fuel is recommended but not suitable for electrical fires or running fires, because they are water-based.

When using any of these portable extinguishers, it is necessary to aim at the base of the flames, rather than at the flames themselves.

A small tabletop fire can be put out with some sand in the fire sand bucket.

If your clothes catch fire, remember: stop, drop and roll until the flames have been extinguished. Don't run around, you'll make the flames worse.

2. Explosion Prevention

In order to prevent accidents due to explosions, following things should be noted:

(1) **NEVER** heat a closed system or conduct a reaction in a closed system.

(2) Before starting a distillation or chemical reaction, make certain that the system is vented. Be sure to use glassware designed for vacuum or pressure operations in a vacuum distillation, do not use flat-bottomed or thin-walled glassware. Always check for star cracks in the flasks before beginning a vacuum distillation. **NEVER** distill to dryness.

(3) Ensure to keep the flammable substance away from the source of ignition.

(4) Ethers have to be tested for formation of peroxides before being used. If peroxides are detected, they must be removed with some reagents, such as ferrous sulfate.

(5) Be extremely careful when using explosive materials, such as acetylide of heavy metals, polynitro compounds (picric acid or heavy metal picrates), trinitrobenzene (TNB), etc. **NEVER** strike explosive materials or allow them to be hit. **NEVER** subject explosive materials to excessive impact or friction. Unused explosives and explosive waste **MUST** be removed and disposed of under the supervision.

3. Poison Prevention

(1) **DO NOT** allow laboratory chemicals to come in contact with skin. Appropriate type of gloves, such as chemical resistant gloves, heat resistant gloves, etc., should be worn for handling hazardous materials. Wash your hands before and immediately after working with any laboratory chemicals, even if gloves have been worn. **NEVER** taste any chemical.

(2) When handling fuming acids or volatile irritants, work in a chemical fume hood, wearing appropriate personal protective equipment. Always wash glass labware immediately after use.

(3) If a harmful substance has splashed onto one's skin or mouth, remove any contaminated items and wash the affected area thoroughly with cool water for at least 10 min. If swallowed, try to spit out anything that is remaining in the mouth, wash out mouth with plenty of water, and then drink some fresh water. If necessary, go to the hospital for treatment. If toxic vapors are inhaled, move the victim at once to fresh air and get medical care as soon as possible.

4. Prevention of Burns

A burn occurs when a corrosive, hot or very cold material contacts the skin. All students are required to wear a lab coat at all times. Wear safety goggles and gloves when handling hazardous chemicals to avoid burns. When corrosive chemical contaminates the skin, wipe it off quickly and immediately rinse it with plenty of water and make forward treatments for different burns as follows.

(1) In case of burns caused by alkalis, use 1% acetic acid solution or 3% boric acid solution to neutralise, and then wash with soap and water.

(2) In case of burns caused by acids, use 3% sodium-hydrogen carbonate (sodium bicarbonate).

(3) In case of burns caused by bromine, use the same procedure as that used for treating an acid burn.

(4) Small burns from touching hot objects should be placed under running cold water for at least 20 min. Major burns need immediate medical attention.

(5) If acid squirts into the eye, wash it by dropping with 2% borax solution, then rinse with physiological saline. In the case of alkali, use 2% boric acid solution, and then rinse with physiological saline, and get medical care as soon as possible.

5. Prevention of Cuts

To avoid cuts, first **NEVER** place glassware under pressure that it is not designed to withstand. Inspect glassware for defects or cracks before use. Do not handle broken glass with bare hands.

In the event of an injury involving a broken glass or sharp instrument, the following steps should be followed. Small cuts should be rinsed thoroughly and examined for traces of foreign materials such as glass. If there is broken glass in the wound, the pieces should be remove carefully first. Bandages are available in the stockroom. Severe cuts and bleeding must be treated by applying direct pressure on the wound to control the rate of bleeding. Seek medical assistance immediately. If the incident involves potential exposure to infectious materials, then immediate medical attention is required.

Each lab should have a first aid kit that is properly stocked for emergency first aid purposes. The first aid kit in the laboratories should contain the following items:

(1) Tweezers, scissors, gauze, cotton wool, bandages, plasters, etc.

(2) Vaseline, band-aid, scald cream, eucalyptus oil, boric acid ointment.

(3) 1% acetic acid or boric acid solution, 1% sodium bicarbonate solution, 2% sodium thiosulfate solution, glycerin, hemostatic powder, medical alcohol, red syrup, gentian violet, and the like.

6. Electrical Safety

Conduct an inspection of laboratory electrical equipment before using it to ensure it is in good condition. Make sure the plug fits snugly into the outlet. Outlets that have loose-fitting plugs can overheat and damage the cord and the plug. **NEVER** force a plug into an outlet if it doesn't fit.

Make sure your hands are dry before plugging or unplugging a power cord. Prevent the human body from contact with electrical conductors. **NEVER** put a metal object in live parts of appliances or outlets.

When the experiment is completed, turn off all power first and unplug the equipment.

1.3 Commonly Used Equipment and Apparatus of Organic Chemistry Experiments

Laboratory glass equipment can be divided into those with ground-glass joints and those without. Standard-taper, ground-glass joint equipment is more convenient and faster to be used. Therefore it is strongly recommended. However, sometimes it is also inevitable to use ordinary glassware.

1.3.1 Standard-taper, Ground-glass Joint Equipment

The standard-taper glassware is manufactured in accordance with international requirements. Pieces of glassware in the same size fit together freely to conveniently set up an apparatus due to the ground glass joints incorporated into each other by an inner (male) joint held above an outer (female) joint. This means it doesn't need to equip plug and drill, also prevent product dirtied by cork or rubber plug. Furthermore, the standard glassware is more favorable to be used in a leak-tight apparatus especially in vacuum distillation because of its tight connections between all parts. What's more, it's safer to use the standard glassware in an experiment involving hazardous or volatile liquids.

The ground glass joints are often labeled in different sizes with different numbers. These numbers represent the outer diameter (OD) in millimeters (mm) at the widest point of the inner joint. The often used standard tapers are 10, 12, 14, 16, 19, 24, 29, 34, and 40 in size.

The ground glass joints can also be labeled with a number, a slash, and another number. The first number represents the OD at the widest point of the inner joint. The second number represents the ground glass length of the joint in millimeters. For example, a 14/20 joint is 14 mm wide at the top (the widest part) and is 20 mm in length. The organic chemistry teaching labs often use 19/22 sized standard taper. Any two joints (male and female) with the same outside diameter can fit into each other, no matter how length the joints are. When connection for two glass apparatuses with different joint size is needed, the connecting adapter (reducing adapter or enlarging adapter) is available to place between to connect them. For example, for enlarging the size of standard-taper glass joints from 14/20, outer to 19/22 inner, the socket size is 14/20, cone size is 19/22.

1. Commonly-used Standard-taper Ground Glassware in Organic Chemistry Experiments

Commonly-used standard-taper ground glassware in organic chemistry Experiments is listed in Fig. 1.1.

2. Tips for Using of Standard-taper Ground Glassware

(1) Before use, all glassware should be checked to ensure that it is free from cracks, flaws or scratches that may cause it to fail in use.

(2) Always keep the ground glass joints of the glassware clean. Wipe sticky dust before connecting the ground joints.

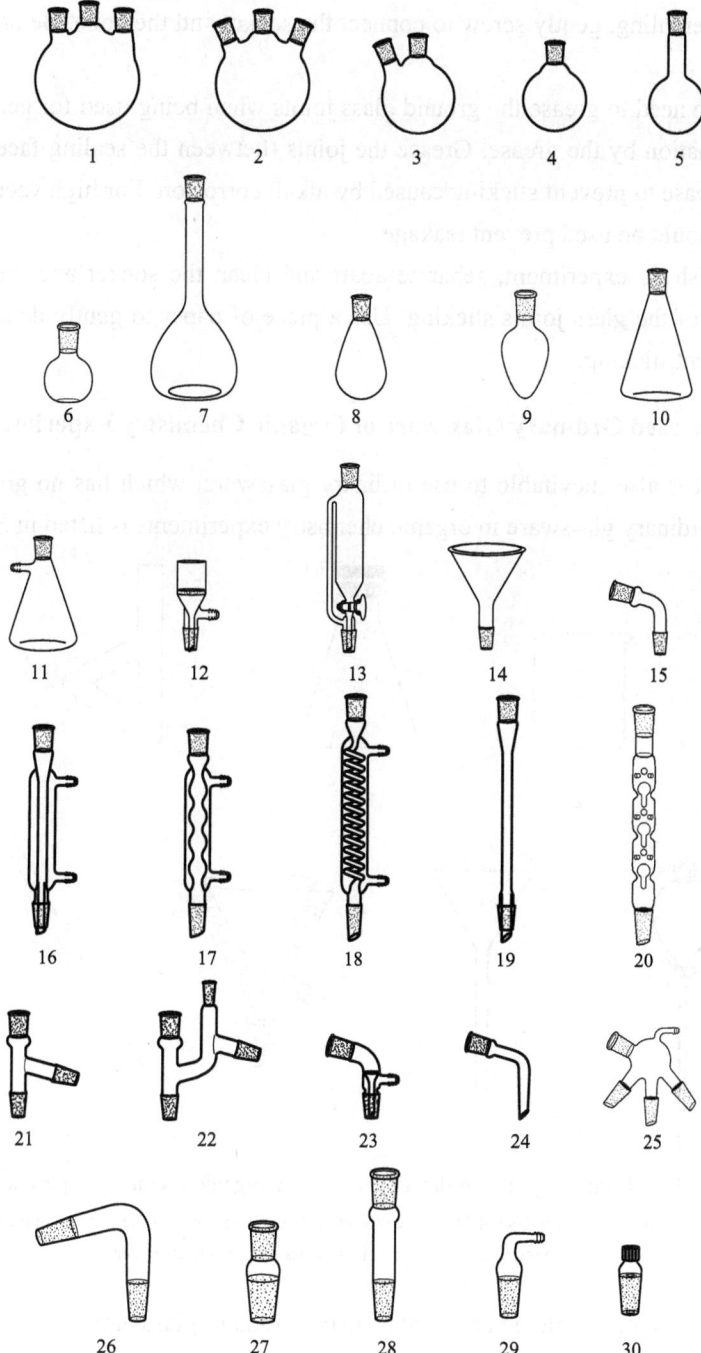

Fig. 1.1 Commonly-used standard-taper ground glassware in organic chemistry experiments

1.three-vertical-neck flask; 2. three-angled-neck flask; 3. two-neck flask; 4. round-bottom flask; 5. long-neck round-bottom flask; 6. flat-bottom flask; 7. long-neck flat-bottom flask; 8. eggplant shaped flask; 9. pear shaped flask; 10. Erlenmeyer flask; 11. suction flask; 12. sand core funnel; 13. constant pressure dropping funnel; 14. funnel; 15. drying tube; 16. straight-shaped condenser; 17. Allihn condenser; 18. graham condenser; 19. air condenser; 20. distilling tube with thorn tube; 21. distillation head; 22. Claisen distillation head; 23. vacuum adapter; 24. distillation adapter; 25. pig receiver; 26. distillation bent; 27. reducing adapter; 28. enlarging adapter; 29. angled cone adapter; 30. thermometer adapter

(3) When assembling, gently screw to connect the socket and the cone. Be careful not to force too much.

(4) There's no need to grease the ground glass joints when being used for general use to avoid potential contamination by the grease. Grease the joints (between the sealing faces) if the reaction involves a strong base to prevent sticking caused by alkali corrosion. For high vacuum applications, vacuum greases should be used prevent leakage.

(5) After finish an experiment, separate apart and clean the socket and the cone rapidly to minimize the risk of the glass joints sticking. Use a piece of paper to gently detach the cone from the socket to prevent sticking.

1.3.2 Commonly-used Ordinary Glassware of Organic Chemistry Experiments

Sometimes, it is also inevitable to use ordinary glassware, which has no ground-glass joints. Commonly-used ordinary glassware in organic chemistry experiments is listed in Fig. 1.2.

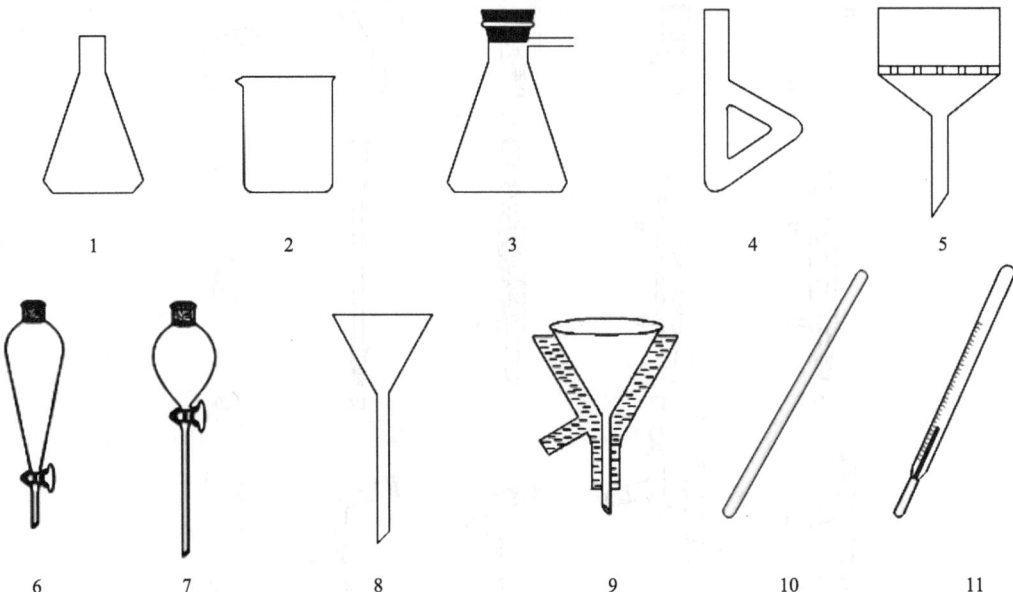

Fig. 1.2 Commonly-used ordinary glassware in organic chemistry experiments

1. Erlenmeyer flask; 2. beaker; 3. suction flask; 4 Thiele tube (b-tube); 5. Büchner funnel; 6. separatory funnel; 7. dropping funnel; 8. funnel; 9. hot water funnel; 10. glass rod; 11. thermometer

The following points should be noted when using ordinary glassware:

(1) No ordinary glassware can be heated directly by fire except a few, such as test tube.

(2) Glassware with thick walls (e.g. suction flask) should **NEVER** be heated over a direct flame. Erlenmeyer flasks are in general not suitable for use under pressure or in a vacuum; wide-mouthed containers such as beakers cannot be used for the storage of flammable solvents.

(3) When storing glass stopcocks and joints, insert a thin strip of paper between joint surfaces to prevent sticking.

(4) **NEVER** use a thermometer to stir, after the thermometer has been used, cool it down

slowly instead of immersing it immediately in cold water. Measured temperature of a thermometer cannot exceed the scale range. If a thermometer has been broken by accident, sprinkle fine powder of sulfur or zinc on the spill site to bind any remaining mercury.

1.3.3 Commonly-used Apparatus of Organic Chemistry Experiments

The apparatus for reflux, distillation, stirring, gas absorption, etc., is commonly applied in organic chemistry experiments.

1. Reflux Apparatus

The reflux apparatus allows an organic reaction to be carried out at the boiling point of the solvent and yet prevents loss of solvent or reagent due to evaporation. Different kinds of reflux apparatus in organic chemistry experiments are shown in Fig.1.3. Fig. 1.3 (a) shows an ordinary apparatus for reflux; (b) shows an apparatus for reflux with a drying tube; (c) shows an apparatus for reflux with gas absorption; (d) shows an apparatus for reflux with a Dean-stark trap; and (e) shows an apparatus for reflux with a dropping funnel.

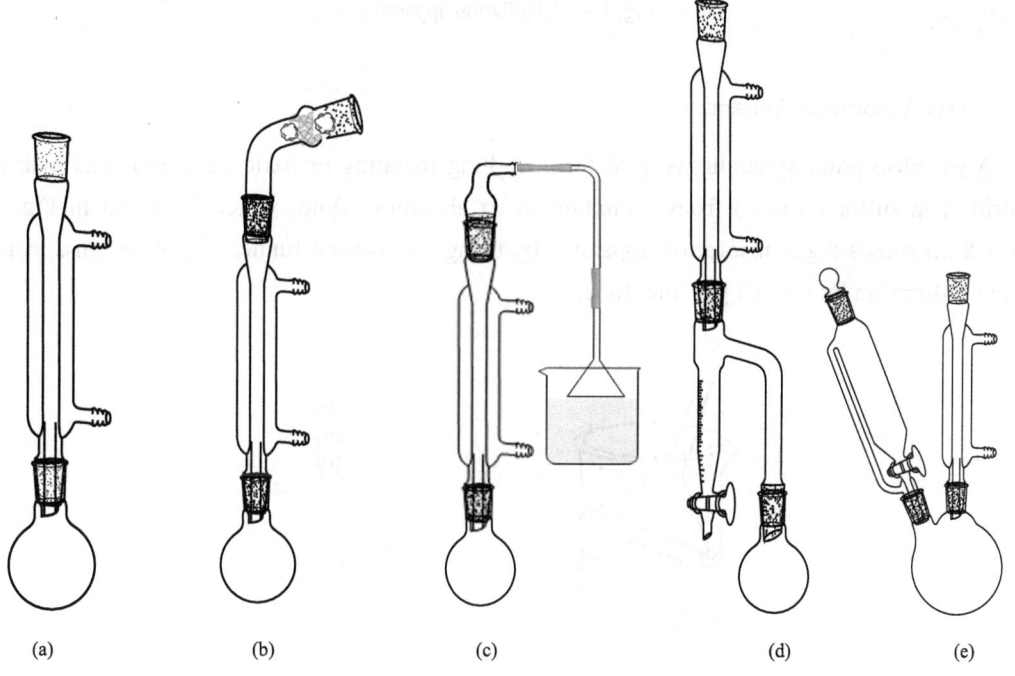

(a)　　　(b)　　　(c)　　　(d)　　　(e)

Fig. 1.3　Reflux apparatus

2. Distillation Apparatus

Distillation is used for the purification of a liquid by heating it to its boiling point, causing vaporization, and then condensing the vapors into the liquid state and collecting the liquid. Separation of two or more liquids requires that they have different boiling points with difference of more than 30℃. Different types of distillation apparatus are shown in Fig. 1.4. Fig. 1.4 (a) shows an ordinary apparatus for distillation using a water condenser; (b) shows an apparatus for distillation using an air condenser; and (c) shows an apparatus for distillation with a dropping funnel.

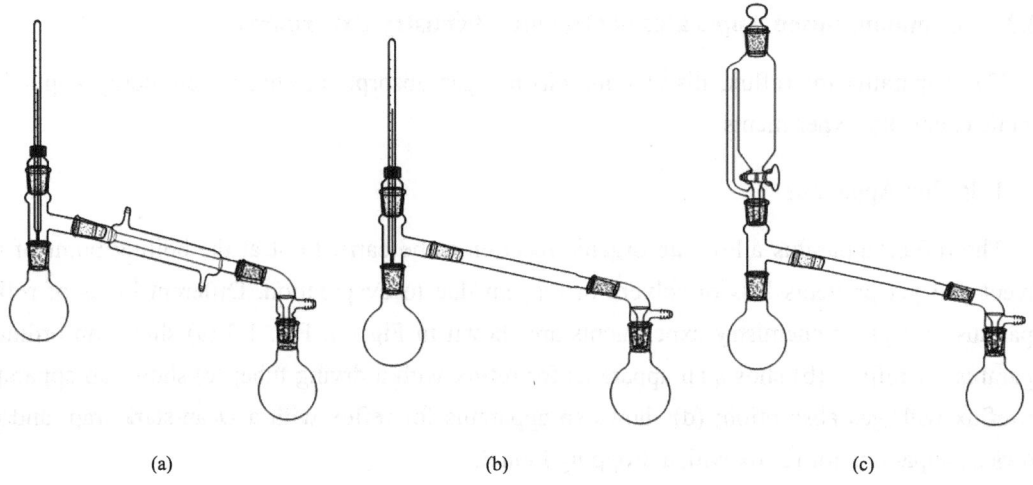

Fig. 1.4　Distillation apparatus

3. Gas Absorption Apparatus

A gas absorption apparatus is used for absorbing irritating or toxic gases (such as hydrogen chloride and sulfur dioxide) from a mixture in an absorbing fluid, which is shown in Fig. 1.5. Fig. 1.5 (a) shows a gas absorption apparatus by using an inverted funnel; (b) shows an apparatus for gas absorption by using a gas inlet tube.

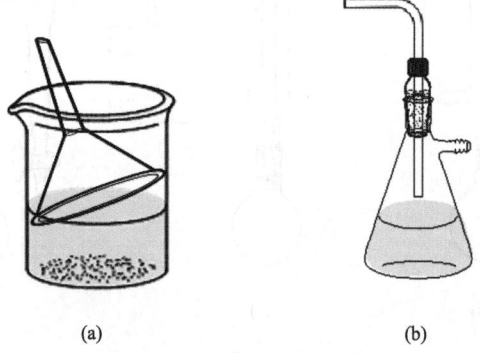

Fig.1.5　Gas absorption apparatus

4. Stirring Apparatus

When a reaction is heterogeneous or a reaction in which certain reactants need to be continuously added, it must be stirred very vigorously to ensure efficient mixing.

Apparatuses for reflux with mechanical stirring are shown in Fig 1.6. Fig. 1.6 (a) shows an apparatus for reflux with heating, overhead stirring and temperature monitoring; (b) shows an apparatus for reflux with heating, overhead stirring and dropping; and (c) is an apparatus for reflux with heating, overhead stirring, dropping and temperature monitoring. Fig. 1.7 are apparatuses for reflux with magnetic stirring, where (a) is an apparatus for reflux with heating, magnetic stirring, dropping and temperature monitoring; (b) is an apparatus for reflux with heating magnetic stirring and water separating.

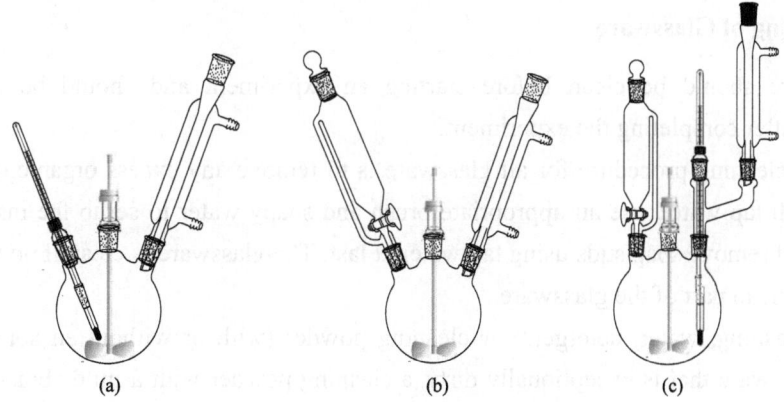

Fig. 1.6 Apparatuses for reflux with mechanical stirring

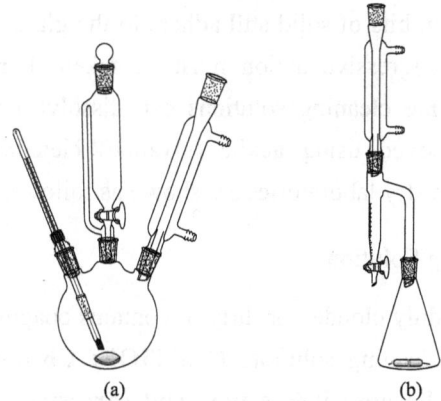

Fig. 1.7 Apparatuses for reflux with magnetic stirring

1.3.4 Assembling Apparatus

Pay attention to the following points when assembling an apparatus:

(1) The glassware and its accessories should be clean before being used.

(2) Always use suitably sized glassware. For example, when choosing a round-bottom flask, it is suitable to make the distilling flask 1/3~1/2 full with sample, the flask should be no more than 2/3 full.

(3) Performing the assembling of apparatus from bottom to top and from left to right, and

disassembling in reverse order.

(4) Reactions should **NEVER** be carried out in, nor applied to, an apparatus that is a closed system unless it is designed and tested to withstand pressure.

(5) The experimental apparatus is required to be correctly and securely installed, and easy to operate.

1.4 Cleaning and Drying of Glassware

1.4.1 Cleaning of Glassware

Glassware should be clean before starting an experiment and should be cleaned again immediately after completing the experiment.

General cleaning procedure for all glassware is to remove any excess organic oils first, rinse glassware with tap water, use an appropriate brush and soapy water to scrub the inside of curved glassware, and remove soap suds using tap water at last. The glassware is clean if no water droplets adhere to the clean part of the glassware.

When washing, water, detergent, or cleaning powder (with or without an abrasive) may be used. For glassware that is exceptionally dirty, a cleaning powder with a mild abrasive action will give more satisfactory results. Highly contaminated glassware can be cleaned by some other proven method such as ultrasonic cleaning.

If, after a second cleaning, bits of solid still adhere to the glass, or if there is clearly a greasy residue on the glass, more aggressive action must be taken. For example, suitable cleaning solutions can be used. Alkaline cleaning solutions can dissolve organic oils. Alkaline (acidic) contaminations are best removed using acidic (alkaline) cleaning solutions. Some cleaning solutions used in organic chemistry laboratories are shown as follows.

1. Chromic Acid Cleaning Solution

If glassware becomes unduly clouded or dirty or contains coagulated organic matter, it can be cleansed with chromic acid cleaning solution. **CAUTION: Chromic acid solution should be handled with extreme care because it is a powerful corrosive and carcinogen.** When using chromic acid cleaning solution, pour off the water in the glassware, add some of the acid cleaning solution to the glassware and swirl the solution to cover the dirty areas for a few minutes. After that, pour the cleaning solution back into the stocking bottle, rinse the glassware with tap water. If carbonized organic debris is present, it may be necessary to let the glassware be soaked with the cleaning solution for a while and then heated mildly until bubbles appear in the solution. The chromic acid cleaning solution is expired and should be discarded when it turns green. Extra care must be taken to be sure chromic acid solution is disposed of properly. Make sure the chromic acid solution is disposed of in accordance with appropriate regulations.

2. Concentrated Hydrochloric Acid

Concentrated hydrochloric acid (conc. HCl) will wash off the attached manganese dioxide or carbonate and some other dirts.

3. Alkaline Cleaning Solutions

Alkaline cleaning solutions contain strong bases like sodium hydroxide or potassium hydroxide, can dissolve fats (including grease), oils, and protein-based substances.

4. Organic Solvents

Organic residues that can't be removed by detergent and water will often dissolve in organic solvents such as technical-grade acetone (**NEVER** use reagent grade solvents for washing). Use acetone sparingly and recycle it after use (don't pour it down the drain). If used for organic analysis, glassware must be washed with distilled water as a final rinse. Clean glassware is hydrophilic and will have a uniformly wetted surface when distilled water is used as a final rinse.

However, try to reduce the use of highly corrosive acids, bases or organic solvents as cleaning agents to avoid environment pollution.

1.4.2 Drying of Glassware

There are several methods for drying glassware:

(1) Natural air drying: when the glassware is not required quickly, natural air drying can offer the simplest solution. A range of drying or draining racks are available to hang glassware with the open neck facing down.

(2) Electric drying: for faster results, an electric glassware dryer can be used. These force warmed air (60~70℃) into each piece of glassware, accelerating the drying.

(3) Oven drying: place items into the drying oven with mouth facing upward (for items that are water-wet only, no flammable solvents). Maintain the temperature of the drying oven at 100~105℃. For drying of glassware with ground-in stopcock, stopcock has to be removed before drying.

(4) Solvent drying: rinse with a solvent such as acetone or ethanol and then gently blowing compressed air into the vessel until it is dry.

Volumetric glassware and cuvettes are **NEVER** to be placed in drying ovens, they should be dried by using method (1) or (4).

1.5 Heating, Cooling and Drying of Reagents in Organic Chemistry Experiments

1.5.1 Heating

Some reaction goes pretty slowly or very difficultly at room temperature, it may be necessary to heat the mixture for the reaction to occur at an economically-viable rate. Sources of heat used in laboratory can be burn kerosene, alcohol, or electric heaters. In the organic chemistry lab, to ensure uniform heating and to prevent decomposition or spontaneous combustion caused by overheating, indirect heating instead of direct contact heating is generally used. Indirect heating is slower, but allows a more even, consistent warming up. In this case, several heating types are used as follows.

1. Air Bath

Air bath can be produced with hot air. For heating a liquid with a boiling point of or exceeding 80℃, an air bath can be used. The simplest air bath can be produced by heating the flask on an asbestos net. However, this heating device is not suitable for liquids with a low boiling point or liquids in a vacuum distillation system for the uneven heat transfer. Heating mantles are good choices for generating air bath, which are relatively safe and easy to handle and will provide stable and even heat over a range of temperatures up to 400℃ controlled by a regulating transformer.

2. Water Bath

When the desired temperature is below 100℃, a water bath can be used. A water bath is laboratory equipment made from a container filled with hot water. When using a water bath, the extremity of the flask should be immersed in water without touching the container's side or bottom. For prolonged heating, maintain the level of water in the bath container slightly higher than that of the liquid in the flask by adding water in the container occasionally to supplement evaporated water. In this case, a flat stainless steel cover with concentric ring sets can be equipped on top of a water bath to minimize evaporation of water.

3. Oil Bath

Temperatures between 100~250℃ can be obtained by using an oil bath. In most cases, the reaction temperature carried out should be about 20℃ less than the maximum temperature attainable for the oil bath. Materials used frequently in oil baths include:

(1) Glycerol: it may be used for temperatures up to 140~150℃. It will decompose when the temperature is too high. Because anhydrous glycerol has a strong water absorption property, long stored glycerol should be dried off water by distillation before using in an oil bath.

(2) Plant oil: plant oils such as cottonseed oil, castor oil and peanut oil, can be heated up to 220℃. To enhance the thermal stability of the oil, 1% of hydroquinone or other antioxidants can be added in.

(3) Liquid paraffin: it may be used at temperatures up to about 220℃. Combustion may occur when it is heated to a higher temperature.

(4) Solid paraffin: it can be used at temperatures up to about 200℃, which is convenient to be stored because it is in a solid state at room temperature.

(5) Silicon oil: it can be heated up to 250℃, which can maintain stable and transparent at a high temperature and therefore is safe to handle and is used most frequently as oil baths in

laboratories.

When heating with an oil bath, use a temperature sensor or thermometer to indicate and control the temperature of the bath. Suspend the thermometer's indicator bulb in the oil bath. Do not allow the thermometer to touch the side or bottom of the container. When removal a reaction vessel from the oil bath after a heating procedure, continue to hold the flask with the clamp and leave the container hanging over the surface of the oil for a while to let the oil adhesive drain off the outside surface of the vessel, hereafter wipe clean with a piece of tissue or cloth.

Make sure to prevent water from leaking into hot oil baths, which can cause hazardous popping and splattering of the oil. Pay close attention to the prevention of fire while using a hot oil bath. If an oil bath starts to produce smoke, turn off the heat immediately.

4. Sand Bath

Sand bath can be used for heating liquids with boiling points above 80℃, it is especially a good choice for temperatures of 220℃ or above. When making a sand bath, a beaker is filled with sand or metal pellets (called shot) and is placed on a hot plate or a heating mantle. The reaction vessel is then partially covered by sand or metal pellets. The sand or shot then conducts the heat from the plate to all sides of the reaction vessel. The sand bath is not used widely now because its temperature rises at a slow rate and is not so easy to be controlled.

1.5.2 Cooling

Some organic chemistry reactions require the use of coolant for cooling operation in the processes of reaction, separation or purification. For example:

(1) Some reactions need to be carried out under defined low temperatures, e.g. diazotization reaction is generally carried out at 0~5℃.

(2) Some reactions release a lot of heat, and are in need of cooling to control the reaction rate.

(3) In cases of accelerating the precipitation of crystals, cooling is required.

(4) When using low boiling point organic chemicals, cooling is needed in order to reduce the loss caused by evaporating.

(5) High vacuum distillation apparatus requires the use of cold trap for cooling.

There are many methods for generating cooling, in which an appropriate cooling method and coolant can be chosen based on different requirements.

1. Cold Water

Use cold water flowing on the outer wall of the container, or immerse the reactor vessel in cold water to exchange heat. A mixture of water and crushed ice can also be used as a coolant, which can provide a cooling temperature at a range of 0~5℃. If water has no influence on the reaction results, the ice can be put into the reactor directly for cooling.

2. Ice-salt Mixture

A coolant can be made by mixing crushed ice and inorganic salts in different proportions. Cooling temperatures provided by different ice-salt mixtures are shown in Table 1.1.

Table 1.1 Mass fractions of ice-salt mixtures and cooling temperatures provided

Salts	Mass fractions of salts	Mass fractions of ice	Cooling temperatures/℃
$CaCl_2 \cdot 6H_2O$	100	246	−9
$CaCl_2 \cdot 6H_2O$	100	123	−21.5
$CaCl_2 \cdot 6H_2O$	100	70	−55
$CaCl_2 \cdot 6H_2O$	100	81	−40.3
NH_4NO_3	45	100	−16.8
NH_4NO_3	50	100	−17.8
NaBr	66	100	−28
NH_4Cl	25	100	−15
NaCl	33	100	−21

3. Dry Ice or the Mixtures of Dry Ice and Organic Solvents

Dry ice (solid carbon dioxide) can be used to achieve a temperature below −60℃. If mixing the dry ice with methanol, acetone or chloroform and other solvents, a temperature down to −78℃ can be generated. Fierce foaming will be observed during this mixing.

4. Liquid Nitrogen

Liquid nitrogen can be used to make a cooling temperature of −196℃ (77 K) or above, which is generally used in scientific researches.

Liquid nitrogen and dry ice are two kinds of good refrigerants for their convenience in use and low prices, which should be stored in Dewar flasks (wide mouth thermos) or other well insulated containers to maintain their cooling effect.

5. Low-temperature Bath

Low-temperature bath can be looked as a mini refrigerator with an upward opening door assembly, in which a cylindrical stainless steel tank filled with alcohol is used as the evaporator and a peripheral compressor is designed for refrigerant circulation of Freon. A circulation pump is always installed outside the low-temperature bath for pumping the cold alcohol into a condenser. A thermometers or some other indicator can be installed in a low-temperature bath. When being used for a reaction, the low-temperature bath is suitable for providing a temperature range of −30℃ to 30℃, wherein the reaction flask should be immersed in the alcoholic liquid.

The above mentioned cooling methods can be selected for using as needed. **CAUTION**: the mercury thermometers cannot be used if temperatures get lower than $-38\,^\circ\text{C}$ because mercury freezes at $-38\,^\circ\text{C}$. Thermometers containing a mixture of organic solvents (alcohol, toluene, and n-pentane) can be used down to $-200\,^\circ\text{C}$.

1.5.3 Drying

Drying is a commonly used method for removing small amounts of water or organic solvents from solids, liquids or gases. In organic chemistry experiments, drying of reagent and products is of great significance.

1. Principles

Drying methods can be divided into two groups: physical drying and chemical drying.

Physical drying: physical drying methods include oven-drying, air-drying, absorption drying and freeze-drying, etc. In recent years, ion exchange resins and molecular sieves are frequently used for drying solvents.

Chemical drying: chemical drying methods use dry agents (also called desiccants) to remove water, which are separated into two classes according to the principle of water removal.

(1) Drying methods in which the dry agents can combine with water reversibly to generate hydrates. For example:

$$CaCl_2 + nH_2O \longrightarrow CaCl_2 \cdot nH_2O$$

(2) Drying methods in which the dry agents react with water irreversibly to produce new compounds. For example:

$$2Na + 2H_2O \longrightarrow 2NaOH + H_2 \uparrow$$

Pay attention to the following points when using dry agents:

(1) When the reaction between the dry agent and water is reversible, it takes time for the reaction to reach equilibrium. Therefore, after adding the dry agent, it will take two hours or more time to dry effectively. In addition, enough drying agent should be added to make effective drying for a reversible dehydration, which is about 5% of the solution by volume. As the temperature rises, the equilibrium position of this reversible reaction moves to dehydration. Therefore, the drying agent has to be removed (by filtration or decanting) from the dry solution prior to removal of the solvent.

(2) The solvent can be distilled without removal of the drying agent in cases where water removal is irreversible.

(3) A drying agent is only suitable for removing small amounts of water. When the water content is too high, the drying effect of dry agents is not good. To this end, remove the aqueous layer as completely as possible in the extraction process, so as to afford a good drying effect and reduce the loss of the product.

2. Drying for Liquid Organic Compounds

1) Selection of Dry Agents

The drying agents for liquid organic compounds should not make chemical reactions with the liquids themselves, including reactions of dissolution, complexation, association and catalysis. For example, acidic compounds cannot be dried by alkaline desiccants. Some commonly-used desiccants for liquid organic compounds are listed in Table 1.2.

Table 1.2 Commonly-used drying agents for liquid organic compounds

Liquid organic compounds	Suitable drying agents
ethers, alkanes, aromatics	$CaCl_2$, Na, P_2O_5
alcohols	K_2CO_3, $MgSO_4$, Na_2SO_4, CaO
aldehydes	$MgSO_4$, Na_2SO_4
ketones	$MgSO_4$, Na_2SO_4, K_2CO_3
carboxylic acids	$MgSO_4$, Na_2SO_4
esters	$MgSO_4$, Na_2SO_4, K_2CO_3
alkyl halides	$CaCl_2$, $MgSO_4$, Na_2SO_4, P_2O_5
organic bases (amines)	NaOH, KOH

2) Drying Efficiencies and Capacities

Capacity refers to the maximum numbers of moles of water that the drying agent can bind. Efficiency refers to the amount of water left in the organic solution after the drying process is completed. The efficiency of an inorganic salt drying agent, which can form hydrates with water, is indicated by measuring the water vapor pressure of the hydrates formed after the drying process. For example, each molecule of sodium sulphate is bonded to 10 water molecules, the drying capacity is 1.27 and the vapor pressure of the hydrated salt is 260 Pa; each molecule of calcium chloride is bonded to at most 6 water molecules, the drying capacity is 0.97 and the vapor pressure of the hydrated salt is 39 Pa at 25 ℃. Therefore, sodium sulfate has a larger drying capacity, but its drying efficiency is weaker. However, calcium chloride has a smaller drying capacity, but its drying efficiency is stronger. When drying compounds with larger water content, a desiccant with a larger capacity is often used first to remove most of the water, and another desiccant with stronger drying efficiency will then be used for a forward drying.

3) Drying Agent Dosage

Minimum required amount of the desiccant can be calculated according to the total moisture to be absorbed and the drying capacity of the desiccant. However, the actual amount of desiccant used greatly exceeds the calculated amount. Generally, the amount of desiccant for 10 mL liquid is 0.5～1 g. However, in actual operation, the drying agent dosage is mainly estimated by observation.

(1) Observation on the liquid to be dried: the liquid is cloudy before drying, and becomes clear after drying, which is a basic sign for the removal of most of the moisture.

(2) Observation on the desiccant: after being added in, if the desiccant is glued to the wall and difficult to be swirled away because of its stickiness resulted by water-absorbing, the dosage of the desiccant is not enough and more desiccant should be added in until the newly added desiccant is not caked, nor sticky, but angular and can be swirled away freely.

As the suspended desiccant (especially $MgSO_4$ and other desiccants in small crystals) can absorb some of the organic liquid, reducing the yield of the product, the amount of desiccant added should be moderate. A small amount of desiccant can be added first and then keep the solution standing still for a while to observe and estimate the deficiency of the desiccant, add some more desiccants when necessary.

4) Drying Temperature

For hydrate-forming desiccant, although the heating speeds up the drying rate, the drying rate speeded-up is far less than the water-releasing rate of the hydrates. Therefore, the drying procedure is usually carried out at room temperature.

5) Steps and Points for Drying Operation

(1) Remove the water in the liquid as much as possible until there is no visible water or suspended water droplets observed in the liquid.

(2) Place the liquid to be dried into an Erlenmeyer flask, put some particles of desiccant in appropriate size (such as anhydrous calcium chloride, should be in bean-size particles with no powder) into the liquid, stopper the flask, swirl the flask gently, and then observe to determine whether the desiccant is adequate or not. In the end, let the flask stand still (for half an hour, preferably overnight).

(3) Then filter the dried liquid into a distillation flask and distill out the desired liquid.

3. Drying of Solid Organic Compounds

Drying of solid organic compounds is mainly for the removal of a small amount of low boiling point solvents (such as water, ether, ethanol, acetone, benzene and the like) remaining in the solids. Since the solid organic compounds are less volatile than the solvents, the methods of evaporation and adsorption are always adopted to achieve the drying. The common drying methods are as follows:

(1) Air-drying.

(2) Oven-drying: ①dry with a constant-temperature oven or constant-temperature vacuum oven; ②dry with infrared light.

(3) Freeze-drying.

(4) Absorption-drying: if it is difficult to dry the solid by vacuum filtration, the solid can be transferred from the Büchner funnel to a filter paper, and 2 to 3 layers of filter paper are placed on both sides of the solid, the solvent can be absorbed by the filter paper by squeezing the solid.

(5) Drier drying: ①ordinary drier; ②vacuum drier; ③thermostatic vacuum drier (drying gun).

4. Drying of Gases

Commonly-used gases in organic chemistry experiments are N_2, O_2, H_2, Cl_2, NH_3 and CO_2. Sometimes, the gas is required with little or no CO_2, H_2O, etc. Therefore, the gases are needed to be dried. Common equipment for drying gases are drying tube, drying tower, U-shaped tube, a variety of scrub bottles (commonly used to hold liquid desiccant) and so on. Commonly-used desiccants for gases are listed in Table 1.3.

Table 1.3 Commonly-used desiccants for gases

Desiccants	Gases
CaO, soda lime, NaOH, KOH	NH_3
anhydrous $CaCl_2$	H_2, HCl, CO_2, CO, SO_2, N_2, O_2, lower alkyl ethers, alkenes, halogenated hydrocarbons
P_2O_5	H_2, N_2, O_2, CO_2, SO_2, alkanes, alkenes
concentrated H_2SO_4	H_2, N_2, HCl, CO_2, Cl_2, alkanes
$CaBr_2$, $ZnBr_2$	HBr

1.6 Basic Requirements for Pre-lab Work, Lab Record and Lab Report

Learning to write up lab report is an important aspect of students' scientific training. A minimum requirement for any record of laboratory work is that it should contain enough information for any practising chemist to replicate the work exactly.

1.6.1 Pre-lab

Preparation of the pre-lab prior to carrying out the experiment helps to familiarize you with the objectives of the experiment and any hazards associated with that experiment. The pre-lab usually includes objectives, table of reagents and an experimental plan, references and answers to pre-lab questions. The students are expected to complete the pre-lab **BEFORE** coming to the lab and be adequately prepared for carrying out the experiment. If a student does not complete a pre-lab, he (or she) will **NOT** be allowed to start the experiment.

The pre-lab must legible and contain the following:

(1) Objectives.

(2) Reactions or isolated substances: for reactions, give the overall reaction with structures and names of reactants and products and the reaction conditions. For isolations, give the structure(s) and name(s) of the compound(s) to be isolated.

(3) Table of reagents: this is a list of all the chemicals being handled (used or produced) during a particular experiment, that includes amount needed (or theoretically produced, for products), density (for liquid reagents), molecular weights, melting or boiling points and hazardous properties.

This will help familiarize the students with potential hazards of the experiment.

(4) Clear accurate diagrams of apparatus.

(5) The questions served to focus on the main concepts of the experiment including questions about operations, key procedure, difficulty, safety, etc.

(6) Experimental plan: formulate a flowchart to summarize the key points of your experiment. For example, a flowchart for the purification of product in the experiment of preparation of 1-bromobutane can be given as Fig. 1.8.

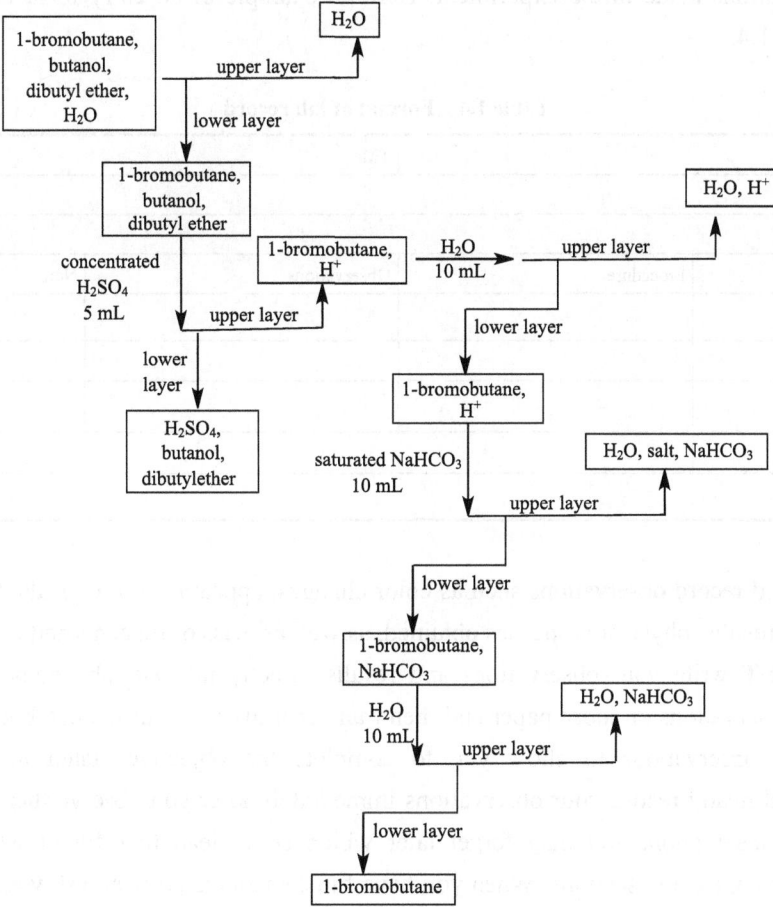

Fig. 1.8 A flowchart for the purification of 1-bromobutane

(7) References: state the origin of all the resource/reference. List the references in the following order: author, year, book/journal name, and page numbers.

1.6.2 Lab Record

Lab record is critical to experimental research. Use a bound notebook rather than a loose-leaf binder or loose paper. Record all observations and date as your experiment progresses. It is important to note any changes in the experimental plan as well.

The first 4~5 pages of the lab notebook should be left for a table of contents. Although formats vary for tables of contents, the important information remains the same. The table of contents should include the date the entry was made, the subject of that entry, and where in the lab notebook that entry can be found.

Each entry starts at a new page and should contain the following items: the date it was made, a title for the experiment, a goal for the experiment, the background for the experiment (including the protocol that was followed, all calculations performed, a list of reagents and equipment used), and all the observations made in the experiment. Here an example of an entry for a lab notebook is listed in Table 1.4.

Table 1.4 Format of lab record

Date		Title	
Objectives			
Background			
Time	Procedure	Observations	Note
Analysis			
Signature			

You should record observations such as color changes, appearance of crystals, temperature of reactions, test results, physical properties obtained, as well as mass of reagents and products used.

You **MUST** write your observations and results directly in your lab notebook. **DO NOT** record your observations on loose paper and then transfer it over to your lab notebook. You should make enough observations to allow you to complete the objective stated in your pre-lab. Accurately and neatly record your observations immediately after you observe them. If you delay recording an observation, you may forget later which could lead to difficult writing up your discussion and conclusion sections. When you have finished an experiment, ask your supervisor or the lab manager to sign on your lab notebook for a leaving.

1.6.3 Lab Report

When you have completed an experiment and have collected all of the necessary information in your lab notebook, you need to write up your results in a lab report. The purpose of writing a lab report is to communicate exactly what occurred in an experiment or observation and to clearly discuss the results. The lab report is an essential element of laboratory work.

The following components should be contained for each experiment, along with any additional material required by the experiment.

(1) Title.

(2) Objectives and principles (reaction formulas), including reaction mechanisms where applicable.

(3) Physical properties of reagents and products (including calculations).

(4) Diagrams of apparatus.

(5) Procedure outline.

(6) Data and observations (e.g. R_f values, m.p., spectral data, appearance, mass, yield, etc.).

(7) Discussion of results (conclusions): in this section, you analyse and interpret the results of the experiment. It is the most important part of the lab report, because here, you show that you understand the experiment beyond the simple level of completing it. For instance, you should indicate the amount of purified compound obtained and how the purity and identity of the compound was assessed. In a preparative experiment, state the percent yield. In this section, you can state whether or not you met your objective; if not, try to make suggestions to improve the method. Be sure to include a discussion of possible sources of error, and how that error would affect the overall yield.

1.6.4 Sample Lab Report

Here's a sample preparative organic chemistry lab report.

Preparation of 1-Bromobutane

Objectives

(1) To prepare 1-bromobutane from 1-butanol.

(2) To practice setting up and operating apparatus for refluxing, distillation and gas absorption.

(3) To further practice the operations for the purification of liquid products, including washing, drying, distillation, etc.

Principles

Main reaction: [*omitted*]

Side reaction: [*omitted*]

Reagents and Physical Properties

Reagents	Molecular weight	Consumption	Relative density (d_4^{20})	Solubility in water /(g/100 mL)	b.p./℃
1-butanol	74	11.2 mL	0.81	7.9	117.7
1-bromobutane	137		1.28	insoluble	101.6
sodium bromide	103	15 g		slightly soluble	
conc. H_2SO_4	98	18 mL	1.84	soluble	

Other Reagents: 10% Na_2CO_3 solution, anhydrous calcium chloride.

Apparatus

Diagram of refluxing apparatus [*omitted*]

Diagram of distillation apparatus [*omitted*]

Procedures

(1) 15 mL of H_2O was placed into a 100 mL round-bottom flask. A magnetic stir bar was added in the flask and the liquid was cooled by an ice-water bath under stirring. 18 mL (33 g) of concentrated sulfuric acid (conc. H_2SO_4) was added in 2~3 mL portions, being thoroughly mixed and cooled after each addition. The mixture was then cooled to room temperature. After that, 11.2 mL (9.0 g) of 1-butanol and then 15 g of sodium bromide was added in the mixture under stirring. A water condenser was attached to the flask and an inverted funnel was connected onto the top of the condenser with long rubber tubing, the funnel was positioned on the top surface of water in a beaker (as depicted in Fig. ××).

(2) The mixture was heated to boiling and maintained reflux for 30 min. The temperature should be lowered down if white fumes were visible out the top of the condenser. After the flask had been air-cooled for a few minutes, 10 mL of warm water was added in the flask and the apparatus was reassembled for a simple distillation. The flask was heated to distill out the liquid in the flask until the distillate is clear.

(3) The distillate was transferred into a 125 mL separatory funnel. About 10 mL of water was added to the funnel. The funnel was then stoppered and inverted **GENTLY** for several times, with venting. Note that two layers had formed; test to determine which layer was the organic layer and separate the two layers. The organic layer was transferred into a new dry separatory funnel, and washed with 5 mL of conc. H_2SO_4 carefully, and then separated from another layer. Next, the organic layer was washed with 10 mL of water, saturated $NaHCO_3$ and water separately in sequence. The cloudy organic layer was transferred to a 50 mL dry Erlenmeyer flask and enough anhydrous $CaCl_2$ was added to cover the bottom of the flask. The flask was swirled occasionally for a period of 5 min. The liquid was decanted into a dry distillation flask after the drying agent was settled down. In the end, the liquid was distilled out and the fraction that boiled in the ranged of 99~103 ℃ was collected as the product.

Results

Yield: 4.2 g

Theoretical yield $= \dfrac{0.9}{74} \times 137 = 16.4 (g)$

Percentage yield $= \dfrac{4.2}{16.4} \times 100\% = 25\%$

Discussion

[*omitted*]

Chapter 2 Basic Operations of Organic Chemistry Experiments

Exp.1 Determination of Melting Point

Objectives

(1) To understand the term "melting point" and know the importance of determining the melting point of an organic compound.

(2) To learn how to obtain an accurate melting point using a Thiele Mel-Temp apparatus.

Principles

The melting point of a pure crystalline substance is the temperature at which the material changes from a solid to a liquid state. Most crystalline organic compounds have characteristic melting points that are sufficiently low (50~300℃) to be conveniently determined with simple equipment. A pure crystalline substance has a constant melting point, and the melting range (the difference between the temperature where the sample starts to melt and the temperature where melting is complete) is not exceeding 0.5~1℃.

A phase diagram combining plots of vapor pressure versus temperature for the solid-liquid phase-transition equilibrium of a substance is shown in Fig. 2.1. The diagram indicates that the solid vapor pressure of the sample varies with temperature at a higher rate than its liquid vapor pressure. The two curves intersect at the point where the solid and liquid phases coexist, the temperature corresponding to the intersection point is the melting point, and therefore a pure organic solid has a sharp and precise melting point.

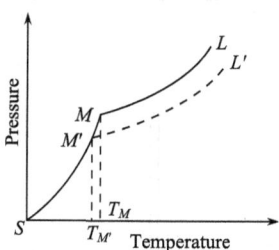

Fig. 2.1 Pressure-temperature phase diagram

According to the Raoult's law, at a certain temperature and pressure, adding solute to a solvent will essentially dilute the solvent molecules, and this leads to a decrease in vapor pressure. Considering the fact that the vapor pressure of the solid and liquid forms must be the same at melting point, the lowering of the vapor pressure leads to the lowering of the melting point.

Impurities are to cause the melting range to become lower and broader. Therefore, we can use melting points to determine the purity of organic compounds.

Materials

Equipment: Thiele tube, mercury thermometer (250℃), alcohol burner, capillary melting point tube (ID 1.0 mm, length 60～70 mm, one closed end).

Reagents: urea, cinnamic acid, a mixture of urea and cinnamic acid (4 : 1 in mass), glycerol (heating oil).

Procedures

In this experiment, capillary tube method using Thiele tube is applied to determine the melting point of the samples, the procedures are as follows.

1. Preparation of the Melting Point Tubes

Capillaries with an inner diameter of about 1.0 mm, a length of 60～70 mm, and one closed end are used[1].

2. Loading Capillary Tube

Place 0.1～0.2 g of sample in a watch glass. The sample may need to be ground into a fine powder with a mortar and pestle. Push the open end of the capillary tube into the compound to load the sample into the tube (as shown in Fig. 2.2). Load only 2～3 mm of sample into the tube. Then, drop the capillary tube down a length of a hollow glass tube, so that the bottom of the capillary tube can jump on a hard surface and the crystals can be packed well at the bottom of the capillary tube. Repeat this process so as to pack the solid well in the capillary tube. This operation should be done quickly as possible to prevent the absorption of moisture from the air. Remove excess sample powder from the outside of the capillary tube so as to avoid introducing contaminants into the bath oil.

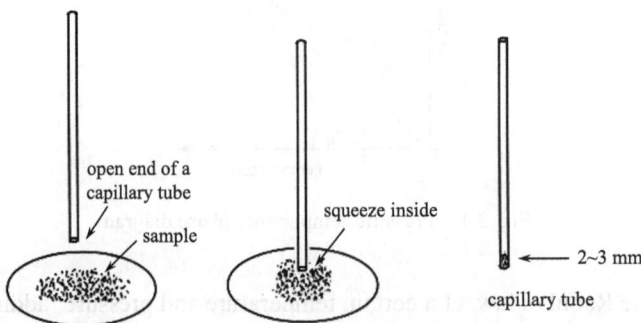

Fig. 2.2 Loading capillary tube

3. Setting Up Apparatus

The simplest equipment for capillary method is Thiele tube. As shown in Fig. 2.3, a Thiele tube is a glass instrument filled with oil that is heated by using an open flame. The side arm of the tube is designed to generate these convection currents without stirring by utilizing the temperature difference of the liquid in the tube and thus transfer the heat from the flame evenly and rapidly throughout the heating oil.

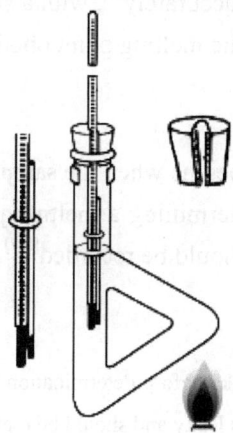

Fig. 2.3 Thiele Mel-Temp apparatus

The choice of liquid for the heating bath depends on several factors: temperature required, toxicity, flammability, expense and ease of handling-particularly removing it from the glassware. The heating oil should be a stable and high boiling liquid. Liquid paraffin can be safely used when the required temperature is below 140℃. Concentrated sulfuric acid can be employed which may be heated to reach but not exceed a temperature of 250℃, owing to excessive fuming above this temperature. Silicone oil provides another alternative for a temperature range of 250~365℃ but it is not economically feasible.

Fill a Thiele tube with heating oil until the level of the oil is just above the top of the side-arm neck. Place the thermometer into a split one-holed stopper that will fit into the top of the Thiele tube. Place this stopper into the Thiele tube and position the thermometer bulb so that it is slightly below the bottom of the side-arm neck of the tube. The sample along with the capillary tube is attached to the thermometer with the help of a rubber band[2]. The sample must be positioned so that it is at the same height as the bulb of the thermometer.

4. Determination of the Melting Point

Use an alcohol burner to gently heat the oil in the arm of the tube. To get accurate results, the melting point should be kept as close to the thermometer as possible. Therefore, the rate of heating should be controlled carefully.

(1) A known sample is heated at a fast rate at which the temperature increases 5~8℃ every

minute at the beginning. Greater rates (1~2℃/min) can be used until the temperature is 10~15℃ below the melting range. And an even slower rate of 0.5~1℃/min can be used when the temperature is close to the melting point.

(2) When the melting point of an unknown substance is to be determined, put off the alcohol burner to allow it to cool to about 50℃ before trying to determine. Use a hairdryer set on the cool setting to lower the temperature of the Mel-Temp faster. Prepare two sample tubes. First tube is used to obtain its approximate melting point using a rapid heating rate (5℃/min); second sample is applied to determine the melting point accurately[3], with a slow heating rate of 0.5~1℃/min when the temperature is nearly 10℃ below the melting point obtained in the first determination.

5. Record the Melting Points

Record the melting range, which begins when the sample first starts to melt and ends when the sample is completely melted while determining a melting point[4], several observations with each readings and the average temperature should be recorded[5-7].

Notes

[1] Check the melting point tube for leaks before determination by dipping its closed end into the oil bath. If there is liquid observed in the tube, the tube is leaky and should be replaced.

[2] Adjust the thermometer before the determination.

[3] Allow the heating oil to cool to about 30℃ and replace the melting point tube by a new one before trying a new determination.

[4] If the sample begins to shrink (collapse) at 120℃, forms droplets at 121℃ and becomes liquid at 122℃, the melting range should be 121~122℃, and the temperature at 120℃ should not be taken as the starting melting point, because the melting starts when the first drop of liquid appears.

[5] The melting point cannot be recorded as the average value of the starting and ending melting point. For example, the melting range of 123~125℃ should not be recorded as a melting point of 124℃.

[6] There are some compounds that are heat sensitive and decompose either before or at the melting point. If a sample becomes severely discolored or evolves a gas, decomposition is taking place. And the decomposition temperature is used as a physical property in place of melting point.

[7] After completing the experiment, do not wash the thermometer with cold water immediately until it is naturally cooled to room temperature, so as to avoid getting a burst. Heating oil should be fully cooled before being poured back into a stock bottle.

Safety Notes

(1) Always wear safety goggles in the lab.
(2) Do not touch the Thiele tube while it is still HOT. Otherwise you will get burned!
(3) Capillary tubes break very easily, handle them with caution.
(4) Wash hands after performing experiment.

Questions

(1) Is the experimental result reliable if the sample is not finely ground in a melting point determination?

(2) Most errors in a melting point determination come from heating the sample too fast, why?

(3) A sample should be used only once for melting point determination, why?

Exp.2 Simple Distillation and Determination of Boiling Point

Objectives

(1) To understand the theory, technique, and use of simple distillation and boiling point determination.

(2) To obtain the essential skills for the operations of heating under reflux and the assembly of a simple distillation apparatus.

(3) To purify an impure liquid by distillation and determine its boiling point.

Principles

The boiling point of a substance is the temperature at which the vapor pressure of the liquid equals the pressure surrounding the liquid and the liquid changes into a vapor. The normal boiling point (also called the atmospheric boiling point or the atmospheric pressure boiling point) of a liquid is the special case in which the vapor pressure of the liquid equals the defined atmospheric pressure at sea level, 1 atmosphere. A liquid has a specific boiling temperature at standard pressure conditions. And the boiling point is useful in the identification of liquids.

Distillation is the process of heating a substance until it is vaporized, cooling the vapors, and collecting the condensed liquid. Since a substance condenses at the same temperature that it boils, the temperature at which the vapor condenses on the thermometer will be the boiling point if a thermometer is added to the distillation apparatus. Therefore, the distillation can be used to determine the boiling point of a liquid.

If there are two liquids in a solution, the liquid with a lower boiling point can be boiled out of the solution without removing a significant amount of the other liquid. By vaporizing the liquid and condensing it in another container, one can obtain a pure sample of the liquid. Any impurities and liquids with higher boiling points remain in the original container. Distillation can be used to remove a volatile solvent from a nonvolatile product; to separate a volatile product from nonvolatile impurities; or to separate two or more volatile products that have sufficiently different boiling points. However, boiling liquid at a constant boiling point is not an indication of purity. Occasionally, a constant boiling mixture with two or more liquids called azeotrope can be encountered a constant-boiling behavior of pure liquids.

Simple distillation is a procedure which is usually used only to separate liquids whose boiling points differ greatly (rule of thumb is 30 ℃). When a reasonably large volume (more than 10 mL) of sample is available, then the boiling point of the liquid can be determined by a macro-scale method using a simple distillation apparatus setup. However, if only a small amount of liquid is available, a micro-scale method should be applied.

The temperature indicated by the distillation thermometer at the instant the first drop of condensate leaves the distillation adapter is the initial boiling point; the temperature indicated at the instant the last drop of liquid evaporates from the lowest point in the distillation flask is the dry point. And the difference between these two temperatures is the distillation range of a liquid. The boiling point of a substance (at a particular pressure) is definite (vary within the range of 0.5~1.5℃) when it is pure but will be lowered and broadened when impurities are added in. Therefore, the purity of a separated liquid can be checked by measuring its boiling point.

Materials

Equipment: round-bottom flask, distillation head, thermometer adapter, condenser, distillation adapter, Erlenmeyer flask, thermometer (150℃), clamp, boiling chips, hot plate, ring stand.

Reagents: industrial alcohol.

Procedures

1. Setting Up the Distillation Apparatus

Set up the distillation apparatus as Fig. 2.4 indicates[1, 2], set these parts of apparatus up from the bottom to the top and from the left to the right. General procedure for set-up of distillation apparatus is as follows:

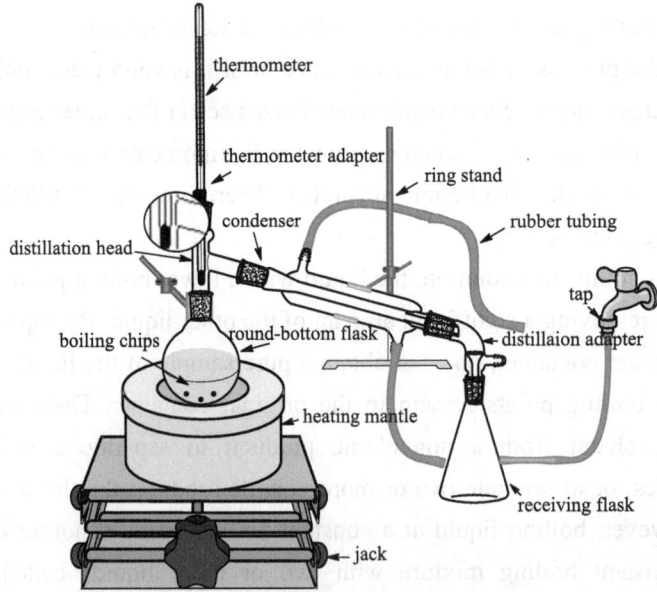

Fig. 2.4 Apparatus of simple distillation

(1) Select a suitable distillation flask. The size of the distillation flask depends on the total quantity of the liquid to be distilled. In simple distillation, the distillation flask should only be 1/3

to 2/3 full of liquid being distilled. A distillation flask should not be filled to more than two-thirds of its capacity. Add quantitive liquid to be distilled into the flask.

(2) Then set a heating mantle (or a hot plate) and ring stand in the hood[3]. Insert the round-bottom flask (distillation flask) into the heating mantle and clamp the neck of the flask to the ring stand so that the flask is touching the bottom of the heating mantle (**NEVER** clamp anywhere except at the joints! It will crack the glassware).

(3) Add 2~3 bean-sized boiling chips to the flask and attach a distillation head[4], thermometer adapter and thermometer to the flask[5]. To make boiling smoothest, boiling chips or capillary tubes should be added to the liquid in the distillation flask before heating has begun. Without a boiling chip, bubbles of vapour do not escape easily from the body of the liquid, with the result that the liquid becomes superheated and the vapour is expelled periodically in a sudden uncontrollable burst that causes the apparatus to bump. It is very important not to add chips or capillary tubes to a heated liquid as it may suddenly begin to boil and eject hot liquid out onto the operator. The chips are generally made of sharp pieces of broken ceramic; capillary tubes are usually of the capillary type with both ends open. Correct positioning of the thermometer is critical to getting accurate measurements of the boiling point. Position the height of the thermometer so that the bulb is just below the side arm of the distillation head (as shown in the zoom part of Fig. 2.4).

(4) Set up a second ring stand. Attach a condenser using a clamp. Attach a distillation adapter to the end of the condenser using a rubber band or a clamp[6].

(5) Attach the receiving flask (a round-bottom flask or an Erlenmeyer flask) and place another clamp (clamped to a second ring stand) at the joint between the distillation adapter and the receiving flask. Be sure all the joints fit snuggly together; otherwise the apparatus will leak and reduce the efficiency of the distillation.

(6) Connect the water hoses to the condenser, with water "in" at the bottom end of the condenser from the tap, and "out" at the top end to the sink. Connect the heating mantle to the variac, and set the variac at approximate values. Check the set-up before finally turning on the rheostat.

(7) Carefully turn the water on, turn on the variac (heat). Only a slow flow of water is required. Adjust the tap to make just a trickle of water coming out of the exit tubing. Do not heat the distilling flask too quickly. The proper rate of distillation is about one or two drops of distillate every second.

(8) As the apparatus is heated, the liquid in the distillation flask begins to bubble; the vapor rises from the liquid and gets into contact with the thermometer bulb, the temperature rises. When the thermometer bulb is completely surrounded by the vapor, the temperature stays fairly constant at the boiling point of the liquid. Distillate should be collected when the thermometer reaches the boiling point temperature. Record the thermometer temperature as the distillate drips into the receiving flask[7].

(9) Continue heating gently until the majority of the liquid has evaporated. **CAUTION: DO NOT LET THE FLASK BOIL DRY**[8]!

(10) When the distillation is completed, unplug the power cord, remove the heating source, and then stop the water supply. After these, disassemble the apparatus in the reverse order from setting up.

2. Simple Distillation of Alcohol

Add 20 mL of industrial alcohol in the distillation flask, set up an apparatus of simple distillation as up mentioned procedures, distill out the alcohol and determine its boiling point.

Notes

[1] Wear safety goggles at all times in the lab.

[2] Be sure all the joints fit snuggly together. Clamp both distillation flask and receiving flask securely but not tight enough to break the glassware.

[3] Heat with a heating mantle or hot water bath. **DO NOT** use an open flame! The liquids and their vapors are flammable.

[4] Remember to add boiling chips or tubes before heating the liquid. **NEVER** add a boiling chip to a liquid which is already hot. If you forget to add a boiling chip before you begin, you must cool the solution before adding one. Boiling chips cannot be re-used since the pores inside these stones become filled with liquid on cooling.

[5] Handle thermometer with care, mercury is very toxic!

[6] Leave the apparatus open to the air at the adapter-receiver end. **NEVER HEAT AN ENTIRELY CLOSED SYSTEM!**

[7] When a simple distillation is used for purification, use a new flask to collect the distillate when the thermometer reaches the boiling point of the desired liquid and the front fraction should be collected and discarded. Stop collecting when the reading of the thermometer changes. The distillate collected in this experiment is 95% ethanol, which has a boiling point of 78 ℃.

[8] **NEVER** distill the distillation flask to dryness as this creates a potential risk of explosion. When the volume remaining in the pot runs down to only 2~3 mL, discontinue heating.

Questions

(1) What is the the definition of boiling point? What is the relationship between pressure and boiling point?

(2) Why is it necessary to add boiling chips to the distillation flask? Can you add a boiling chip to a solvent which is already hot? Can the boiling chips be re-used?

(3) The proper rate of distillation is about one or two drops of distillate every second, why?

(4) Can the liquid with a constant boiling point be considered as a pure substance?

Exp.3 Recrystallization

Objectives

(1) To purify samples of organic compounds which are solids at room temperature.

(2) To learn the principle and the technology of recrystallization.

(3) To be skilled in the operations of hot filtration and vacuum filtration.

Principles

Recrystallization is one of the most common methods used to purify solid organic compounds. The method of purification is based on the principle that the solubility of most solids increases with increased temperature, and that the impurities present will have different solubilities from the desired compounds. The process requires suitable solvent. A suitable solvent is one which readily dissolves the solid (solute) when the solvent is hot but not when it is cold. The procedure involves: selecting a suitable solvent; dissolving the impure material in a minimum amount of hot solvent; filtering the hot solution at its boiling point to remove insoluble impurities when present; allowing the solution to cool and deposit crystals of the compound; filtering the crystals from the solution (called the mother liquor); washing the crystals with a little cold solvent to remove the mother liquor; drying the crystals to remove the last traces of solvent.

Repeated recrystallization will result in an even purer crystalline precipitate.

There are seven major steps in the recrystallization process: selecting solvent; dissolving the sample; decolorizing with activated charcoal; hot filtration; cooling the hot filtrate; separating the crystallized product from mother liquor; washing the filtered product and drying it.

1. Selecting Solvent

If a recrystallization is to be effective, the solvent must be properly selected. A good recrystallization solvent should:

(1) Do not react with the substances to be purified.

(2) Dissolve a moderate quantity of the substance to be purified at an elevated temperature, but only a small quantity at low temperatures.

(3) Dissolve impurities readily at a low temperature or not dissolve them at all.

(4) Be readily removable from the purified product, the solvent must have a relatively low boiling point and evaporate readily.

(5) Be good for producing quality crystals.

(6) Be cheap, low-toxic and safe.

Assume that there is an unknown compound to be purified by crystallization. To find a good recrystallizing solvent, you can follow the following steps.

First, its solubility properties in simple available solvents should be tested.

1) Single-solvent Method

Add about 0.1g of material to 1 mL of cold solvent in a test tube. Shake up to 3 min to help dissolution. If a good portion of the solid dissolves at room temperature, the solvent **IS NOT** suitable. If insoluble, warm the solvent to boiling and some more solvent with a maximum amount of 3 mL can be added at the boiling temperature to make a complete dissolution. If still insoluble, the solvent **IS NOT** suitable. If soluble, cool down the solution to observe the precipitation. If a large amount of crystals can be precipitated from the cold solution, the solvent will probably be suitable for recrystallization.

2) Mixed-solvent Method

If no single solvent is found suitable, then a mixed solvent for recrystallization is in order. For this, two miscible solvents are required, such as ethanol-water, acetone-water, acetic acid-water, ether-methanol, ether-petroleum ether, benzene-petroleum ether, and the like. For mixed-solvent recrystallization, the material should be relatively soluble in one solvent and relatively insoluble in another solvent.

In this experiment, recrystallization from a single solvent (water) will be illustrated.

2. Dissolving the sample

Add crude solid to an Erlenmeyer flask. Add a minimum amount of hot solvent and one boiling chip to the flask and heat the mixture to boiling on a hot plate, while stirring the mixture and boiling gently to dissolve the solid completely. If there are particles of sample still undissolved, then add an additional amount of hot solvent in small increments while continue heating the solution until the entire solid dissolves in only as much as hot or near boiling solvent as is necessary. There may be some solid material that will not dissolve at all. If additional solvent can be added with no appreciable change in the amount of solute present, the particulate matter is probably insoluble impurities. These impurities will be removed by hot filtration. Do not add too much solvent or the solution will not be saturated and the yield of purified product will be reduced.

When using a mixed solvent, for example, the material is soluble in ethanol and insoluble in water. The correct procedure is to dissolve the solid in the minimum amount of boiling alcohol and add warm water dropwise. Each drop will produce a cloudiness which at first clears on mixing. When the warmed solution just fails to clear on shaking, a few drops of alcohol are added; the mixture is re-heated and set aside for crystallization. Any insoluble impurities are best removed by filtering the hot alcohol solution before adding water. When the sample has dissolved, remove the flask from the hot plate and add an additional small volume of solvent (about 20% of the amount required for solution).

3. Decolorizing with Activated Charcoal

If the solution is not the expected color, activated charcoal can be used. Be sure the solution is slightly below the boiling point and add a small amount of activated charcoal (1%~5% by mass of the sample)[1]. An excessive quantity of activated charcoal should be avoided since it may adsorb the product. Heat the mixture to boiling for 5~10 min, stirring occasionally. **NOTE**: If the solution is not cooled somewhat when you add the activated charcoal, it will froth badly and the solution may splash out of the flask!

4. Hot Filtration

This step is optional if there is no visible particulate matter and the solution is the expected color.

The two types of filtration commonly used in organic chemistry experiments are gravity filtration [shown in Fig. 2.5(a)] and vacuum or suction filtration [shown in Fig. 2.5(b)].

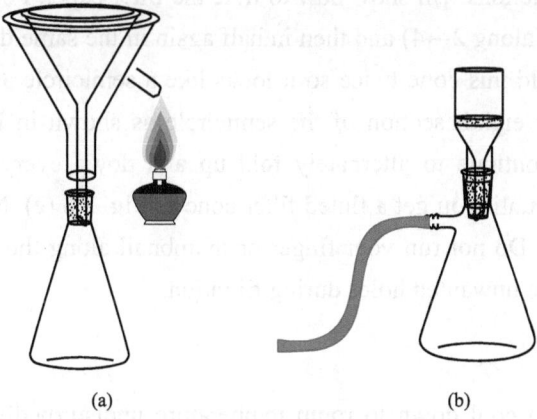

Fig. 2.5 Apparatus for hot filtration

Gravity filtration is generally used for larger-scale operations and usually employs a device for keeping the filter funnel hot in recrystallization. Set up a hot filtration assembly as in Fig. 2.5(a). Fit a fluted filter paper (a fluted filter paper has many indentations and high surface area, which allows for a fast filtration. The fluted filter paper is made as the way shown in Fig. 2.6) into a stemless glass funnel and place the funnel in a Brass hot water funnel, which is placed on top of a clean Erlenmeyer flask. The hot water funnel is filled with hot water beforehand and being heated at the bottom of the side arm by an alcohol burner in the process of hot filtration. Prior to filtering the hot solution, the fluted filter paper, glass funnel and Erlenmeyer flask should be warmed to reduce the risk of crystals separating out on the filter paper.

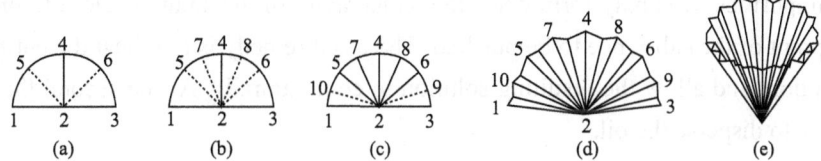

Fig. 2.6 Creating a fluted filter paper (the dotted lines represent locations to crease and fold the filter paper)

Filter the mixture through the fluted filter paper in the hot filtration assembly. Filter as rapidly as possible, returning the flask to the hot plate after each "pour" to keep it hot. Do not fill the filter paper in the funnel more than 3/4 full and pour the mixture into the center of the filter so that the unfiltered mixture does not go over the sides of the filter paper. If the filtrate is not clear of solids, it will have to be heated and filtered again. Use a little hot solvent to wash down any crystals which may be retained on the filter or return the crystals back into the original flask to make a new hot filtration when there are many crystals retained. If only a small to medium amount solution (up to 100 mL) is being filtered, a vacuum filtration using a pre-heated Büchner funnel is preferred. To avoid puncturing the filter paper, two layers of filter paper are suggested in the vacuum filtration. After the suction filtration, transfer the hot filtrate to a clean beaker.

The following instructions will show how to flute the filter paper. Fold paper in half along line 1→3, then in half again (along 2→4) and then in half again in the same direction. You should have a 1/8 section cone. Unfold this cone twice so it looks like a semicircle as Fig. 2.6 (a). Alternately fold up and down every eighth section of the semicircle as shown in Fig. 2.6 (b), (c), and (d). Unfold the paper and continue to alternately fold up and down every sixteenth section of the semicircle. Open the fan until you get a fluted filter cone as Fig. 2.6 (e). Note: To make all creases, fold and press the paper. Do not run your finger or thumbnail along the folds. It may weaken the paper enough to introduce unwanted holes during filtration.

5. Crystallization

Allow the filtrate to cool down to room temperature undisturbedly, and crystallization will start. Generally the solution should not be disturbed as it cools, since this also leads to production of small crystals. Do not make cooling too fast or stir the filtrate when cooling, because these will lead to production of small crystals. Very small crystals can absorb impurities from solution. The formation of crystals larger than about 2 mm should be avoided because some of the solution may become occluded or trapped within the crystals. When overly large crystals begin to form, agitate the solution gently to produce smaller crystals.

When failure of crystallization occurs after the solution has cooled, a seed crystal of the original solid is added to the solution to induce crystallization. Alternatively, crystallization can often be induced by using a glass rod to rub the inside surface of the crystallization vessel.

Occasionally the solute will separate from solution as oil rather than a solid. This type of separation is undesirable for purification of solutes because the oils usually contain significant amounts of impurities. Oils may form when the temperature of the solute is close to or higher than the melting point of the substance to be purified. The usual remedy is to reheat the entire mixture to affect dissolution, and allow the resulting solution to cool naturally. At the second the oil appears, stir vigorously to disperse the oil.

If crystallization fails, other purification ways should be applied instead.

6. Vacuum Filtration

Set up a Büchner funnel in the manner shown in Fig. 2.5(b)[2]. Place a circle of filter paper over the holes in the Büchner funnel, wet it with a little water, and "seat it" by applying the vacuum. Swirl the crystallization flask to suspend the crystals, and pour the contents into the Büchner funnel while applying vacuum. To recover crystals still adhering to the flak, add a small amount of ice-solvent to the flask, swirl and pour this through filter. It is best to turn off the vacuum before adding the wash liquid; let the wash liquid stand for a few seconds and then turn on the vacuum to suck the liquid through. To discontinue the vacuum, always open the stopcock on the safety bottle and then turn off the water vacuum pump, otherwise water may be sucked back into the filter flask.

7. Drying the Crystals

A spatula may be needed to aid in the transfer of the crystals. Draw air through the crystals for a few minutes, then transfer them to a clean, weighed watch glass and let the crystals air dry at room temperature. To speed up drying, slightly damp crystals can be dried more quickly in an oven. Determine the weight and melting point of the dry product. Take the melting point of the impure sample at the same time and compare the value with the melting point of the purified sample.

Materials

Equipment: digital scale, water vacuum pump, hot plate, Erlenmeyer flask, measuring cylinder, beaker, glass rod, spatula, hot water funnel, glass funnel (stemless), vacuum flask, Büchner funnel, alcohol burner, watch glass, weighing paper, rubber tubing (hose), boiling chips, filter paper.

Reagents: impure acetanilide, distilled water, activated charcoal.

Procedures

Place 2 g of crude acetanilide into a 100 mL Erlenmeyer flask and add 40 mL of water into the flask. Heat the mixture on a hot plate under stirring to dissolve the solids. If the solution is near boiling and there are still undissolved solids, continue to add some hot water slowly with stirring until all the solids are dissolved. When the sample has dissolved, remove the flask from the hot plate and add an additional small volume of solvent (about 20% of the minimum amount required for dissolution).

If the solution is not the expected color, cool down the solution to be slightly below the boiling point and add a little (0.2~0.4 g) of activated charcoal and heat the mixture to boiling for 5~10 min. Set up an apparatus for a suction filtration using a preheated Büchner funnel and a warm suction flask. Two layers of filter paper are used to prevent the filter paper from being punctuated. Turn on the water vacuum pump; pour the hot mixture into the Büchner funnel rapidly to make a hot suction filtration[3].

Transfer the filtrate to a clean 250 mL beaker while hot. Make the filtrate stand undisturbed to be cooled down naturally to form crystals.

Set up an apparatus for a suction filtration again. Swirl the crystallization flask to suspend the crystals, and pour the contents into the Büchner funnel while applying vacuum, use a little cold water to wash the solid and suck it dry[4, 5]. Scrape off the crystals on the filter paper onto a watch glass. Air dry the crystals or dry the crystals in an oven which is set at 80~100℃. Weigh the product and record the yield.

Acetanilide has a melting point of 114℃.

Recovery rate/% = crystal quality / crude product quality × 100

Notes

[1] Do not add the activated charcoal to the boiling solution, or the hot solution will bump out.

[2] The filter paper should be smaller in diameter than the base of Büchner funnel, but it must cover all the holes.

[3] Make the hot filtration as rapid as possible, otherwise the crystals will precipitate on the filter paper and in the funnel neck.

[4] Use minimum amount of solvent for washing the crystals to avoid large loss of crystals.

[5] Before turning off vacuum pump in the suction filtration, disconnect the pump and the suction flask first (or open the piston on the safety bottle) to prevent back-flowing of water into the suction flask.

Questions

(1) What major steps are usually involved in the recrystallization process? What are the basic requirements of a good solvent for recrystallization?

(2) Will it be certainly pure when an impure substance is purified after one recrystallization? How to decide if a second recrystallization is needed or not?

(3) You should not use too much or too little solvent in recrystallization, give the reasons? How to control the amount of solvent required for recrystallization?

(4) How to wash crystals in a Büchner funnel?

Exp.4   Extraction

Objectives

(1) To learn the principles and methods of extraction.

(2) To be skilled in the operations of extraction and washing.

(3) To separate a two-component mixture using extraction techniques.

Principles

Extraction is a very important method used for purification and separation of organic compounds, in which the processes of separation, extraction or purification are achieved by using the different solubility or distribution ratio of a solute in two immiscible (or slightly miscible) solvents. The solute to be extracted can be solid, liquid or gas. Extraction used for the purpose of removal a small amount of impurities from a mixture is often referred to as "washing".

The principle of extraction can be described as follows: a solution is assumed to be formed by dissolving organic compound **X** in solvent **A**, and **X** is to be extracted from this solution. Solvent **B**, which has excellent solubility for **X**, is immiscible with solvent **A** and has no chemical reactions with **X** or **A**, can be chosen as a good extraction solvent. Put the solution into a separatory funnel, add solvent **B** and shake the funnel vigorously. After shaking the separatory funnel for a moment, the solution is kept still for several minutes to allow stratification. At this time, as long as the temperature remains unchanged, the ratio of concentrations of compound **X** in the two immiscible phases of solvent **A** and **B** at equilibrium defines a constant, **K**, which is called "distribution coefficient". This relationship is called the distribution law and can be expressed as the following equation:

$$\frac{\text{concentration of X in solvent A}}{\text{concentration of X in solvent B}} = K \text{ (distribution coefficient)}$$

(Note: The distribution law is suitable when solvent **B** is assumed to have no chemical reactions with **X**.)

When extracting compound **X** with a certain amount of solvent **B**, it is highly desirable that the amount of **X** remaining in solvent **A** is minimized. Therefore, one question is developed: which method is more favorable in the extraction, one single extraction with large volume of solvent or multiple extractions with small amounts of solvent each time? This can be explained by the following derivation:

In the first extraction: If V is the volume (mL) of the extracted solution, which is approximately the same as that of solvent **A** (the volume of solute is negligible); W_0 is the total amount (g) of solute (**X**) in the extracted solution; S is the volume (mL) of solvent **B** used in the extraction; W_1 is the amount (g) of solute **X** left in solvent **A** after the first extraction; W_2 is the amount (g) of solute **X** remaining in solvent **A** after the second extraction; W_n is the amount (g) of solute **X** remaining in solvent **A** after n times of extraction.

$W_0 - W_1 =$ the amount (g) of solute **X** extracted in solvent **B** after the first extraction
$W_1 - W_2 =$ the amount (g) of solute **X** extracted in solvent **B** after the second extraction
Then

$$\frac{W_1/V}{(W_0 - W_1)/S} = K$$

Sort and categorize it to give

$$W_1 = \frac{KV}{KV + S} W_0$$

Similarly, we can prove that

$$\frac{W_2/V}{(W_1 - W_2)/S} = K$$

Sort and categorize it to give

$$W_2 = \frac{KV}{KV + S} W_1 = \left(\frac{KV}{KV + S}\right)^2 W_0$$

W_n is the amount (g) of solute **X** left in solvent **A** after the n times of extraction:

$$W_n = \left(\frac{KV}{KV + S}\right)^n W_0$$

It should be noticed that the above equations apply only to solvents that are almost immiscible with solvent **A**.

For example, at 15℃, the partition coefficient of *n*-butyric acid in the hydrated benzene, K, is 1/3. If 4 g of *n*-butyric acid in 100 mL of aqueous solution are extracted with 100 mL of benzene each time, we can deduce from the above equations that the amount of *n*-butyric acid remaining in the aqueous solution after the first, second, third, fourth or fifth extraction would be

$$W_1 = 4 \times \frac{\frac{1}{3} \times 100}{\frac{1}{3} \times 100 + 100} = 4 \times \frac{1}{4} = 1.0 \text{(g)}$$

$W_2 = 4 \times (1/4)^2 = 0.250 \text{ (g)}$ $W_3 = 4 \times (1/4)^3 = 0.0625 \text{ (g)}$
$W_4 = 4 \times (1/4)^4 = 0.016 \text{ (g)}$ $W_5 = 4 \times (1/4)^5 = 0.004 \text{ (g)}$

If 100 mL of benzene is divided into 3 equal portions, each portion is used in one time of extraction of the aqueous solution of *n*-butyric acid mentioned above, the amount of *n*-butyric acid remaining in the aqueous solution after three times of extraction is

$$W_3 = 4 \times \left(\frac{\frac{1}{3} \times 100}{\frac{1}{3} \times 100 + \frac{100}{3}} \right)^3 = 4 \times \left(\frac{1}{2} \right)^3 = 0.5 \text{(g)}$$

Calculation proves that:

(1) It is always more efficient to carry out several extractions using a small volume of solvent each time than to carry out a single extraction using a large volume of solvent.

(2) However, when the total amount of extraction solvent is kept unchanged, the volume of solvent used is reduced with the increase of the number of extraction times. If too little solvent is used at each time, the increased amount of extracted solute is not worth the time and troublesome operation that it requires. Therefore, 3-5 times of extraction is enough in normal cases.

1. Selection of an Extraction Solvent

An ideal extraction solvent should have the following desirable features:

(1) It should be insoluble or slightly soluble with the solvent of the solution being extracted and form no emulsions when being mixed with the solution being extracted.

(2) It should be chemically inert to the extracted substance, other components in the mixture, and the solvent to the solution being extracted.

(3) It should have a favorable distribution coefficient for the substance being extracted.

(4) It should have a low boiling point so that the solvent can be easy to recycle.

(5) It should be reasonably safe to work with and relatively inexpensive.

2. How to Use a Separatory Funnel

A separatory funnel is usually used for the liquid-liquid separation, which is used in an extraction process according to the steps involved below:

(1) Choose an appropriate separatory funnel of suitable size so as to make the total volume of liquid be 1/2~3/4 the capacity of the funnel.

(2) Before use, make sure the stopper on the top and the stopcock at the bottom fit properly and are liquid tight[1]. If the funnel is fitted with a glass stopcock, make sure the plug is greased adequately[2]. If the plug turns smoothly in the stopcock (which should have a clip to help hold the plug in place) then it is ready to test. Shaking the funnel 50% full of water should not cause any leakage at either the stopper or the stopcock.

(3) Wash and dry the separatory funnel. A separatory funnel with vaseline attached on the stopcock should not be placed into an oven for drying. When a separatory funnel with Teflon stopcock is used, the stopcock should not be overtightened because it will warp; and should not be exposed to sudden heat, since Teflon expands much faster than glass.

(4) Support the separatory funnel in a ring on ring stand. Close the stopcock and add the solution to be extracted and the extraction solvent to the funnel.

(5) Insert the stopper, holding the funnel horizontally, shake the funnel two or three times and invert the funnel. Point the barrel away from your face and that of your neighbors[3]. **Open the stopcock to release the pressure**, which may have accumulated inside the funnel (volatile solvents such as ether develop considerable pressure) (Fig. 2.7).

(6) Replace the funnel in the holder (ring on ring stand) and **remove the stopper**[4]. Allow the liquids to stand until the layers have completely separated. Draw the lower layer into a flask or beaker of proper size. When the bottom layer has been removed, close the stopcock and pour off the upper layer from the top of the funnel[5].

3. Liquid Delamination

Sometimes, delamination may be difficult in the extraction process (especially when the material to be extracted is alkaline, the two phases are emulsified and it is difficult to delaminate them). At this time, the causes should be carefully analyzed and appropriate measures should be taken.

(1) If the organic extraction solvent is to have a density close to that of water, it may be hard to delaminate them. In this case, delamination can be accelerated by adding some water-soluble inorganic salts in the solution to increase the density of the water. What's more, the formation of saturated aqueous solution by adding inorganic salts (usually with NaCl) in the water phase can significantly reduce the solubility of organic materials in water and thereby improve the extraction effect greatly. This method is known as "salting-out".

(2) If an emulsification is produced because of the partial miscibility of extraction solvent and water, delamination can be achieved as long as the mixture is allowed to stand still for a long time.

(3) If the delamination is not obvious when particles of suspended solids present at the two-phase interface in the extraction, filter the mixture to solve the problem.

(4) If the emulsification in the extraction is due to an alkali solution or surfactant, lower the

pH of the solution with a little dilute sulfuric acid and shake it gently to achieve delamination.

In addition, some other methods can be used to reduce emulsions. Such as, emulsions can be destroyed by heating up to necessary temperature, or by decreasing the surface tension of the solutions using alcohol compounds, and so on.

Acetic acid is extracted from acetic acid aqueous solution by ether in the following two methods:

(1) One-time extraction of acetic acid with 30 mL ether.

(2) Multiple extraction of acetic acid by ether for 3 times, using 10 mL each time.

Materials

Equipment: separatory funnel, Erlenmeyer flask, base burette.

Reagents: mixed solution of glacial acetic acid and water (glacial acetic acid : water=1 : 19), ether, 0.2 mol/L NaOH, phenolphthalein indicator.

Procedures

1. One-time Extraction

(1) Pipette exactly 10 mL of aqueous solution of acetic acid into a separatory funnel and extract the solution with 30 mL of ether.

(2) Hold the funnel firmly but gently in both hands, keeping the stopper tightly seated with one hand at all times, fitting palm of another hand against the stopcock and cover the funnel piston, shake the solution vigorously, occasionally **unscrew the cork** to release the pressure that may be built up from solvent vapor or evolved gases (Fig. 2.7). Close the stopcock and, holding the funnel horizontally; shake the funnel two or three times. Invert the funnel and release the pressure as before. Repeat this process until opening the stopcock causes no further pressure release. Close the stopcock and shake the funnel 15~20 times.

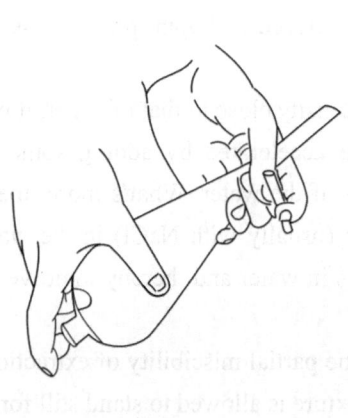 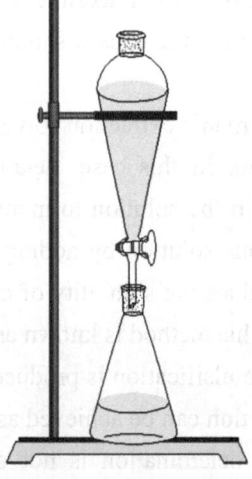

(a) shake and vent   (b) stand and delaminate

Fig. 2.7   Operation of the separatory funnel

(3) Replace the funnel in the holder (ring on ring stand) and **remove the stopper** [Fig. 2.7 (b)]. Allow the liquids to stand until the layers have completely separated. Drain the lower layer into a 50 mL Erlenmeyer flask [see Fig. 2.7 (b)]. Do not draw the liquid through the stopcock too rapidly. Slow the flow carefully as the boundary between the two layers approaches the stopcock. Stop the flow of liquid completely just as the upper layer enters the hole in the stopcock. Pour the upper layer through the neck of the funnel into a second flask. **NEVER discard either layer until you are absolutely certain which is the proper layer to keep**[6].

(4) Add 3~4 drops of phenolphthalein as an indicator, titrate the solution already extracted with 0.2 mol/L NaOH solution, and record the volume of the NaOH solution being used.

(5) Calculate: ①the amount and mass fraction of acetic acid left in the water; ②the amount and mass fraction of acetic acid left in the ether.

2. Multiple Extraction

(1) Pipette exactly 10 mL of aqueous solution of acetic acid into a separatory funnel and extract the solution with 10 mL of ether as above. Separate and set aside the upper, organic layer.

(2) The lower, aqueous layer is returned to the separatory funnel, extracted with 10 mL of ether and the ether solution is separated and set aside.

(3) The lower, aqueous layer in the second extraction is then extracted with 10 mL of ether for the third time and the ether solution is separated.

(4) Titrate the aqueous solution already extracted with 0.2 mol/L NaOH solution.

(5) Calculate: ①the amount and mass fraction of acetic acid left in the water; ②the amount and mass fraction of acetic acid left in the ether.

Compare the effect of the two extraction methods.

Notes

[1] Before extraction, be sure to check the separatory funnel for leaks and check the stopcock for flexibility.

[2] The stopcock plugs can be made from glass or Teflon. A very small amount of grease can be applied to the glass plugs to improve the seal and allow for better movement of the plug. **The grease should be used sparingly, because it will clog up the hole in the plug!**

[3] When you vent the funnel, **NEVER** point the stem of the funnel towards anyone!

[4] **The stopper has to be removed when draining the lower layer**. If the stopper were not removed, a vacuum will build up above the liquid upon draining. This vacuum will reduce the rate of draining and ultimately stop it completely. After some time, the vacuum will suck air in (from the stem) and the phases will mix again.

[5] When separating the liquid phases, drain off the lower layer from the bottom of the funnel, and pour the upper layer out the top.

[6] **NEVER throw any layer away, until you are absolutely sure that you isolated your final product.** It is easier to isolate it from a small amount of solution than from the waste container.

Questions

(1) What are the factors affecting on the extraction efficiency of the extraction? How to choose an appropriate solvent for an extraction?

(2) What is the purpose of using a separatory funnel? What are the rules for using separatory funnels?

(3) When making the liquid separation of two liquids which are incompatible in a separatory funnel, where is the lower layer released from, and where should the upper layer be transferred from?

Exp.5　Vacuum Distillation

Objectives

(1) To learn the basic principles and applications of vacuum distillation.

(2) To be familiar with the main equipment of vacuum distillation.

(3) To distill a high boiling point liquid that would decompose at its normal boiling point by performing a vacuum distillation.

Principles

Vacuum distillation is one common method for the purification of organic compounds, especially some high boiling point liquids, which will be decomposed, oxidized or polymerized under a normal distillation.

The distillation process that occurs below one atmosphere and in vacuum is called as vacuum distillation. This distillation method works on the principle that boiling occurs when the vapor pressure of a liquid reaches or exceeds the ambient pressure, therefore, the boiling point of the liquid changes with the change of the external pressure. If the external pressure above a substance is reduced by mean of a mechanical pump, the boiling point will be reduced. Low pressure permits vaporization at low temperatures.

Fig. 2.8 illustrates that vapor pressure increases exponentially with temperature. If we follow the temperature dependence of vapor pressure for a substance left out in an open container, we would find that the equilibrium vapor pressure of the liquid would increase until it reached 1 atm (101.3 kPa, 760 mmHg) or 101325 Pa. At this temperature and pressure, the liquid would begin to boil and would continue to do so until all of the liquid distilled or boiled off.

It illustrates an important principle that decreasing the external pressure would decrease the boiling point of the liquid. In addition, many compounds cannot be distilled at atmospheric pressure because their boiling points are so high. At their normal boiling points, the compounds decompose. Some of these materials can be distilled under reduced pressure because the required temperature to boil the substance can be lowered significantly.

A nomograph (Fig. 2.9) is a useful device that can be used to estimate the boiling point of a liquid at reduced pressure under any conditions providing either the normal boiling point or the boiling point at a given pressure. To use the nomograph given the normal boiling point, simply place a straight edge at the temperature in the central column of the nomograph (*B*). Rotating the straight edge about this temperature will afford the expected boiling point for any number of external pressures. Simply read the temperature and the corresponding pressure from where the straight edge intersects the column *A* and *C*.

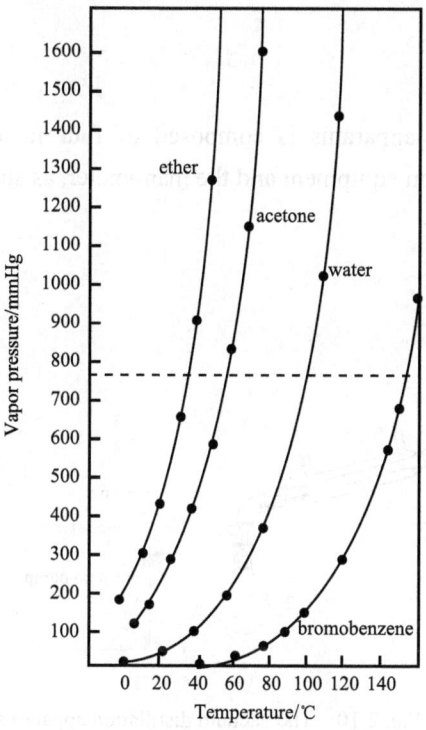

Fig. 2.8 Vapor pressures of liquids versus temperature

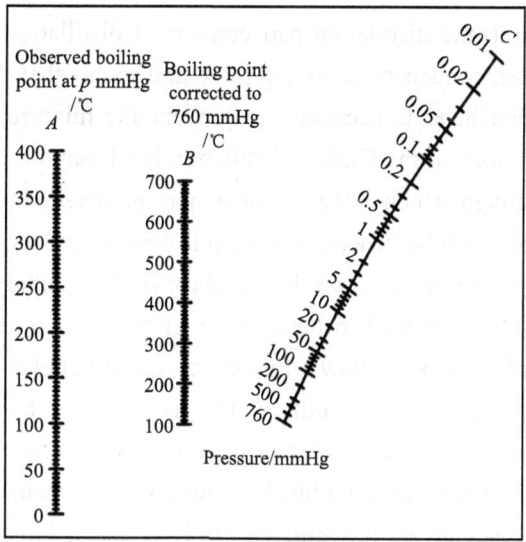

Fig. 2.9 A nomograph used to estimate boiling points at reduced pressures (1 mmHg≈133 Pa)

For example, ethyl salicylate has a normal boiling point of 234℃, and its boiling point at a reduced pressure of 15 mmHg can be found by using the nomograph in Fig. 2.9. Use the nomograph in Fig. 2.9, place a straight edge at 234℃ in column B, rotate the straight edge about 234 to intersect column C at 15 mmHg, read the temperature from where the straight edge intersects column A, and the expected boiling point of ethyl salicylate at 15 mmHg is found to be at about 113℃.

Apparatus

The vacuum distillation apparatus is composed of four major parts: the distillation unit, vacuum pump, safety protection equipment and the manometer, as shown in Fig. 2.10.

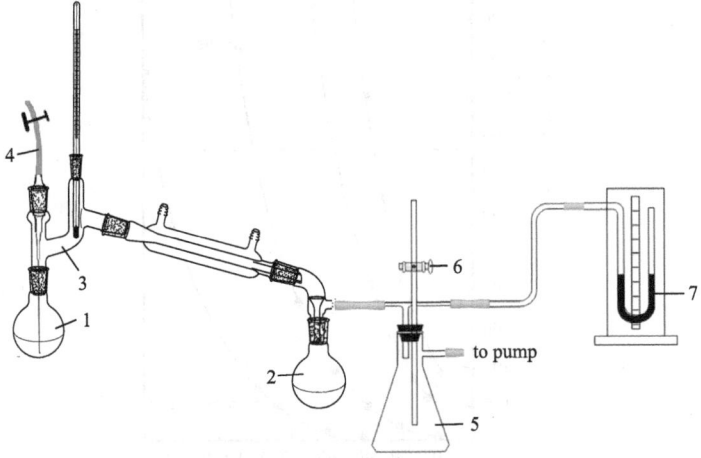

Fig. 2.10 The vacuum distillation apparatus

1. distillation flask; 2. receiving flask; 3. Claisen distillation head; 4. capillary; 5. vacuum trap; 6. stopcock; 7. manometer

(1) The distillation unit: the distillation part consists of distillation flask, Claisen distillation head, capillary, thermometer, condenser, receiving flask, and some other components.

The Claisen distillation head is necessary to prevent the mixture from splattering into the condenser. The right-hand neck of the Claisen distillation head may be fitted with a screw-capped adapter (see Fig. 2.10) through which a thermometer may be inserted, while the left-hand socket accommodates a stout capillary tube, whose upper end is closed by a short piece of pressure tubing and a screw clip to regulate the amount of air passing through the capillary.

At reduced pressures, liquids tend to bump. One of the most reliable ways used to control bumping in a vacuum distillation is to allow a fine stream of air bubbles drawn through a very fine capillary into the boiling liquid. The capillary tube is prepared by drawing out a piece of thick-walled capillary tube of 3~5 mm external bore and then the length of the fine drawn-out capillary is adjusted so that it reaches to within 1~2 mm of the bottom of the flask. If the organic compound is likely to oxidize with air, the capillary can be connected to a nitrogen supply instead.

When running a vacuum distillation, it's good to use a pig receiver. The pig receiver rotates to allow the collection of three fractions into three separate receiver flasks, without having to stop the distillation. Attach the receiving flasks to the pig receiver and the pig receiver to the condenser. Use a water condenser for distillation below 140℃ and replace it by an air condenser for above 140℃.

(2) Safety protection equipment: placing some traps between an apparatus and vacuum source establishes a safe and controlled working system. A vacuum trap controls the vacuum applied to the distillation system by turning a stopcock at the top and prevents vapors from being emitted back into the system; a cold trap prevents volatiles from entering the oil pump; and absorption towers are installed between the cold trap and the oil pump in order to prevent acidic or water vapors from getting back into the oil pump. A mixture with a moderate amount of solvent and cooling agent such as ice-water, ice-salts or dry ice-acetone can be selected as needed and used to create a cold trap.

Three absorption towers (also known as drying towers) are usually set up for absorbing different vapors: the first tower is loaded with anhydrous $CaCl_2$ or silica gel for absorbing water vapors; the second is packed with granular NaOH for absorption of acidic gases; and the third is filled with paraffin for absorption of hydrocarbon gases.

Cold trap and absorption towers are shown in Fig. 2.11.

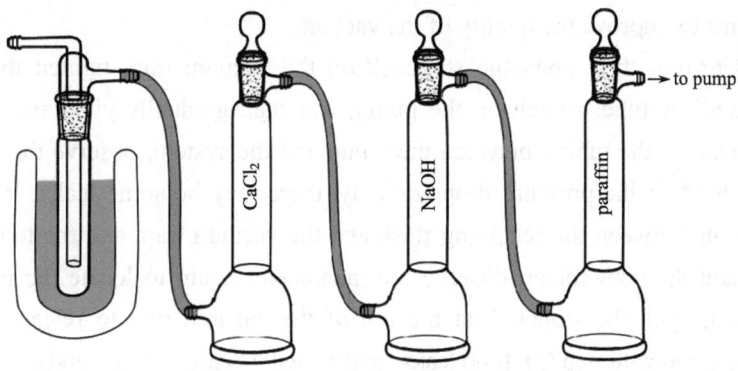

Fig. 2.11 Cold trap and absorption towers

(3) Pumping equipment: water or oil pumps are frequently used in the laboratories. A water pump can produce pressure range at 1.333~100 kPa (10~760 mmHg); an oil pump can generate a vacuum pressure down to 0.133~133.3 Pa (0.001~1 mmHg); a diffusion pump can provide a pressure below 0.133 Pa (10~3 mmHg).

When using an oil pump in a vacuum system, a cold trap should be added to the vacuum pumping system either to remove volatile impurities (e.g. water, solvents, acidic or alkaline compounds) from the gas stream or to prevent pump backstreaming. These conditions can cause loss of efficiency or damage when introduced into or emanating from the vacuum pumping system. Additionally, the oil in an oil pump should be changed on a regular basis.

In a typical pump down sequence, a water pump would be used for a "rough" vacuum to remove most of the volatile fraction from the flask. Then an oil pump would be used to bring the pressure down to a "second high" vacuum, a diffusion pump would be used to bring the pressure further down to a "high" vacuum.

(4) Manometer: a closed-end U-shaped manometer is frequently used in the laboratory to measure the pressure in a vacuum system, and the difference between the heights of the columns of mercury in the manometer is the pressure of the system. For example, in a closed-end manometer, if the mercury level was 690 mm higher on the closed end than on the gas side, the gas pressure equals 690 mmHg. Great care should be exercised when using this manometer: if air is allowed to enter the exhausted apparatus rapidly, the mercury may rise to the top of the closed end with sufficient velocity to break it. Therefore, when the pumping operation is completed, be careful to unscrew the cock on the vacuum trap to let the gas get into the system very slowly.

Procedures

1. Operating Method

(1) Install the apparatus as shown in Fig. 2.10, after all connections are made, check that the entire system is airtight[1]. Heavy-walled glassware as well as thick-walled tubing should be used[2]. Normal tubings will collapse when a vacuum is applied! Paint a little vacuum grease on the ground-glass joints to improve the quality of the vacuum.

(2) Leak detection: first open the stopcock on the vacuum trap, tighten the capillary tube clamp on the capillary tube, switch on the pump, and then gradually close the stopcock on the vacuum trap. Pinch off the tubing between the pump and the system, observe the manometer. The pressure should hold. If the pressure drops quickly, there may be some leaks. If there are leaks, pinch off the tubing between the receiving flask and the vacuum trap, and the tubing between the receiving flask and the manometer, observe the manometer again to locate the possible leaks. If leaks are detected, open the stopcock at the top of the vacuum trap to release pressure, repair leak(s), and begin again with step (2). If no leaks are detected, release pressure and continue to step (3).

(3) Fill the distillation flask (no more than 1/2 full) with liquid to be distilled[3], open the vacuum pump, close the stopcock of the vacuum trap gradually, and tighten or loosen the screw clamp on the capillary tube until a small steady stream of bubbles continuously emerges from the capillary tube. Start running cooling water through the condenser and heat the distillation flask slowly[4]. Collect the low-boiling fraction with one receiving flask, and when a stable reading is observed in the thermometer, rotate the multi-outlet distillation receiver rapidly to receive the desired fraction with a new receiving flask. Then adjust the heat until the distillate drops at a regular rate of 1~2 drops per second.

(4) After the distillation is complete, perform operations in the following order: ①remove the heat source, after the distillation flask is cooled to room temperature, unscrew the clamp on the capillary tube[5]; ②slowly open the stopcock on the vacuum trap to release the vacuum and allow the mercury column in the manometer restore slowly; ③switch off the pump, wait until the pressure inside and outside the system is equalized and then close the stopcock on the vacuum trap[6]; ④disassemble the apparatus.

2. Purification of Crude Ethyl Acetoacetate

In a 50 mL round-bottom flask, 20 mL of crude ethyl acetoacetate is added and distilled under reduced pressure.

The boiling point of ethyl acetoacetate is shown in Table 2.1.

Table 2.1 The boiling point of ethyl acetoacetate

Pressure/mmHg	760	80	60	40	30	20	18	14	12
b.p. /℃	181	100	97	92	88	82	78	74	71

Notes

[1] Safety goggles must be worn during the vacuum distillation, especially when reading the temperatures.

[2] All glassware used must be thick-walled, flat-bottomed glassware such as Erlenmeyer flasks, flat-bottomed flasks and the like cannot be used.

[3] Before an oil pump is applied for a vacuum distillation, a simple distillation or a vacuum distillation using a water pump should be performed to remove the most volatile fraction of the mixture to be distilled.

[4] The pressure of the system must be stabilized before the heating of the distillation flask. Otherwise, the temperature of the mixture may exceed the boiling point at the desired vacuum before the desired vacuum is reached, and bumping may occur.

[5] The heat source should be removed before stopping the distillation; the distillation flask should be cooled to room temperature before slowly opening the stopcock on the vacuum trap. Otherwise, a large amount of air is possible to be sucked into the flask at a high temperature, and the remaining residual liquid will be easily oxidized, which will lead to explosions.

[6] Before turning off the pump, be sure to connect the system with atmosphere first (by opening the stopcock on the vacuum trap), or the oil or water in the pump will be sucked backwards.

Questions

(1) What kind of compounds need to be purified by vacuum distillation?

(2) What precautions should be taken when using a water pump for a vacuum distillation?

(3) It is necessary to reduce the pressure before heating when performing a vacuum distillation, why?

(4) How to stop the vacuum distillation after the desired chemicals are collected? Why?

Exp.6   Steam Distillation

Objectives

(1) To learn the principles and applications of steam distillation.

(2) To gain participant understanding of the equipment and operations of steam distillation.

(3) To isolate temperature sensitive materials using a steam distillation set-up.

Principles

Steam distillation is a special type of distillation in which steam is introduced into the distillation apparatus. The water vapor carries small amounts of the vaporized compounds to the condensation flask, where the condensed liquid phase separates, allowing for easy collection. This process effectively allows for distillation at lower temperatures for purification of temperature sensitive organic compounds, reducing the deterioration of the desired products.

Dalton's law of partial pressure states that the total pressure of a gaseous mixture is equal to the sum of partial pressures for all its components. As a result, the total pressure of a solution, which is formed by dissolving organic compound A in water, can be given by: $p_{total} = p_{H_2O} + p_A$. When the total vapor pressure (p_{total}) equals the ambient atmospheric pressure, the solution boils. This means that such a solution would boil at a temperature lower than the boiling point of pure water (100℃) and lower than that of organic compound A. Therefore, the steam distillation can allow for distillation of organic compounds at lower temperatures.

Theoretical calculation on the composition of the distillate in a steam distillation is fully explained as follows: assume that the two components are ideal gases, and take an introductory look at the Ideal Gas Law:

$$pV = nRT = WRT/M$$

The composition of the mixture is calculated with the following equation:

$$W_A / W_{H_2O} = M_A p_A / M_{H_2O} p_{H_2O}$$

For example, a mixture of benzaldehyde (b.p. 178℃) and water at normal pressure boils at 97.9℃. In this case,

$$p_{H_2O} = 703.5 \text{ mmHg}$$

$$p_{C_6H_5CHO} = 760 - 703.5 = 56.5 \text{ (mmHg)}$$

$$M_{C_6H_5CHO} = 106 \text{ g/mol} \qquad M_{H_2O} = 18 \text{ g/mol}$$

Substitute these results into the above equation, and the composition of the distillate is obtained as follows:

$$\frac{W_{C_6H_5CHO}}{W_{H_2O}} = \frac{106 \times 56.5}{18 \times 703.5} = 0.473 \text{ (g)}$$

That is, 1 g of water is theoretically required to distill 0.473 g of C_6H_5CHO by a steam distillation. If 10 mL of C_6H_5CHO (ρ=1.041 g/mL) is distilled by means of steam distillation, the volume of water contained in the distillate can be derived as follows:

$$10 \times 1.041 / 0.473 = 10.41 / 0.473 = 22 \text{(mL)}$$

Attention should be paid to the point that these values, which are derived theoretically, are approximate. In a real experiment, the volume of water in the distillate often exceeds the calculated value because a considerable part of the steam leaves the distillation flask before getting full contact with the material to be distilled; what's more, benzaldehyde is slightly soluble in water.

A substance which can be purified by steam distillation must exhibit the following features:
(1) It must be insoluble or slightly insoluble in water.
(2) There should be no chemical changes when it is boiled with water.
(3) Its vapor pressure around 100℃ must be above 10 mmHg (1.33 kPa).

Steam distillation can be effectively used in the following cases:
(1) A liquid substance is isolated from a mixture containing a large portion of solids.
(2) A liquid substance is separated from a mixture containing tarry materials.
(3) A high-boiling point organic liquid, which will decompose under normal distillation conditions, is to be purified.

Materials

Equipment: water vapor generator, round-bottomed flask, Claisen distillation head, straight condenser, Erlenmeyer flask, measuring cylinder, separatory funnel, beaker, receiver adapter.

Reagents: benzaldehyde, anhydrous $MgSO_4$.

Physical properties of benzaldehyde are shown in Table 2.2.

Table 2.2 Physical properties of benzaldehyde

Reagent	M_r	m.p./℃	b.p./℃	d_4^{20}	S_{H_2O}
benzaldehyde	106	−26	179	1.04~1.046	lightly soluble

Apparatus

A set up for steam distillation commonly used in the laboratories is shown in Fig. 2.12, which includes two parts: steam generator and distillation unit.

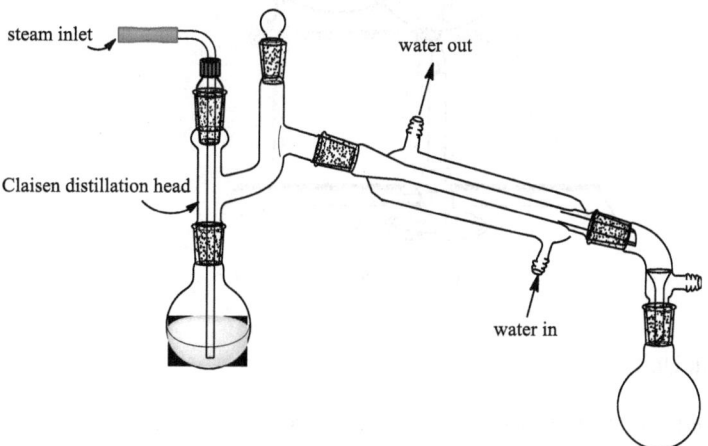

Fig. 2.12 Apparatus of steam distillation

1. Steam Generator

Steam generator can be made of metal, or be replaced with a short-neck flask. The apparatus is installed as Fig. 2.13. A long glass tube, which has an internal diameter of about 5 mm and is 1 m in length, is inserted into the steam generator as a safety tube. The bottom of the safety tube should be well below the water surface in the steam generator, but not touching the bottom of the flask. The steam generator is then connected to a T-shaped tube. One end of the T-shaped tube is connected to a short length of rubber tubing, which is fitted with a screw clamp; the other end of the T-shaped tube is attached to the distillation flask with a bent inlet tube to pass steam into the distillation flask. The length of rubber tubing connecting the inlet pipe should be as short as possible to minimize the condensation of water vapor. The additional end of the T-shaped tube is used to drain the condensed water, and to protect the steam system from over pressure. If steam pressure continues to rise, the safety clamp installed on the T-shaped tube will be screwed off to ensure the steam generator open with atmosphere.

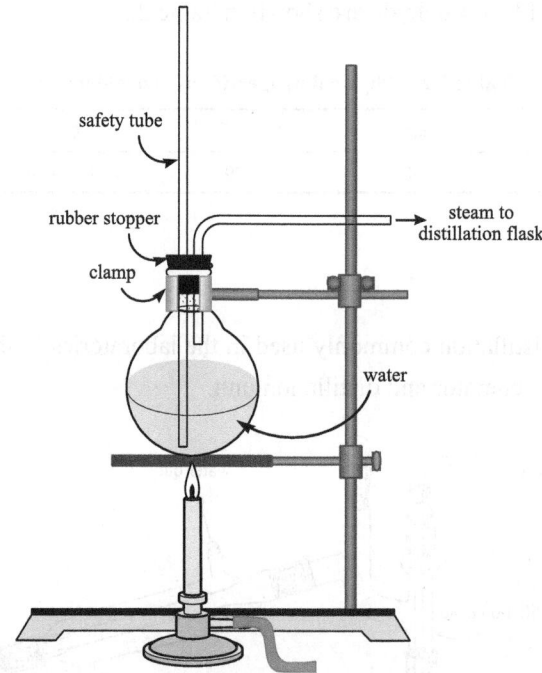

Fig. 2.13 Steam generator

2. Distillation Unit

A long-necked round-bottomed flask is often used in the distillation flask, and the volume of the liquid to be distilled cannot exceed 1/3 of its capacity. The long-necked steam distillation flask is inclined at an angle so as to prevent the solution in the flask from being splashed into the entrance of the outlet tube and thus being blown over mechanically into the condenser. One end of the steam inlet tube is bent to face the middle bottom of the distillation flask, and be 8~10 mm away from the bottom. Or a short-necked round-bottomed flask with a Claisen distillation head can be applied to help stop solution in the flask from splashing over into the distillation receiver (shown in Fig. 2.12).

The steam generator is filled with hot water to 3/4 capacity. After the whole apparatus is checked to ensure no leaks, loosen the clamp on the T-shaped tube, and heat the steam generator to boil the water. When there is a large amount of steam emitted from the T-shaped tube, tighten the clamp to lead the steam into the distillation flask, and then start the distillation.

Detect whether the steam flowing is smooth by observing the height of water column in the safety tube. When a clogging is indicated by an unreasonable high water column in the safety tube, unscrew the clamp on the T-tube, remove the heat source, detach the equipment and then check and repair the equipment to prevent danger.

Keep the distillate dropping at a regular rate of 2~3 drops per second.

When no more oily droplets are noticed coming over with the steam, unscrew the clamp on the T-tube, remove the heat source and stop the distillation.

Procedures

(1) Connect the device as shown in Fig. 2.12[1].

(2) Add 15 mL of benzaldehyde in the round-bottom flask.

(3) Heat to distill: before heating, keep the clamp on the T-shaped tube open. Heat the steam generator until steam is let off the tube, and then close the clamp to make the steam introduced into the round-bottomed flask. If necessary, the round-bottom flask may be heated to accelerate the distillation so as not to avoid condensation of the water steam[2].

(4) When the distillate contains no obvious oil beads, unscrew the clamp on the T-shaped tube, remove the heat source, and stop the distillation (to prevent reverse flow)[3].

(5) After the water in the distillate is separated off with a separatory funnel, dry the distillate with a desiccant. The desiccant is then filtered off to give the product. Weigh the product and calculate its yield.

Notes

[1] Ensure proper and tight installation of the apparatus.

[2] Adjust heat to keep the distillate dropping at a regular rate of 2~3 drops per second. Always pay attention to the safety tube.

[3] Before removing the heat source, make sure to open the clamp on the T-shaped tube to avoid reverse flow.

Questions

(1) When a steam distillation is performed, why should the end of the steam inlet pipe be inserted close to the bottom of the vessel?

(2) What are the items to be checked frequently during steam distillation? How to solve the problem when the water in the safety pipe rises too high?

Exp.7　Fractional Distillation

Simple distillation is not effective in separating closely boiling components of a mixture. Fractional distillation is a technique used to separate miscible liquids that have boiling point difference of less than 25 ℃, with which it is even possible to separate mixtures with boiling points differing only by 1 to 2 ℃.

Objectives

(1) To understand the principle of fractional distillation.

(2) To obtain pure components from a mixture of organic compounds using fractional distillation.

Principles

Fractional distillation involves heating the mixture and partial condensation of the vapors along a fractionating column, which is set up such that components with lower boiling points pass through the column and are collected earlier than components with higher boiling points. Fractional distillation is essentially equivalent to multiple simple distillations in a single run. In a typical fractional distillation, a liquid mixture is heated in the distilling flask, and the resulting vapor rises up the fractionating column, and condenses in the column. The condensed liquid trickling down the column is constantly reboiled by up-coming vapour. As each reboiling makes the condensed liquid trickling down richer and richer in the less volatile component, by the time the liquid drips back into the flask, it will be very rich in less volatile component. At the same time, each reboiling makes the moving-up vapor richer in the more volatile component. By the time the vapor is condensed and collected in the receiving flask, the condensate is enriched in the more volatile component. These successions of boiling-condensing-reboiling operations result in the separation of the components of the mixture. The units for measuring how many times the vapor goes through this cycle of condensation and re-evaporation are called "number of theoretical plates".

It should be noted that azeotropes presenting at specific concentrations at a constant temperature usually distill at a constant boiling temperature and cannot be separated by simple or fractional distillation. This happens because when an azeotrope is boiled, the vapour has the same proportions of constituents as the unboiled mixture. For example, an ethanol-water mixture on fractional distillation yields a solution containing approximately 95.5% by volume of ethanol. Once this composition has been achieved, the liquid and vapour have the same composition, and no further separation occurs.

There are five parts containing in the apparatus of fractional distillation: heat source, round-bottomed flask, fractionating column, condenser and receiving flask. The apparatus of fractional distillation is essentially the same as simple distillation except that a fractionating column is fitted in between the distilling head and the distillation flask (shown in Fig. 2.14). The efficiency of fractional distillation depends on the use of the fractionating column.

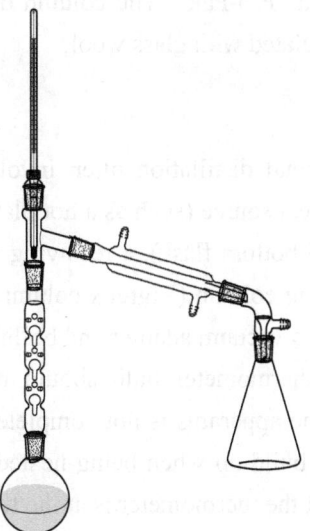

Fig. 2.14 Fractional distillation apparatus

There are many types of fractionating columns that are used in fractional distillation. There are spherical columns that are open, columns with glass indentations called Vigruex columns, and columns which are loosely packed with glass, metal or ceramic material. The spherical columns are poor in separation efficiency; a Vigruex column has lower separation efficiency than a packed column with the same length and is suitable for the separation of small quantities of liquids with large differences in boiling points; and a packed column is suitable for separating liquids with small differences in boiling points for its high separation efficiency. They are all similar in that the surface area, which contacts the distilling vapor, is increased.

The efficiency of a column depends upon column length and the packing. Excellent efficiency can be achieved by filling the column with packing material with large surface and sufficient length of the column. Increasing the surface area over which the vapor must travel increases the purity of the distillate because the number of times that the liquid is distilled is effectively increased. The larger the surface area contacted by the vapor, the more efficient the column is in separating the components.

Different packing materials have different efficiencies. For example, short pieces of glass tubing (6~20 mm in length) can be used as packing materials with high stability but low efficiency; metal pieces are highly efficient as packing materials but they can react with some organic compounds (such as alkyl halides). In packing a column, it is important to pack it loosely enough to leave some free space for the flow of liquid and vapor, but not too loosely or its efficiency will suffer.

The fractionating column is often insulated to keep the temperature of the column nearly constant. If the temperature of the column fluctuates widely, it is difficult to maintain a slow, constant distillation rate. Therefore, a good choice is a vacuum jacketed glass column packed with punched metal pieces, trade-named "Pro-Pak". The column doesn't have to be vacuum jacketed, but in this case must at least be insulated with glass wool.

Apparatus

A laboratory setup for fractional distillation often involves use of the following pieces of apparatus (shown in Fig. 2.14): a heat source (such as a hot plate with a bath or a heating mantle), a distillation flask (typically a round-bottom flask), a receiving flask (often a round-bottom flask or an Erlenmeyer flask), a fractionating column (Vigreux column), a distillation head, a thermometer and adapter if needed, a condenser, a vacuum adapter and boiling chips.

Notice that the top of the thermometer bulb should be just below the side arm of the distillation flask. Make sure that the apparatus is not completely sealed. There has to be a vent in the system otherwise the pressure build-up when being heated will blow the apparatus apart. The mixture is heated at such a rate that the thermometer is at the temperature of the boiling point of the more volatile component.

Procedures

1. General Method of Fractional Distillation

(1) Take the mixture to be separated in a round-bottomed flask[1]. Add 2~3 boiling chips in the flask to avoid bumping[2]. As shown in Fig. 2.14, fit the round-bottomed flask with an adapter and attach a fractionating column, introduce a thermometer, connect a water condenser to the side tube of the adapter[3].

(2) Circulate water from the lower end of the condenser, prior to heating the round-bottomed flask. Gently heat the flask until the liquid begins to boil. Lower the heating rate and heat slowly at a constant rate to permit a good separation. The collection rate of the distillate is better at about 1 drop per 2~3 s[4].

(3) Collect the distillate in a receiving flask. Record the temperature when distillate begins to collect. Continue boiling the liquid until the temperature either rises or drops quickly, record the temperature when the distillate stops to collect. Increase the heat input and change to the second receiving flask. When the liquid boils again, collect the second fraction and record the temperatures when you start and finish collecting this fraction.

(4) Continue collecting the additional fractions until approximately one to two mini liters of liquid remain in the distillation flask. At this point remove the heat source and discontinue the distillation. When the distillation flask is cool, detach the apparatus.

2. Fractional Distillation of an Ethanol-water Mixture

(1) Add 60 mL of 60% aqueous solution of ethanol and 2 boiling chips in a 100 mL round-bottomed flask; set up the apparatus and turn on the circulating tap water.

(2) Gently heat the ethanol solution to boiling and adjust the rate of heating so that the vapor in the fractionating column ascends slowly to the top of the column.

(3) When the temperature reading on the thermometer registers 78℃, start to collect the distillate and adjust the heat so that the distillate drops at the rate of 1 drop per 2～3 s.

(4) Continue boiling the liquid under a constant external condition and then stop heating until the thermometer reading continues to drop.

(5) Record the exact temperature range and volume for each fraction you collected.

(6) Determine alcohol concentration of the distillate by measuring its density using an alcohol hydrometer.

Notes

[1] Remember that the distillation flask should only be 2/3 full of the liquid to be distilled.

[2] A boiling chip must be added to the cold liquid to ensure smooth distillation and prevent "bumping". **NEVER** add a boiling chip to the heated liquid as you will run the risk of releasing large amounts of vapour at once. Since the pores of the boiling chip fill with liquid as soon as boiling ceases the stone cannot be reused.

[3] The fractionating column could be insulated with a material such as glass wool around the outside of the column to reduce heat loss, thereby to keep the temperature of the column nearly constant.

[4] To achieve optimal separation, the distillation must be done at a slow steady rate (no more than one drop every 2～3 s).

Questions

(1) What is the difference between fractional and simple distillation? If a mixture of two liquids with very close boiling points can be purified by a fractional distillation?

(2) The efficiency of the fractional distillation will be significantly reduced if the liquid mixture is heated too fast, why?

Exp.8 Column Chromatography

Chromatography, a group of methods for separating complex mixtures as well as for purifying and identifying chemicals, is one of the most important techniques in organic analysis. It is particularly suitable for the separation, purification, and identification of small quantities of substances with much better resolution than extraction, distillation, fractional distillation and recrystallization.

Chromatography is a physical method of separation in which the components to be separated are distributed between two phases, one of which is stationary while the other is mobile. The mobile phase moves through the stationary phase by capillary suction or by gravity. The separation process occurs because the components of mixture have different affinities for the two phases and thus move through the system at different rates, a component with high affinity for the mobile phase moves relatively quickly through the chromatographic system, whereas one with a high affinity for the solid phase moves more slowly.

There are many types of chromatography. Depending on the sorbents used as the phase, it is divided into partition chromatography (a liquid on the surface of an inert hard support), adsorption chromatography (on a very porous sorbent), and ion-exchange chromatography (on ion-exchange resins) and gel permeation chromatography (on macroporous inert sorbents). Depending on the type of mobile phase, chromatography is divided into gas, liquid chromatography and supercritical fluid chromatography. According to the shape of the solid-phase matrix, liquid chromatographic methods can be roughly divided into column and planar chromatography; and the latter includes thin layer chromatography (TLC) and paper chromatography (PC).

Objectives

(1) To learn the principles and applications of column chromatography.

(2) To be skilled in the operations for separating organic compounds using column chromatography.

Principles

Column chromatography is one of the most useful methods for the separation and purification of both solids and liquids. It is a solid-liquid technique in which the stationary phase is a solid and mobile phase is a liquid. In a typical column, the stationary phase, a solid adsorbent, normally silica gel (SiO_2) or alumina (Al_2O_3) is placed in a vertical glass column. The sample to be purified is added to the top of the column. Hereafter, the mobile phase, a solvent, is added to the top of the column as an eluting solvent and flows down through the column by either gravity or external pressure (flash chromatography). Separation of compounds is achieved through the varying absorption on and interaction between the stationary and mobile phases. The individual components are retained by the stationary phase differently and separate from each other while they are running at different speeds through the column with the eluent. At the end of the column they elute one at a time. During the entire chromatography process the eluent is collected in a series of fractions.

Reagents

A mixture of methyl orange and methylene blue (a solution prepared by dissolving 1 mg of methyl orange and 5 mg of methylene blue in 2.2 mL of 95% ethanol), neutral alumina.

Apparatus

The apparatus for column chromatography is shown in Fig. 2.15.

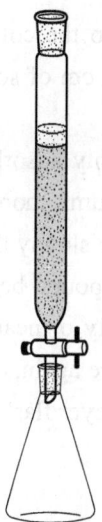

Fig. 2.15 Apparatus for column chromatography

Procedures

1. Packing the Column

Find a clean, empty and dry column of suitable size. Hold the column to a stand in a vertical position. Place a small piece of cotton or glass wool into the column using a long glass rod. Pour a little sea sand into the column (approx. 0.5 cm in height)[1]. Tap on the side of the column with a rubber stopper or tubing to help the sand settle uniformly, and then close the stopcock. Fill the column with 95% ethanol to a height of 3/4 the column, and open and adjust the stopcock to control the eluent drop at a constant flow rate of 1 drop per second. Add neutral alumina to the column slowly using a long-necked funnel. Tap the side of the column gently to help the adsorbent packed evenly and firmly. Fill 3/4 of the column with alumina. Add about a 0.5 cm thick layer of sand to the top of the neutral alumina in the column to create a flat surface[1]. Drain the excess solvent from the column until its level is about 1 cm above the top surface of the sea sand. Close the stopcock.

2. Loading Sample

Apply 2 mL of the sample solution to the top of the column using a long pasteur pipette. Use a clean pasteur pipette to rinse the sides of the column with the minimum amount of eluent. Open the stopcock and drain the solution on the sand until it reaches the top lever of the alumina. Close the stopcock, stopper the column and let it stand for 2 min.

3. Eluting

Keeping the stopcock closed, fill the column with 10 mL of 95% ethanol through a dropping funnel loaded on top of the column. Turn on the stopcock to start the chromatographic elution by collecting the eluate with Erlenmeyer flasks, keeping a rate of flow of approximately 1 drop every second. Add an additional few of ethanol to the column to prevent the alumina from becoming dry[2]. Top up the solvent when more than 2 cm of solvent still remains above the protective sand layer.

Polar compounds usually are most firmly adsorbed and stay on the column longer, whereas less polar compounds move through the column more quickly. In this experiment, the more polar compound is methyl orange and travels more slowly than methylene blue. Collect uncolored eluent in the waste, but as soon as the colored compound begins to emerge, collect this in an Erlenmeyer flask. When the first dye is either completely or nearly emerged from the column, add 10 mL of water carefully to the top of the column. Once again, collect clear eluent in the waste[3]. The second dye should be collected in a separate Erlenmeyer flask.

Notes

[1] Try to pack the column as evenly as possible; air bubbles, cracks, and channels will lead to a poor separation; a flat surface is also needed for efficient distribution of a fluid flow over the adsorbent.

[2] The adsorbent should **NEVER** go dry until the procedure is complete. Otherwise, bubbles or cracks can be generated in the column and a poor separation will be resulted in.

[3] For every column there is an optimal flow rate. Adjust the stopcock to control the flow. If the flow rate is too slow, diffusion processes will lead to band widening and sometimes decomposition of the sample may occur. If it is too fast, there is not enough time for equilibration and the compound will be forced down the column, leaving a long tail behind.

Questions

(1) In the experiment, first a less polar eluting solvent is used and then some eluting solvents with increasing polarity are used. Can this order of eluting be reversed?

(2) Why should the level of the liquid be adjusted just above the top surface of the adsorbent before the sample is added?

Exp.9 Thin Layer Chromatography

Objectives

(1) To learn the principles and applications of thin layer chromatography.

(2) To be skilled in the operations for separating organic compounds using thin layer chromatography.

Principle

Thin-layer chromatography (TLC), as a distribution chromatography, is a quick, inexpensive microscale technique that can be used to separate and qualitatively analyze the components in a mixture, to monitor the progress of a reaction. The most typical method for TLC is to uniformly coat a thin layer of adsorbent (known as the stationary phase) on a glass plate to make a TLC plate, thereafter use a capillary to dot the sample solution on the plate, and then develop in a closed container containing a developing solvent (known as the mobile phase). The components are drawn up the plate via capillary and move across the layer from one side to the opposite at different rates determined by their different distribution coefficients between the stationary and mobile phases, resulting in separation of the components because of their different traveling distances on the plate. The traveling distance of the material on the TLC plate can be expressed by a retardation factor (R_f).

A R_f value is the ratio of the distance the analyte traveled versus distance the mobile phase travelled, which can be calculated by the following formula (shown in Fig 2.16):

$$R_f = \frac{\text{distance traveled by the compound}}{\text{distance traveled by the solvent front}}$$

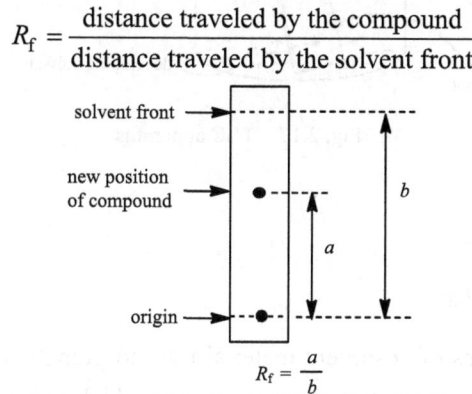

Fig. 2.16 Calculation of retardation factor (R_f)

R_f is connected to the chromatography conditions: solvent system, adsorbent thickness of the adsorbent, amount of material spotted, and temperature, etc. However, a particular compound will have a specific R_f value for a specific solvent (or solvent mixture) when other experimental conditions are kept constant. Therefore, R_f values can be used for identification by comparing R_f values of the unknown sample (or its constituents) with R_f values of known compounds. In addition, spots can be identified by comparing their R_f values with those of standards.

TLC is generally applied in the following four areas:
(1) Separation and identification of micro-scale samples.
(2) Refinement of normal-scale samples.
(3) Monitoring the progress of the reaction.
(4) Determining appropriate conditions for column chromatography.

Reagents

Silica gel GF254, sodium carboxymethylcellulose (CMC) (0.5% in water), cyclohexane, ethyl acetate, petroleum ether, ether, ethanol, *n*-butanol, iodine, azobenzene (saturated in ethanol), resorcinol (1% in ether), *o*-aminophenol (1% in ether), Sudan III (saturated in ethanol), oil-based ballpoint pen ink (1% in ethanol).

Apparatus

TLC apparatus is shown in Fig. 2.17.

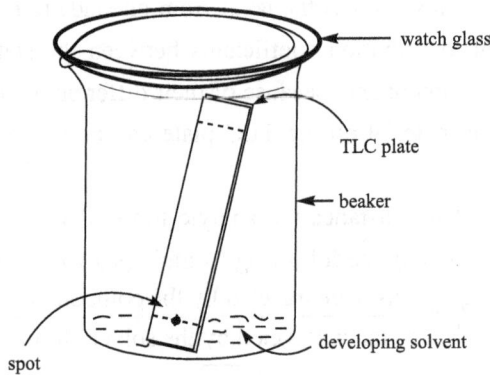

Fig. 2.17　TLC apparatus

Procedures

1. Preparation of TLC Plates

There are different types of adsorbent material used to prepare TLC plates. Commonly used adsorbent include silica gel and neutral alumina, among which silica gel is used for separating a wide variety of substance.

(1) Take four glass plates of dimension 7 cm×3 cm, wash and rinse them until no droplets visible, then dry them in an oven. Take care not to touch the clean plates with bare fingers, which would leave grease spots.

(2) Weigh 3 g silica gel GF254 and mix it with 8 mL 0.5% CMC solution into slurry.

(3) There are various technique used in coating with the slurry. These include pouring, immersion, spreading and spraying. The method used in this experiment is that of pouring. The glass plate is placed on a flat surface. The slurry is then poured at the centre of the plate and a glass rod is used to roll the slurry smoothly over the glass plate. Gently tap the plate on the edge of a table to produce even layers[1]. The plates are air-dried for about 45 min and then heated in an oven set at 105~110℃ for about 30 min; this treatment helps activate the adsorbent.

2. Spotting

As shown in Fig. 2.18, measure 1 cm from the bottom of the plate. Using a pencil, draw a line across the plate at the 1 cm mark. This is the origin line, on which the sample is to be spotted the plate. Take care not to press so hard with the pencil that you disturb the adsorbent. On the line, mark lightly the locations of the samples to be spotted on the plate. Leave enough space (1~1.5 cm) between the samples so that they do not run together. Dip a micro capillary into the sample solution and then gently touch the end of it onto the proper location on the TLC plate. Don't allow the spot to become too large (smaller than 2 mm). If necessary, you can touch the capillary to the plate, lift it off and blow on the spot to make it denser[2].

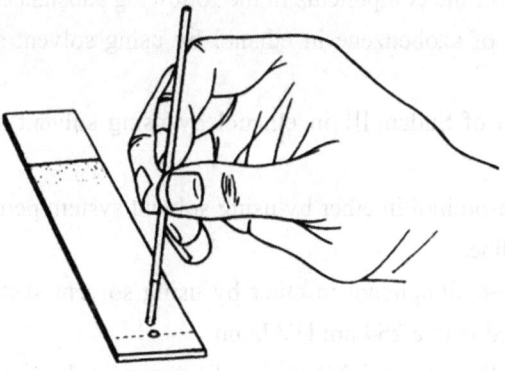

Fig. 2.18 Spotting

3. Developing

The developing container for TLC can be a specially designed chamber, a jar with a lid, or a beaker with a watch glass on the top. Pour solvent into the chamber to a depth of just less than 0.5 cm, and allow it to stand for 5~10 min so as to be saturated with solvent vapors. Place the prepared TLC plate in the developing beaker, cover the beaker with a watch glass, and leave it undisturbed on the bench top. Make sure the solvent does not cover the spot[3]. Allow the plate to develop until the solvent is about 1 cm below the top of the plate. Remove the plate from the beaker and immediately mark the solvent front with a pencil.

Generally a solvent or solvent mixture which gives a R_f of 0.2~0.8 for the sample should be selected.

4. Visualizing

If there are any colored spots, circle them lightly with a pencil. Most samples are not colored and need to be visualized with a UV lamp or by the iodine staining technique. When using a UV lamp, hold the lamp over the plate and circle any spots you see. Beware! UV light is damaging both to your eyes and to your skin! Make sure you are wearing your safety goggles and do not look directly into the lamp. Protect your skin by wearing gloves.

Another way to visualize colorless organic compounds separated on a TLC plate is by placing them in iodide (I_2) vapor to test their absorption of iodide vapor[4]. These TLC plates with colorless marks are placed in a bath of iodine vapor prepared by placing a small amount of iodine crystals in a tightly capped jar. Colorless spots gradually gain a dark brown color after placing the TLC plates in the bath for approximately 10 min.

5. Analyze the Components in the Mixtures Using TLC

(1) Find the R_f values of the components in the following substances:

(i) saturated solution of azobenzene in ethanol by using solvent system cyclohexane : ethyl acetate (9 : 1 in volume).

(ii) saturated solution of Sudan III in ethanol by using solvent system cyclohexane : ethyl acetate (9 : 1 in volume).

(iii) 1% solution of resorcinol in ether by using solvent system petroleum ether : ether (1 : 2 in volume), visualized by iodine.

(iv) 1% solution of o-aminophenol in ether by using solvent system petroleum ether : ether (1 : 1 in volume), visualized with a 254 nm UV lamp.

(2) 1% solution of ballpoint-pen ink in ethanol by using solvent system n-butanol : ethanol : water (8 : 5 : 2 in volume).

6. Cleaning up

After the experiment is completed, wash the used TLC plates with water, rinse twice with distilled water, and then dry them for use.

Notes

[1] The TLC plate should be prepared in a uniform and hard surface without cracks or blisters.

[2] As a rule of thumb, a concentration of 1% usually works well for TLC analysis. If the sample is too concentrated, the spots will be streaked and/or run together; if it is not concentrated enough, it will be invisible on the plate.

[3] The solvent level has to be below the starting line of the TLC, otherwise the spots will dissolve away.

[4] For the reason that the colored spots usually disappear in a short period of time when the plate is visualized by the iodine vapor staining method, the spots should be outlined immediately with a pencil after the TLC plate is taken out of the iodine bath.

Questions

Why do TLC plates need to be activated before being used?

Exp.10 Paper Chromatography

Paper chromatography is a method using filter paper strips as carrier or inert support. It is mainly used for the separation and detection of polyfunctional or highly polar hydrophilic compounds, such as alcohols, hydroxy acids, amino acids, carbohydrates and flavonoids.

Objectives

(1) To learn the principles and applications of paper chromatography.

(2) To be skilled in the operations for separating organic compounds using paper chromatography.

Principles

The principle of paper chromatography is a little complex and is generally considered to be partition chromatography. In paper chromatography, support material consists of a layer of cellulose highly saturated with water. In this method a thick filter paper comprised the support, and water drops settled in its pores made up the stationary "liquid phase." Mobile phase consists of an appropriate fluid placed in a developing tank.

The sample mixture is applied near the bottom edge of a piece of filter paper, the edge of the paper is immersed in a solvent. The solvent moves up the paper by capillary action, carrying the samples with it. During this process the components partition themselves between the two phases to different extents (the lipophilic components are distributed more in the mobile phase and travel faster than hydrophilic ones), and effecting their separation. The sample is distributed between the aqueous phase and the organic phase during the process of advancement, and the partition coefficients of the components in the two phases are different to achieve the purpose of separation.

Materials

Equipment: paper clip, watch glass, beaker, glass rod, filter paper (6 cm×15 cm), hair dryer or oven, capillary, sprayer, ruler, pencil.

Reagents: 0.5% aqueous solution of valine, 0.5% aqueous solution of leucine, 0.5% solution of ninhydrin in ethanol, mixed solution of valine and leucine, developing solvent (n-butanol : acetic acid : ethanol : water=4 : 1 : 1 : 2, V/V).

Procedures

1. Selection of Chromatography Filter Paper

The paper should be highly purified and have appropriate thickness, adsorptive and water holding abilities. The choice of a particular chromatographic paper depends upon the type of separation. Fast speed papers are useful for major applications and when high resolutions are necessary slow speed papers are preferred. Thick paper with increased sample capacity may be used for preparative studies. The most commonly used in the laboratory is Xinhua No. 1 filter paper. In this experiment, the paper strip is cut into 6 cm×15 cm. A set-up for paper chromatography is shown in Fig. 2.19.

2. Spotting

Draw a thin line across the paper 2~3 cm from the edge with a pencil[1]. On the line, mark lightly the locations and the symbols of the three samples to be spotted with a minimum separation of 1.5 cm. Dip a capillary into the sample solution and then gently touch the end of it onto the proper

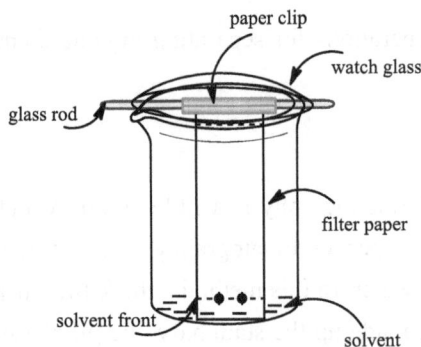

Fig. 2.19   Paper chromatography

location on the paper to make the sample spots[2]. Three spots correspond to the sample of 0.5% valine solution, 0.5% leucine and the mixed solution of the former two are thereby made across the line. The spot must be as small as possible without any spread for better separation. It is best done by placing the sample dropwise, drying it with hot air blower (or hairdryer) to evaporate the solvent. After drying, another drop may be put and dried. Make sure to change the capillary when making spots for another sample.

3. Saturation and Developing

Hold the top edge of the spotted paper with a glass rod or strip as shown in Fig. 2.19. After the development chamber is presaturated with the solvent system, hung the paper with its spotted end dipping in the solvent as shown in Fig. 2.19. The spotting line should be about 1 cm above the solvent[3]. The solvent starts rising slowly and then it stops after some time. It may take an hour or even longer.

The development time will depend on the complexity of the mixture being separated, solvent system and the quality of paper and the ambient temperature. For good resolution, reasonable R_f values must be in the range of 0.4~0.8.

When the front edge of the solvent rises close to the top of the filter paper, take out the paper, mark the solvent front immediately and dry the paper[4,5].

4. Visualizing

Spray the paper evenly under a fume hood with 0.5% ninhydrin solution with a sprayer[6], avoiding excess use of the spray. The paper should appear translucent but should not be saturated with the reagent. Then use a hair dryer to dry the filter slowly until purple spots appear (or transfer the sheet to a chromatography drying oven which has been previously adjusted to 90℃).

5. Measurement of R_f Value and Identification of Samples

Circle all the sample spots lightly with a pencil. Measure the distance moved by each of the samples and the solvent. Calculate the R_f values of all the samples. Determine the samples by comparing their R_f values.

Notes

[1] The filter paper should **NEVER** be handled by bare hands since the skin's oils show up on the developed chromatogram.

[2] Use different capillary tubes for each amino acid and unknowns so that cross contamination can be avoided.

[3] The sample spots must be above the solvent level.

[4] Dispose of any leftover developing solvent (butanol/acetic acid/water) in reclaim container provided.

[5] Butanol is a flammable liquid. Make sure there are no open flames and/or heat near the developing solvent.

[6] Wear gloves when using ninhydrin solvent, because this material will form colored compounds with all amino acids, including those in your skin.

Questions

Why should a baseline be drawn on the chromatography paper using a pencil? Can a ballpoint pen or a pen be used to draw the baseline?

Chapter 3 Preparation Experiments of Organic Compounds

Exp.11 Preparation of Cyclohexene

Objectives

(1) To synthesize cyclohexene by the intramolecular dehydration of cyclohexanol.

(2) To learn the basic operations such as fractionating reaction, salting-out method, drying method and liquid separation.

(3) To be familiar with the operations such as water-bath heating, distilling and washing.

Principle

The main and side reactions occur in this experiment are given as follows.

Main reaction:

$$\text{C}_6\text{H}_{11}\text{OH} \xrightleftharpoons[\Delta]{85\%\text{H}_3\text{PO}_4} \text{C}_6\text{H}_{10} + \text{H}_2\text{O}$$

Side reaction:

$$2\,\text{C}_6\text{H}_{11}\text{OH} \xrightleftharpoons[\Delta]{85\%\text{H}_3\text{PO}_4} (\text{C}_6\text{H}_{11})_2\text{O} + \text{H}_2\text{O}$$

This reaction is reversible, therefore, we utilize the method that distills the binary azeotropic mixture of cyclohexene and water (b.p. 70.8℃, containing 10% of water) out from the system during the reaction in order to enhance its conversion. However, it is noteworthy that the starting material cyclohexanol can be mixed with water to form another binary azeotrope (b.p. 97.8℃, containing 80% of water). In order to distill the product out together with water without the loss of cyclohexanol, the fractionating-reaction method should be used. By applying this method, the continuous removal of reaction products from the reaction system can raise the conversion far beyond what is expected by the equilibrium. By controlling the distillation temperature at the top of the fractionating column, the loss of the reaction material can be reduced.

Materials

Equipment: round-bottom flask, fractionating column, Erlenmeyer flask, straight condenser, thermometer, thermometer adapter, distillation head, adapter, heating mantle, separatory funnel.

Reagents: 85% aqueous H_3PO_4, 5% aqueous Na_2CO_3, saturated aqueous NaCl, anhydrous $CaCl_2$, cyclohexanol.

Procedures

1. Preparation of Cyclohexene

In an oven-dried 50 mL round-bottom flask, place 10 mL of cyclohexanol (9.6 g, 0.096 mol)[1] and 4 mL of 85% aqueous H_3PO_4 (or 1 mL of conc. H_2SO_4[2]). Shake the flask vigorously to mix the materials well[3]. Then add a few boiling chips. Set up a fractionating-reaction apparatus as shown in Fig. 3.1. Connect a short fractionating column onto the round-bottom flask; place a distillation head on top of the fractionating column. To the distillation head, attach the thermometer adapter and thermometer. Then connect a water condenser to the outlet of the distillation head. Attach a vacuum adapter to the water condenser and an Erlenmeyer flask to the vacuum adapter. At last, clamp the Erlenmeyer flask and place it in a beaker containing cold water.

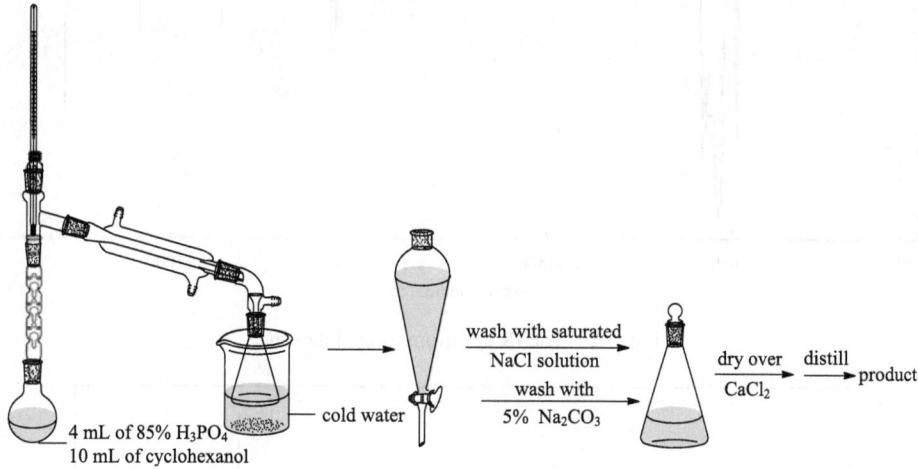

Fig. 3.1 Procedures for the preparation of cyclohexene from cyclohexanol

Heat the mixture in the round-bottom flask gently over a heating mantle until it boils, and then control the temperature at the top of the fractionating column to be no higher than 90 ℃ by tuning the heating power[4]. The distillate is a cloudy liquid (as a mixture of cyclohexene and small amount of water). When there is no distillate coming out, increase the heating temperature appropriately to facilitate the completion of the reaction. Stop the distillation when white fumes appear in the round-bottom flask and only very small amount of liquid remained in the flask[5]. This reaction process may take about 40 min.

2. Work-up Process

Transfer the distillate into a separatory funnel. Run off the aqueous layer, add in an equal volume of saturated NaCl solution, stopper the funnel and shake the separatory funnel vigorously. Allow the layers to separate and then run off the aqueous layer. Wash the remained organic layer with 3~4 mL 5% aqueous Na_2CO_3[6]. Allow the layers to separate and run off the aqueous layer completely. The aqueous phase is removed thoroughly. Transfer the organic phase into a dried Erlenmeyer flask, and dry the solution over anhydrous $CaCl_2$[6]. Allow the solution to stand over the drying agent for about 15 min, then, filter the dry liquid into a dried round-bottom flask. Place a couple of boiling chips in the flask, heat the mixture in a water bath and collect the distillate that boils at 80～85℃ with a dry pre-weighed Erlenmeyer flask. The isolated yield is 4～5 g.

The IR and ^{1}H NMR spectra of cyclohexene are shown in Fig. 3.2 and Fig. 3.3, respectively.

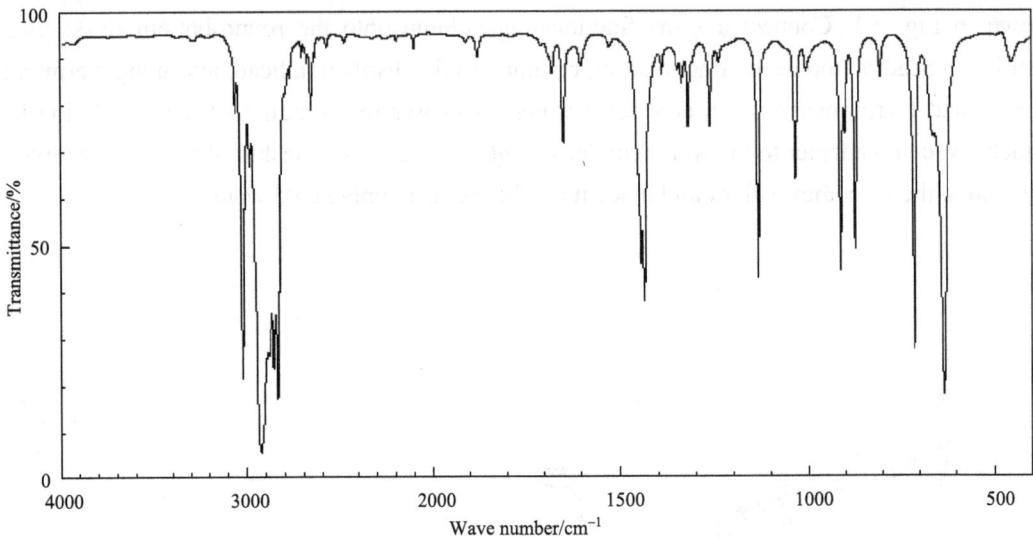

Fig. 3.2 The IR spectrum of cyclohexene

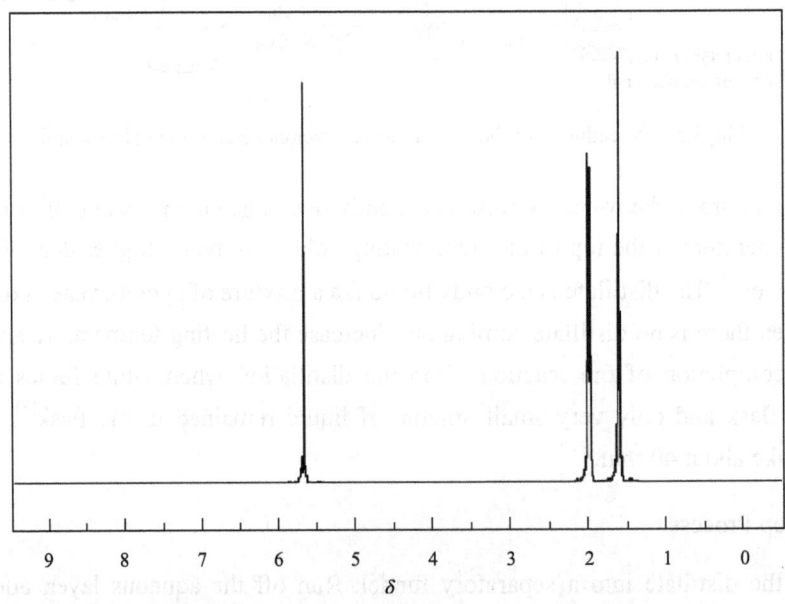

Fig. 3.3 The ^{1}H NMR spectrum of cyclohexene

Notes

[1] Cyclohexene is a viscous liquid at room temperature. Therefore, take care to minimize the loss of product in the transfer process.

[2] Sulfuric acid (1 mL, 95%) could also be utilized as the catalyst and the procedures are the same. However, the side reactions such as oxidation could be reduced when H_3PO_4 is used because of its weaker oxidizing ability.

[3] Cyclohexanol and H_3PO_4 should be mixed thoroughly before being heated, otherwise, carbonization might occur partially and the color would be dark when the mixture is heated.

[4] It would be better to heat the distilling flask using an air bath or an oil bath. Since not only cyclohexene can form an azeotrope with water, but also cyclohexanol can form an azeotrope with water, the distillation temperature should not be too high, otherwise cyclohexanol would also be distilled out. Some reports say that the distillation temperature at the top of the fractionating column should be no higher than 73 ℃ to reduce the loss of the material. However, the reaction would be very slow and it may take a long time. In order to facilitate the distilling process, we can control the temperature no higher than 90 ℃.

[5] We can determine the end point of reaction according to the following factors:

(i) The fractionating reaction has taken about 40 min.

(ii) The amount of the distilled azeotropic mixture of cyclohexane and water reaches the theoretical amount.

(iii) There is a fitful white fume in the flask.

(iv) The temperature at the top of the fractionating column drops and then goes up to be above 85 ℃ again.

[6] The residue of aqueous phase should be removed thoroughly before drying it. Otherwise, the amount of $CaCl_2$ would have to be increased, and more products may be absorbed and lost. $CaCl_2$ is a proper desiccant for this experiment because it can also remove trace amount of cyclohexanol. The amount of $CaCl_2$ used depends on the moisture content of the crude product.

Questions

(1) Why should we control the temperature at the top of the fractionating column no higher than 90 ℃ during the distillation?

(2) What are the contents in the fitful white fumes in the flask before ending the distillation?

(3) During the work-up process, why should the aqueous phase be saturated with NaCl?

(4) During the work-up process, why should the aqueous layer be removed thoroughly before being dried?

Exp.12 Preparation of 1-Bromobutane

Objectives

(1) To study the principles and method for the preparation of alkyl halides from alcohols.

(2) To master the skills for the operations for the experiments that produce toxic gases; to be familiar with the procedures of the refluxing reaction utilizing equipment for exhaust gas absorption.

(3) To consolidate the skills for the basic operations such as washing, drying, distilling and separating.

Principle

Main reactions:

$$NaBr + H_2SO_4 \xrightarrow{\Delta} HBr + NaHSO_4$$

$$n\text{-}C_4H_9OH + HBr \rightleftharpoons n\text{-}C_4H_9Br + H_2O$$

Side reactions:

$$2n\text{-}C_4H_9OH \xrightarrow[\Delta]{conc.H_2SO_4} n\text{-}C_4H_9OC_4H_9\text{-}n + H_2O$$

$$n\text{-}C_4H_9OH \xrightarrow[\Delta]{conc.H_2SO_4} CH_2{=}CHCH_2CH_3 + CH_3CH{=}CHCH_3 + H_2O$$

$$2HBr + H_2SO_4 \longrightarrow Br_2 + SO_2 + 2H_2O$$

Materials

Equipment: a refluxing unit with an exhaust gas absorber, separatory funnel, 75° bent adapter, Erlenmeyer flask, measuring cylinder.

Reagents: 1-butanol, NaBr (anhydrous), conc. H_2SO_4, 10% aqueous $NaHCO_3$, anhydrous $CaCl_2$, 5% aqueous NaOH.

Procedures

1. Preparation of 1-Bromobutane

As shown in Fig. 3.4, in a 100 mL round-bottom flask charged with 10 mL of water, slowly add 12 mL (0.22 mol) of conc. H_2SO_4 while shaking the flask (take care!). When the mixture is cooled to room temperature, add 7.5 mL (0.08 mol) of 1-butanol and 10 g (0.10 mol) of finely grinded NaBr powder to the flask (note: do not allow particles remain in contact with the ground joint surface of flask)[1]. After the starting materials have been well mixed together, add a few boiling chips. Set up a refluxing apparatus with an exhaust gas absorber using 5% aqueous NaOH as the absorbent (note: only part of the funnel should be immersed into water)[2]. Heat the mixture under gently reflux over a low flame using a heating mantle until all the solid materials have been resolved. Then tune the temperature properly to keep it refluxing steadily. This reaction process may take 30~40 min. When the mixture is cooled, add in 10 mL of water. Then set up a distillation apparatus, and distill out the crude product using a simple steam distillation until no more water-insoluble droplets come over.

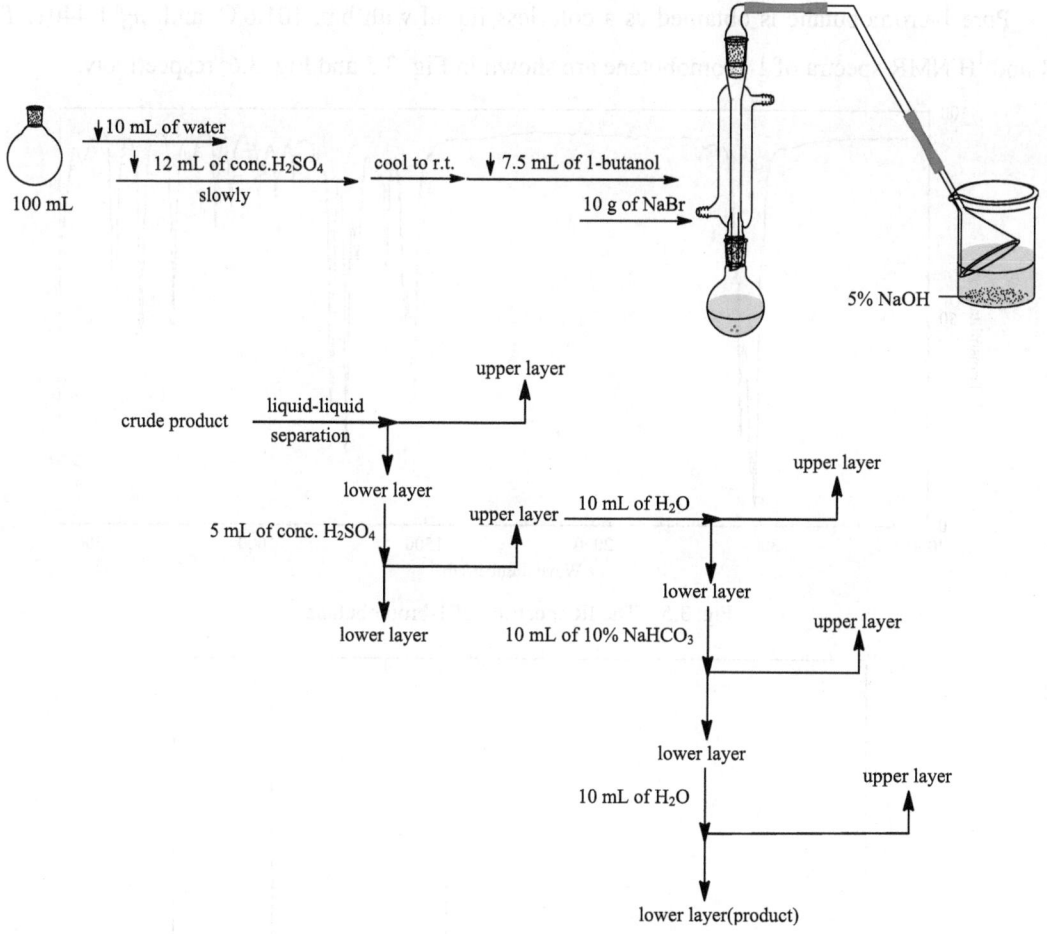

Fig. 3.4 Procedures for the preparation of 1-bromobutane from 1-butanol

2. The Purification of the Crude Product

Transfer the obtained distillate into a separatory funnel. Then add 10 mL of water to wash the distillate. The mixture separates into two obvious layers. Note that 1-bromobutane now is mainly contained in the lower layer. Carefully drain the organic phase into another dry separatory funnel through the stopcock. Then wash the organic phase with 5 mL of conc. H_2SO_4 to remove unreacted starting materials such as any alkenyl or ethereal by-products. Allow the two layers to separate completely, and then remove the sulfuric acid layer. Identify the two layers carefully. Separate the layers and wash the 1-bromobutane layer with 10 mL of water, 10 mL of 10% aqueous $NaHCO_3$, and 10 mL of water, respectively[3]. Then transfer the organic phase into a dry Erlenmeyer flask and dry the cloudy liquid by adding 1~2 g of anhydrous calcium chloride pellets and mixing until the liquid clears[4].

Decant the dried liquid into a dry 50 mL round-bottom flask and distill, collecting material boiling in the range 99~103 ℃.

Pure 1-bromobutane is obtained as a colorless liquid with b.p. 101.6℃ and n_D^{20} 1.4401. The IR and ^{1}H NMR spectra of 1-bromobutane are shown in Fig. 3.5 and Fig. 3.6, respectively.

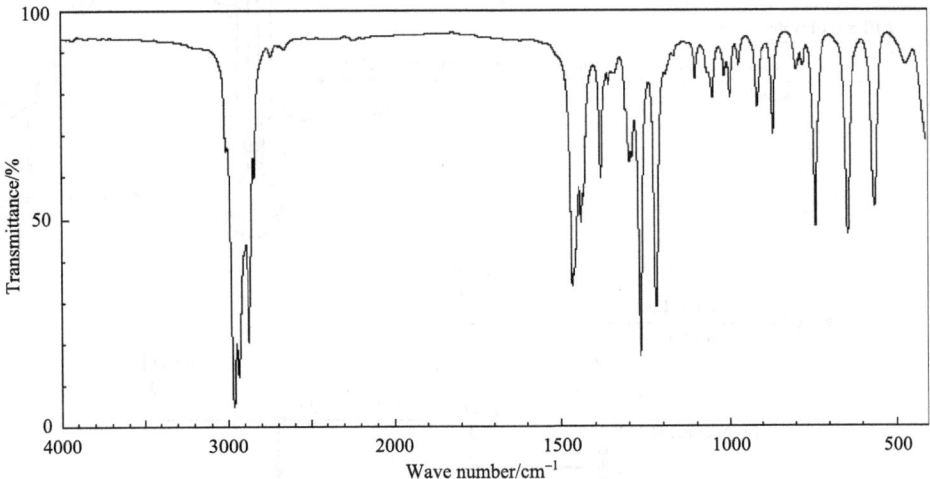

Fig. 3.5　The IR spectrum of 1-bromobutane

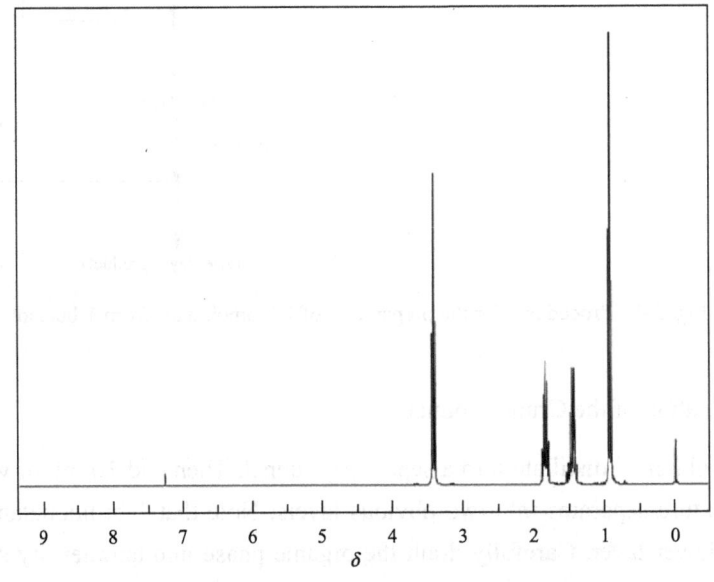

Fig. 3.6　The ^{1}H NMR spectrum of 1-bromobutane (source: Bio-Rad Laboratories)

Notes

[1] Strictly follow the order given in the textbook when adding the reactants. And the materials should be well mixed before the reaction starts.

[2] The reaction mixture should be kept refluxing steadily. And the back suction should be avoided.

[3] Be very careful when determining which layer will be on the top and which layer is at the bottom in the process of liquid-liquid separation.

[4] The amount of the desiccant should be reasonable.

Questions

(1) Why should a refluxing-reaction apparatus be utilized for the preparation of 1-bromobutane?

(2) Why should we use the sulfuric acid with proper concentration instead of concentrated sulfuric acid for the preparation of 1-bromobutane?

(3) Why should concentrated H_2SO_4 be used to wash the crude product?

(4) How could we determine the end point of the distillation of the crude product?

Exp.13 Preparation of Bromobenzene

Bromobenzene is colorless oil with a pungent odor. Its m.p. is −30.7℃ and b.p. is 156.2℃. And its n_D^{20} is 1.5590. It is insoluble in water but easily soluble in many organic solvents such as ethyl ether, acetone, benzene, methanol and carbon tetrachloride. Bromobenzene could be used as an organic solvent, analytic agent or organic reagents.

Objectives

(1) To study the principles and method for the preparation of bromobenzene.

(2) To consolidate the skills for some basic operations, such as distillation.

Principles

In the presence of iron powder, the electrophilic substitution reaction between benzene and bromine gives bromobenzene.

$$\text{C}_6\text{H}_6 + \text{Br}_2 \xrightarrow{\text{Fe}} \text{C}_6\text{H}_5\text{Br} + \text{HBr}$$

Strictly speaking iron isn't a catalyst; it reacts with bromine to form $FeBr_3$, which acts as the catalyst in this reaction. The reaction is moisture-sensitive, because $FeBr_3$ hydrolyzes easily, resulting reduced catalytic activity. On another hand, the further bromination of the obtained product might occur to give the benzene dibromides (the main byproduct is *p*-dibromobenzene).

$$2\,\text{C}_6\text{H}_5\text{Br} + 2\text{Br}_2 \xrightarrow{\text{Fe}} p\text{-C}_6\text{H}_4\text{Br}_2 + o\text{-C}_6\text{H}_4\text{Br}_2 + 2\text{HBr}$$

Although there are great differences between the boiling points of benzene, bromobenzene and *p*-dibromobenzene, it is difficult to separate them by simple distillation process. Therefore, we utilize a secondary distillation to further purify the crude product.

Materials

Equipment: a refluxing apparatus with an exhaust gas absorber, separatory funnel.

Reagents: benzene, liquid bromine, iron powder, 10% aqueous NaOH.

Procedures

1. Preparation of Bromobenzene

As shown in Fig. 3.7, place a water condenser and a constant pressure dropping funnel onto each neck of a 100 mL three-necked flask containing 6 mL (5 g, 0.06 mol) of anhydrous benzene and 0.2 g of iron powder, put an exhaust gas absorber on top of the condenser[1]. Add 2.5 mL (8 g, 0.05 mol) of liquid bromine in the dropping funnel (**caution**: liquid bromine is highly volatile and irritant)[2]. First, add 1~2 drops of bromine into the mixture. The reaction would be initiated immediately (heat if necessary), and there is gaseous HBr escaping from the liquid. Then add the rest of bromine slowly into the mixture while keeping the mixture under gently boiling. This feeding process might take 20 min. After that, keep the mixture under gentle boiling until there is no gaseous HBr escaping from the liquid. This reaction process would take about 10 min.

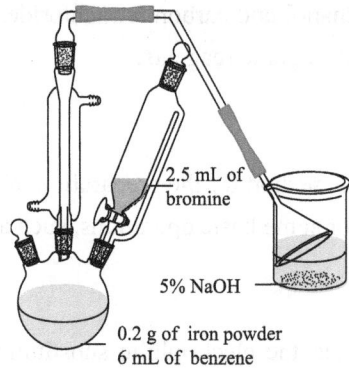

Fig. 3.7 Apparatus for the preparation of bromobenzene from benzene

2. Purification of the Crude Product

Add 15 mL of water in the mixture while shaking the flask vigorously. Filter the mixture under reduced pressure. Transfer the filtrate into a 50 mL separatory funnel and wash the liquid with 25 mL of water, 6 mL of 10% NaOH[3] and 25 mL of water, respectively. Then transfer the organic phase into a dry Erlenmeyer flask and dry it over anhydrous $CaCl_2$, affording the crude product. After that, purify the crude product by distillation. Remove the residual benzene at a relatively lower temperature (<135℃), then use an air condenser and collect the distillate boils at 140~170℃. Finally, make a secondary distillation on the above obtained liquid and collect the distillate boils at 150~160℃[4]. The yield of the pure product might be 4~5 g (55%~60%). This experiment would take 5~6 h.

Pure bromobenzene is obtained as colorless oil with b.p. 156.2℃ and n_D^{20} 1.5590. The IR spectrum of bromobenzene is shown in Fig. 3.8.

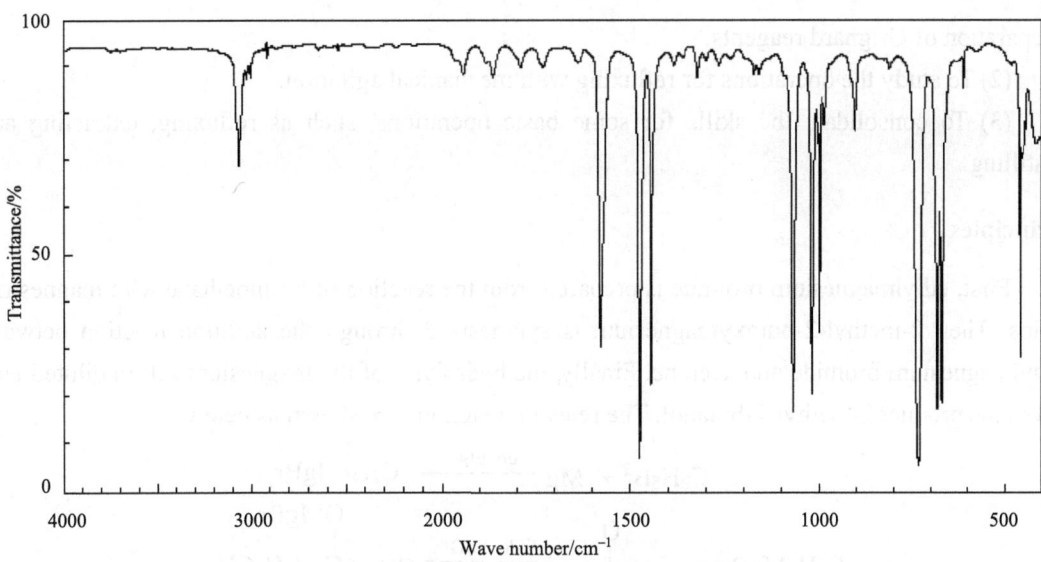

Fig. 3.8 The IR spectrum of bromobenzene

Notes

[1] In this experiment, all the parts of the reaction apparatus should be dried thoroughly. Otherwise, the reaction would be very slow or even not occur. Therefore, at the beginning of the experiment, you should carefully check whether all the pieces of the apparatus are dried. In addition, you should carry out a gas tightness test because liquid bromine, benzene and bromobenzene are all highly volatile.

[2] Take care when you treat liquid bromine, because it's highly irritant and corrosive. The operations of weighing and transferring of this reagent should be carried out in a good fume hood. And the operator should wear protective gloves and goggles. Avoid skin contact with bromine. In case of skin contact, wash with plenty of water immediately, and then massage with some glycerol and ointment.

[3] Because bromine is weakly soluble in water, aqueous NaOH is utilized to remove it.

[4] The crude product could also be further purified by steam distillation.

Questions

(1) What should we pay attention to when we employ the strongly corrosive and irritant chemicals such as liquid bromine?

(2) How can we decrease the generation of the byproduct *p*-dibromobenzene in this experiment?

(3) Why a secondary distillation process should be used in this experiment?

Exp.14　Preparation of 2-Methyl-2-Butanol

Objectives

(1) To study and understand the methods, application scope and reaction conditions for the

preparation of Grignard reagents.

(2) To study the operations for refluxing with mechanical agitation.

(3) To consolidate the skills for some basic operations, such as refluxing, extracting and distilling.

Principles

First, ethylmagnesium bromide is prepared from the reaction of bromoethane with magnesium chips. Then 2-methyl-2-butoxymagnesium is synthesized through the addition reaction between ethylmagnesium bromide and acetone. Finally, the hydrolysis of the magnesium salt in diluted HCl gives the product 2-methyl-2-butanol. The relevant reactions are shown as below.

$$C_2H_5Br + Mg \xrightarrow{dry\ Et_2O} C_2H_5MgBr$$

$$C_2H_5MgBr + CH_3COCH_3 \xrightarrow{dry\ Et_2O} CH_3-\underset{\underset{CH_3}{|}}{\overset{\overset{OMgBr}{|}}{C}}-CH_2CH_3$$

$$CH_3-\underset{\underset{CH_3}{|}}{\overset{\overset{OMgBr}{|}}{C}}-CH_2CH_3 \xrightarrow{H_3O^+} CH_3-\underset{\underset{CH_3}{|}}{\overset{\overset{OH}{|}}{C}}-CH_2CH_3$$

Materials

Equipment: mechanical agitator, constant pressure dropping funnel, three-necked flask, distillation head, Allihn condenser, straight condenser, drying tube, adapter, round-bottom flask.

Reagents: magnesium chips, bromoethane, acetone, anhydrous ether, 5% Na_2CO_3, 20% H_2SO_4, ether, anhydrous K_2CO_3, anhydrous $CaCl_2$.

Procedures

1. Preparation of Ethylmagnesium Bromide

As shown in Fig. 3.9, equip a mechanical agitator, an Allihn condenser and a constant pressure dropping funnel to a thoroughly dried three-necked flask (100 mL). Place a $CaCl_2$ drying tube on the top of the condenser[1]. The funnel is charged with a well mixed solution of 6.5 mL (14.6 g, 0.13 mol) of bromoethane in 10 mL of anhydrous ether. Then add 1.7 g (0.07 mol) of magnesium chips and a small particle of iodine into the flask. To initiate the reaction, add about 1/3 of the mixed liquid from the dropping funnel in the flask[2]. The mixture would become opaque and the color of iodine would disappear subsequently (otherwise the flask should be warmed in a water bath). When the reaction is initiated, start the agitator and stir the mixture while continue adding the rest of the solution. Please note that the dropping rate should be carefully controlled so that the reaction mixture is kept gently boiling. If the reaction mixture becomes viscous, a proper amount of anhydrous ether could be supplied. After complete addition of the solution, continue to stir the mixture under gentle refluxing in a warm water bath for about 30 min until there are almost no observable magnesium chips.

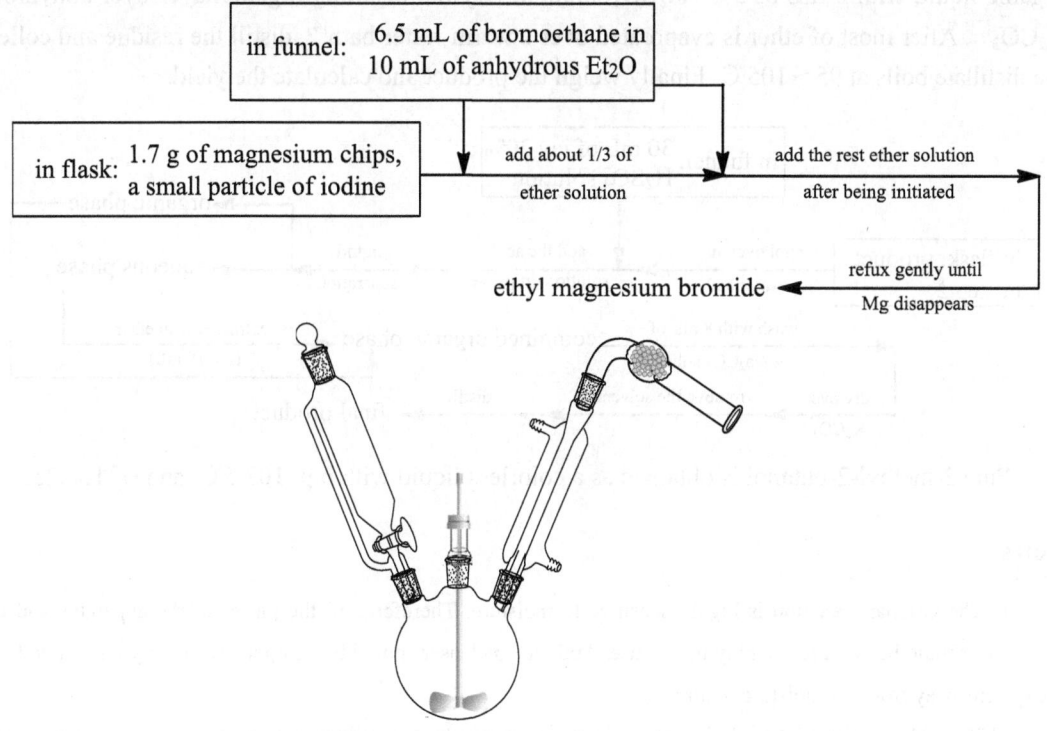

Fig. 3.9 Apparatus for the preparation of ethylmagnesium bromide

2. The Addition Reaction with Acetone

Put the flask into an ice-water bath subsequently. Mix 5 mL (3.95 g, 0.07 mol) of acetone with 5 mL of ether in the dropping funnel and add the solution dropwise into the flask with mechanical stirring. After the addition of the solution is completed, continued to stir the mixture for about 15 min at room temperature. Some grey-white viscous solid would precipitate out during this process.

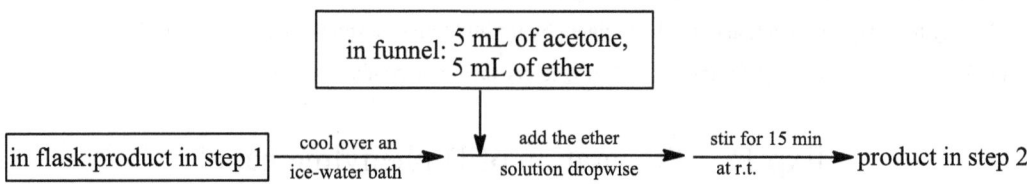

3. The Work-up Procedures

Place the flask into an ice-water bath, add 30 mL of 20% H_2SO_4 solution (which has been cooled thoroughly in an ice-water bath before use) dropwise and carefully from the dropping funnel to hydrolyze the crude product. After the addition, separate the mixture and collect the ether layer. The aqueous phase is extracted by ether (2×10 mL). Then combine the ether layers and wash the organic liquid with 8 mL of 5% Na_2CO_3 solution. Dry the resulting organic layer over anhydrous K_2CO_3[3]. After most of ether is evaporated over a warm water bath[4], distill the residue and collect the distillate boils at 95~105 ℃. Finally weigh the product and calculate the yield.

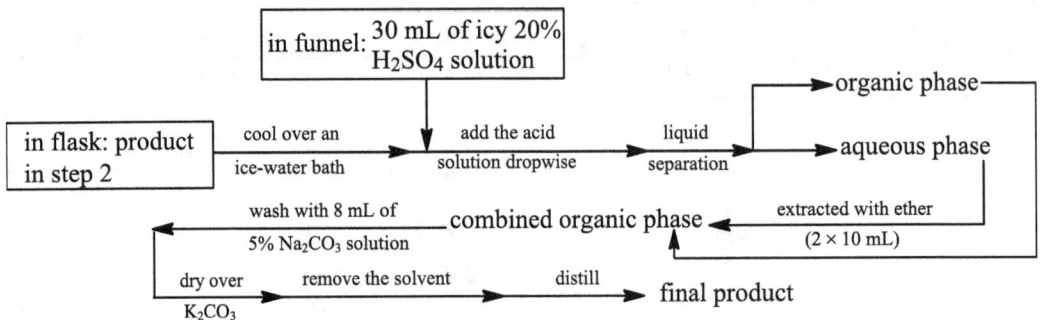

Pure 2-methyl-2-butanol is obtained as a colorless liquid with b.p. 102.5 ℃ and n_D^{20} 1.4025.

Notes

[1] The Grignard reaction is highly sensitive to moisture. Therefore, all the pieces of the apparatus and the chemicals should be dried thoroughly before use. And the condenser should be equipped with a drying tube to keep the system away from the moisture of air.

[2] The addition rate of the solution of bromoethane should be appropriate to keep the reaction mixture gently boiling.

[3] 2-Methyl-2-butanol can be mixed with water to give an azeotrope. Therefore, the crude product should be dried thoroughly using anhydrous K_2CO_3.

[4] Ether is volatile, easily flammable and combustible. Therefore, the joints of the apparatus should be connected tightly and the equipment should be kept far away from fires.

Questions

(1) What are the key points for the success of this experiment?

(2) How could we increase the yield of this experiment?

(3) Why the Grignard reaction could be initiated by the addition of a small amount of iodine?

(4) Why should we use K_2CO_3 instead of $CaCl_2$ to dry the crude product?

Exp.15 Preparation of Diphenylmethanol

Objectives

To learn the principle and method for preparing a secondary alcohol from a ketone by using sodium borohydride as a reducing agent.

Principles

Diphenylmethanol is mainly used in organic synthesis and is an important intermediate for some pesticides, pharmaceuticals and some other products. For example, diphenylmethanol can be used as an intermediate for benaztropine and diphenhydramine.

Reduction reactions are one of the major classes of reactions in organic chemistry. Alcohols are often produced by reduction of the corresponding aldehydes or ketones. Diphenylmethanol can be obtained from benzophenone by using a variety of reducing agents, such as zinc-alkali, sodium borohydride, lithium aluminum hydride, etc. Sodium borohydride is a hydride (H^-) transfer reagent which reduces carbonyl groups selectively and conveniently. The reaction can be carried out in a water-containing alcohol. In a microscale experiment, sodium borohydride is a good choice. Theoretically, 1 mol of sodium borohydride reduces 4 moles of aldehyde or ketone. However, an excessive amount of sodium borohydride is often used in the actual reaction. The reaction formula is as follows.

$$\text{Ph-CO-Ph} \xrightarrow{NaBH_4} (\text{Ph}_2\text{CH-O})_4\text{B}^-\text{Na}^+ \xrightarrow{H_2O, H^+} \text{Ph}_2\text{CHOH} + H_3BO_3$$

Materials

Equipment: beaker, magnetic stirrer, suction filter, Büchner funnel.

Reagents: benzophenone, sodium borohydride, 95% ethanol, petroleum ether (30~60℃), 10% hydrochloric acid.

Procedures

In a 200 mL beaker, add 1 g (5.5 mmol) of benzophenone, 3 mL of water, and 20 mL of 95% ethanol, stir to dissolve all the solids, and carefully add 0.4 g (10.5 mmol) of sodium borohydride in batches with stirring[1]. At this time, bubbles occur in the solution and the solution becomes hot. The sodium borohydride should be added in such a rate that the reaction temperature does not exceed 50℃. After the addition of sodium borohydride is finished, continue stirring for about 20 min until precipitates appear[2]. Cool the mixture to room temperature, add 50 mL of cold water to decompose excess sodium borohydride, then slowly add 10% hydrochloric acid dropwise until the solution is acidic.

Cool the reaction liquid with an ice water bath, filter the mixture under reduced pressure, wash the obtained solid with a little cold water, and dry the solid to afford the crude product. Recrystallize the crude product from petroleum ether (30~60℃) to give pure diphenylmethanol as needle crystals[3]. The melting point of pure diphenylmethanol is 69℃.

Notes

[1] Sodium borohydride is highly caustic and must not come in contact with your skin, eyes, or nasal passages.

[2] If no precipitation occurs, you can distill off most of the ethanol in a water bath, and pour the residue into a mixture of 6 g of crushed ice and 1 mL of 10% hydrochloric acid after the residue is cooled down.

[3] Recrystallization can also be carried out from hexane instead of petroleum ether.

Questions

(1) Explain the difference between the two hydride reducing reagents (sodium borohydride and lithium aluminum hydride) in reactivities and operations.

(2) Give some other methods for synthesizing diphenylmethanol.

(3) Why should 10% hydrochloric acid be added in the reaction mixture after the completion of the reaction?

Exp.16   Preparation of 1, 1′-Binaphthol (a Reference-based Experiment)

Objectives

(1) To learn the method for searching literature in English, and to synthesize the target product according to the references.

(2) To learn and grasp the method of solid-state grinding synthesis.

(3) To learn the synthesis of simple chiral compounds, and the properties of chiral ligands.

Principles

Some reactions can occur in the absence of any solvent. The solvent-free reactions are less energetic consumptive, more efficient, and more selective compared to the reactions carried out in solvents. What's more, they are more environmentally friendly because solvents are not required.

In this experiment, the racemic mixture of 1,1′-binaphthol (BINOL) will be synthesized by the oxidative coupling of β-naphthol in the presence of ferric trichloride by using the solid-state grinding synthesis[1].

$$2 \text{ } \beta\text{-naphthol} + 2FeCl_3 \cdot 6H_2O \xrightarrow{\text{solid-state grinding}} \text{(R)-BINOL} + \text{(S)-BINOL}$$

Materials

Equipment: oven, mortar, pestle, water bath, filtration equipment.

Reagents: β-naphthol, ferric trichloride hexahydrate, 5% aqueous HCl, ethanol.

Procedures

1. The Solid-state Grinding Reaction

Put 1.0 g of β-naphthol and 3.8 g of ferric trichloride hexahydrate into a mortar and mix them well. Then grind the mixture rigorously for about 0.5 h[1,2]. Then place the mortar containing the reaction mixture in an oven set at 50 ℃ for about 2 h.

2. The Work-up Procedures

Take out the mortar from the oven and cool it to room temperature. Add 5% aqueous HCl under stirring to hydrolyze the crude product. Filter the mixture after it is cooled down. And wash the precipitate with small amount of water for 3 times to remove the residual ferrum (Ⅲ) and ferrum (Ⅱ) ions. Dry the crude product thoroughly under reduced pressure. The product could be further purified by the recrystallization in ethanol. The pure product is obtained as white crystals with m.p. 214~217℃.

Notes

[1] The mixture should be grinded vigorously to make the particles as small as possible. Thus the particles would have a larger specific surface and the reaction would proceed more smoothly.

[2] The mixture may be wet and paste-like during the grinding process because of the high hydrophilicity of ferric trichloride. This will not affect the experimental results and the experiment can be continued.

Questions

(1) Comparing with organic reactions with use of solvents, what are the advantages of solid-state organic synthesis?

(2) What are the roles of $FeCl_3$ in this experiment? What are the advantages of using $FeCl_3$?

(3) Why the reaction mixture should be grinded vigorously?

(4) Why the reaction mixture should be treated with diluted HCl after the reaction?

References

Toda F, Tanaka K, Iwata S. 1989. Oxidative coupling reaction of phenols with $FeCl_3$ in solid state. J Org Chem, 54: 3007-3009.

Exp.17   Preparation of Di-*n*-butyl Ether

Objectives

(1) To understand the principles and method for the preparation of ethers by the intermolecular dehydration of alcohols.

(2) To learn the operations for a refluxing reaction using an equipment with a Dean-stark trap.

Principles

Main reaction:

$$2CH_3CH_2CH_2CH_2OH \xrightarrow[134\sim135^{\circ}C]{conc.\ H_2SO_4} CH_3CH_2CH_2CH_2OCH_2CH_2CH_2CH_3 + H_2O$$

Side reaction:

$$CH_3CH_2CH_2CH_2OH \xrightarrow[>135^{\circ}C]{conc.\ H_2SO_4} CH_3CH_2CH_2=CH_2 + CH_3CH=CHCH_3 + H_2O$$

In order to move the position of equilibrium to the right and thereby increase the conversion rate, the generated water from the refluxing reaction mixture will be removed continuously from the system using a Dean-stark trap.

Materials

Equipment: two-necked flask, Allihn condenser, Dean-stark trap, thermometer, separatory funnel, round-bottom flask.

Reagents: 1-butanol, boiling chips, conc. H_2SO_4, saturated aqueous NaCl, 5% aqueous NaOH, saturated aqueous $CaCl_2$, anhydrous $CaCl_2$.

Procedures

1. Preparation of Di-*n*-butyl Ether

Add 15.5 mL (12.5 g, 0.17 mol) of 1-butanol, 2.5 mL of conc. H_2SO_4 and 2~3 pieces of boiling chips to a 50 mL two-necked flask. Shake the flask to make the starting materials mixed well. Then set up an apparatus for a refluxing reaction with a Dean-stark trap (as shown in Fig. 3.10). Fill the Dean-stark trap with water until the water level just reaches the level of its side arm and then drain 2.0 mL of water from the bottom of the trap[1]. Insert an Allihn condenser into one neck of the flask

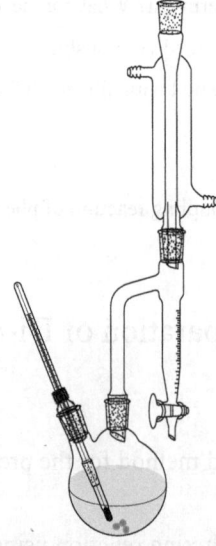

Fig. 3.10 Apparatus for the preparation of di-*n*-butyl ether

and a thermometer into the other with the mercury bulb immersed in the liquid mixture. After that, heat the mixture to boiling and continue heating it under gentle reflux. The water generated during the reaction would be condensed and trapped as the bottom layer in the Dean-stark trap. The vapor containing the reaction solvent and water travels out of reaction flask up into the condenser where the vapor is cooled and drips into the trap. Here, the immiscible liquids separate into layers (water below and solvent above it). When their combined volume reaches the level of the side-arm, the upper, less-dense layer will begin to flow back to the reactor to react continually while the water layer will remain in the trap. Continue to heat until the temperature reaches at 134~136℃[2]. Stop heating when the level of the water in the trap reaches the level of the side-arm. If continuing to heat the flask, the mixture would become darker and more byproducts would be generated. This process might take about 1 h.

2. The Work-up Procedures

Cool the reaction mixture to room temperature and transfer it into a separatory funnel containing 25 mL of water. Shake the funnel vigorously, and then let it stand for a while. Remove the aqueous phase and wash the organic phase containing crude product with 10 mL of water, 8 mL of 5% aqueous NaOH[3], 8 mL of water and 8 mL of saturated aqueous $CaCl_2$, respectively. Dry the organic phase over 1~2 g of anhydrous $CaCl_2$ and transfer the dried liquid into a dry 50 mL round-bottom flask. Make a further purification of the crude product by distillation. Collect the distillate boiling at 140~144℃. The pure product is obtained as a colorless liquid with b.p.142.4℃ and n_D^{20} 1.3992.

Notes

[1] In this experiment, the theoretical volume of water generated from the reaction system is about 1.5 mL. However the actual volume of water separated from the system should be slightly larger than the theoretical one. Otherwise, the yield would be lower.

[2] The appropriate temperature for preparation of di-*n*-butyl ether is 130~140℃. However, at the beginning of the refluxing reaction, it would be difficult to reach at this temperature because the generated di-*n*-butyl ether and water form an azeotrope containing 33.4% by weight of water (b.p. 94.1℃). And di-*n*-butyl ether, 1-butanol and water could also form an azeotrope containing 29.9% of water and 34.6% of 1-butanol (b.p. 90.6℃). In addition, 1-butanol and water form another azeotropic mixture containing 44.5% of water (b.p. 93℃). Therefore, it needs about 0.5 h of refluxing reaction at 100~115℃ before the temperature of the mixture reaches 130℃.

[3] When washing the mixture with aqueous NaOH, the funnel should not be shaken vigorously. Otherwise, the mixture would be emulsified and difficult to separate.

Questions

(1) How can we increase the yield of this reaction in this experiment?

(2) Why should the Dean-stark trap be filled with water with 2.0 mL of space unoccupied instead of being fully filled with water?

(3) Could we synthesize ethyl isopropyl ether from ethanol and isopropanol under acid catalysis? If not, how can we prepare it?

Exp.18 Preparation of Cyclohexanone

Objectives

(1) To learn the principles and method for the preparation of ketones via the chromic acid oxidation of alcohols.

(2) To learn the operations for oxidation reactions.

(3) To learn the methods for controlling the temperature and the rate of the reaction.

Principles

Ketones are a class of important starting materials for many organic syntheses. Ketones can be synthesized by the oxidation or dehydrogenation of secondary alcohols. The most common approach to the synthesis of ketones in industry is the dehydrogenation of secondary alcohols under high temperatures (250~350℃) catalyzed by transition metals such as Ag, Cu, Cu/Cr, etc. In laboratory, oxidation reactions are usually utilized for the synthesis of ketones. And the acidic sodium/potassium bichromate is a commonly used oxidant. In this experiment, cyclohexanone will be synthesized by the oxidation reaction of cyclohexanol using dichromic acid as the oxidant. The equation is shown as below:

$$\text{Cyclohexanol} \xrightarrow[50\sim55℃]{Na_2Cr_2O_7,\ conc.\ H_2SO_4} \text{Cyclohexanone}$$

Ketones are rather difficult to be further oxidized; therefore they are stable in the presence of oxidants. The oxidation of alcohols is an exothermic reaction, whereas it requires certain energy to complete the reaction. Therefore, the reaction temperature needs to be strictly controlled to avoid vigorous reaction.

Materials

Equipment: beaker, round-bottom flask, a distillation unit, thermometer, separatory funnel.

Reagents: cyclohexanol, $Na_2Cr_2O_7 \cdot 2H_2O$, conc. H_2SO_4, ether, refined salt, anhydrous $MgSO_4$.

Procedures

1. Preparation of Cyclohexanone

To a 50 mL beaker containing 20 mL of water, add 3.5 g of $Na_2Cr_2O_7 \cdot 2H_2O$ in batches under stirring. Stir the mixture until the solid is dissolved completely. Then add 3 mL of conc. H_2SO_4 dropwise via a syringe while stirring. An orange solution is obtained. Cool the solution to below 30℃ for the further steps.

To a 50 mL round-bottom flask charged with 3.5 mL of cyclohexanol, place the above obtained dichromic solution. Shake the flask to mix the materials well. In this process, monitor the temperature of the mixture with a thermometer. Keep the reaction temperature at 50~55℃ for about 15 min (once the temperature reaches 55℃, cool the mixture in a cold water bath)[1]. After that, remove the cold water bath and allow the flask stand at room temperature for 15 min. Shake the flask occasionally to facilitate the completion of the reaction. Finally, a deep green mixture would be generated.

2. The Work-up Procedures

Add 20 mL of water and two boiling chips into the above flask. Set up an apparatus for a simple distillation. In the distillation process, cyclohexanone would be distilled out together with water because cyclohexanone and water form an azeotrope (b.p. 95℃). When there are no more oily droplets coming out of the flask, stop distilling. About 15 mL of distillate would be collected.

The distillate is saturated by the addition of about 3 g of refined salt[2]. Then transfer it into a separatory funnel and let it stand for a while. Separate the liquid to collect the organic phase, and extract the aqueous phase with 5 mL of ether. Combine the organic layers and dry the obtained liquid over anhydrous $MgSO_4$. Transfer the organic phase into a round-bottom flask and distill it over a water bath. After the ether has been removed, collect the distillate boiling at 151~155℃. The b.p. of pure cyclohexanone is 155.7℃.

Notes

[1] This reaction is an exothermic process, and the reaction temperature should be controlled strictly.

[2] The addition of refined salt is to saturate the distillate and decrease the solubility of the product in water, leading to a better separation. Water should not be distilled out too much. Otherwise, a small amount of cyclohexanone would be dissolved and lost. The solubility of cyclohexanone in water at 31℃ is 2.4 g.

Questions

(1) The oxidation of cyclohexanol by dichromic acid gives cyclohexanone, but adipic acid is obtained when potassium permanganate is utilized as the oxidant. Please give your reasons.

(2) What would happen if the reaction temperature is too high or too low? What are the possible byproducts? Please give the related reaction equations.

(3) Why should the distillate be treated with salt in the work-up process?

Exp.19 Preparation of Benzalacetone and Dibenzalacetone

Objectives

(1) To study the principles and method for the growth of carbon chain using aldol condensation.

(2) To study the method for the preparation of specific products by controlling the feed ratio of staring materials.

Principles

Aldol condensation is the reaction of two molecules of aldehydes/ketones bearing active α-hydrogen in diluted acidic or basic solution to give corresponding β-hydroxy aldehydes/ketones through an intermolecular condensation; or the reaction followed by a further dehydrolysis to afford corresponding α,β-unsaturated carbonyl compounds. It is an important method for the preparation of α, β-unsaturated carbonyl compounds and a significant strategy for the growth of carbon chain.

There are two classes of aldol condensation: homo aldol condensation and cross aldol condensation. The cross aldol condensation between an aromatic aldehyde without a α-hydrogen and an aldehyde/ketone bearing an active α-hydrogen is called as Claisen-Schmidt reaction. This reaction provides an important method for the synthesis of the aromatic compounds with two types of functional group on the side chain and the aliphatic systems bearing several aryl rings. In the cross aldol condensation of benzaldehyde with acetone, two different products could be generated when different ratios of substrates are applied.

When excessive amount of benzaldehyde is used, the reaction equation is as follows.

$$2\ PhCHO + CH_3COCH_3 \xrightarrow[-2H_2O]{OH^-} PhCH=CHCOCH=CHPh$$

When excessive amount of acetone is used, the reaction equation is as below.

$$PhCHO + CH_3COCH_3 \xrightarrow[-H_2O]{OH^-} PhCH=CHCOCH_3$$

Benzalacetone (also called 5-phenyl-3-buten-2-one) has a flavor of coumarin. It could be utilized as the material for the preparation of perfume, the flavor-tuning agent of floral perfume, the mordant in dye industry and the brightener in electroplating industry.

Materials

Equipment: round-bottom flask, Büchner funnel, suction flask, magnetic stirrer, Erlenmeyer flask.

Reagents: benzaldehyde, acetone, 95% ethanol, anhydrous ethanol, glacial acetic acid, 10% aqueous NaOH, ether, 1 : 1 aqueous HCl, saturated aqueous NaCl, anhydrous $MgSO_4$.

Procedures

1. Preparation of Dibenzylideneacetone

To a 250 mL round-bottom flask is added a stir bar. While magnetically stirring, 5.3 mL (0.05 mol) of freshly distilled benzaldehyde, 1.8 mL (0.025 mol) of acetone, 40 mL of 95% ethanol and 50 mL of 10% NaOH are added successively into the flask. The reaction is stopped after the mixture is stirred for 20 min at room temperature.

The above reaction mixture is filtrated under reduced pressure. And the precipitates are washed by small amount of water. Then the filter cake is soaked and washed by the mixture of 1 mL of glacial acetic acid and 25 mL of 95% ethanol. Finally, the solid is washed by small amount of water[1]. The solid is transferred into a 100 mL Erlenmeyer flask and recrystallized in anhydrous ethanol[2].

Pure dibenzylideneacetone is obtained as a light yellow plate crystal with m.p. 110~111 ℃ (decomposed at 113 ℃).

2. Preparation of Benzalacetone

A stir bar is added to a 100 mL three-necked flask, which is connected with a dropping funnel, an Allihn condenser and a thermometer. While magnetically stirring, 22.5 mL of 10% NaOH and 4.0 mL (0.054 mol) of acetone are added successively into the flask[3]. Then 5.3 mL (0.05 mol) of freshly distilled benzaldehyde is added dropwise into the flask. Note that benzaldehyde should be added in an appropriate rate so that the temperature of the mixture varies between 25 and 30 ℃[4]. The mixture is further stirred for 30 min.

Then proper amount of 1 : 1 aqueous HCl is added into the flask through the funnel, till the pH value of the reaction mixture is almost 7.0. The mixture is transferred into a separatory funnel and the yellow oil is separated. The aqueous phase is extracted with ether (3×10 mL). And the combined organic layers are washed with 10 mL of saturated aqueous NaCl, dried over anhydrous MgSO$_4$ and filtered. The filtrate is distilled using a water bath and the ether is recovered. About 5 g of the product could be obtained.

Pure benzalacetone is obtained as a white or light yellow crystal with m.p. 42 ℃.

Notes

[1] During the preparation of dibenyzlideneacetone, the washing and soaking steps can be preceded on a Büchner funnel.

[2] If the reaction mixture becomes brownish red during the preparation of dibenyzlideneacetone, active charcoal can be used for the decoloration. During the second filtration process of the recrystallization, the mixture should be cooled to sub-zero temperatures in order to reduce the loss of the product.

[3] The amounts of benzaldehyde and acetone should be accurate (transfer pipettes can be utilized).

[4] The reaction temperature should not be too high. Otherwise, the byproduct yield would increase and the desired product yield would decrease.

Questions

(1) What are the possible side reactions in this experiment? How can we avoid or reduce these side reactions?

(2) What would occur if a higher concentration of base is applied?

(3) Why is brownish red product obtained in some cases? What should we do in these cases?

Exp.20 Preparation of Benzoin

Objectives

(1) To learn and understand the principles of benzoin condensation.

(2) To learn and be skilled in the operations for the benzoin condensation using vitamin B_1 as the catalyst.

(3) To be more adept in the basic operations such as cooling, refluxing, recrystallization, etc.

Principles

Benzoin condensation reaction is a coupling reaction that occurs between two aldehydes for the formation of parent benzoin. The reaction is traditionally catalyzed by a cyanide ion, which involves the use of highly poisonous sodium cyanide.

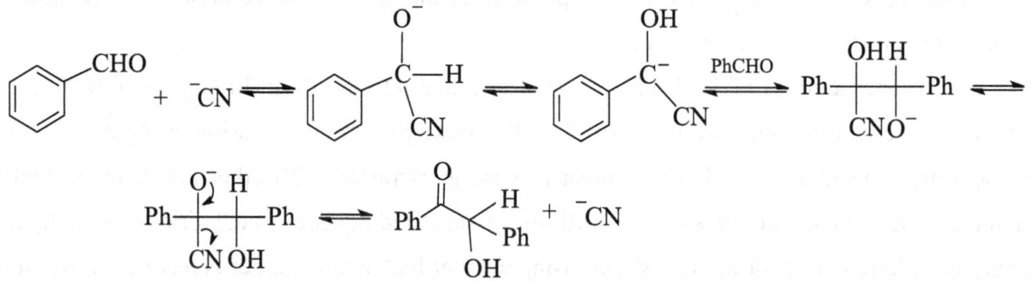

However, the traditional method utilized super toxic cyanides, which are dangerous to human health. A safer alternate catalyst for this reaction is vitamin B_1, namely thiamine, which can function in a manner completely analogous to cyanide ion in promoting benzoin condensation. Vitamin B_1 can act as a coenzyme in the biochemical processes such as the decarboxylation of α-keto acids and formation of acyloin (α-hydroxy ketones). The structure of vitamin B_1 is depicted as below.

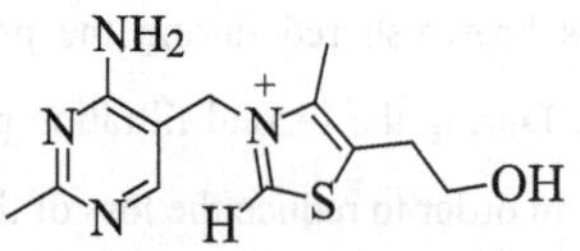

The hydrogen atom on the carbon between S and N of the thiazolium ring of vitamin B_1 is obviously acidic and could be captured by a certain base to form a carbanion, which promotes the benzoin condensation. The mechanism of the benzoin condensation catalyzed by vitamin B_1 is shown below:

Materials

Equipment: round-bottom flask, condenser, Büchner funnel, suction flask, thermometer, melting point apparatus.

Reagents: vitamin B_1, distilled water, freshly distilled benzaldehyde, 95% ethanol, 10% NaOH.

Procedures

In a 50 mL round-bottom flask, place 1.0 g of vitamin B_1[1], 2 mL of distilled water and 8 mL of 95% ethanol. Shake the flask frequently to dissolve vitamin B_1 completely. Then plug the flask tightly and cool the mixture in an ice-water bath. Meanwhile, place a tube containing 2 mL of 10% NaOH in the ice-water bath[2]. Then add the cooled 10% NaOH dropwise into the flask under vigorous shaking. Adjust the pH value of the mixture to be at 9~10 (the mixture turns to yellow). After that, remove the ice-water bath and add 5 mL (5.2 g, 0.05 mol) of freshly distilled benzaldehyde dropwise into the flask[3]. Add a few pieces of boiling chips in the flask and connect a condenser to the flask. Keep the mixture warmed in a water bath (60~75℃) for about 1.5 h (the temperature can be raised to 80~90℃ in the late stage). In this process, shake the flask occasionally and keep the pH value of the mixture at 9~10 (additional 10% NaOH might be needed to maintain the pH value). The mixture is changed into a homogeneous orange or orange-red solution[4]. After that, remove the water bath and cool the mixture to room temperature. Then place the flask in an ice-water bath to precipitate the crystals. Filter the mixture and wash the precipitates with a little cold water for 2 times and dry the crystals in air to afford light yellow solids as the crude product.

Purified the crude product by the recrystallization in 95% ethanol (activated charcoal can be used if necessary)[5]. Dry and weigh the product (about 2 g of the product would be obtained). And the melting point can be measured by a melting point apparatus.

Pure benzoin is obtained as white needles with m.p. 137℃.

Notes

[1] Vitamin B_1 is used as a catalyst in the reaction, and its quality would directly affect the reaction. Vitamin B_1 is stable in acid conditions but it is susceptible in base because its thiazole ring is easily open in basic solution. It is hygroscopic, and it can be easily oxidized in aqueous solution to lose its activity. Meanwhile, light and certain metal (such as copper, ion, and manganese) ions may accelerate the oxidation of vitamin B_1.

[2] The carbanion could be formed when vitamin B_1 is treated with NaOH. Cold NaOH solution is used in the first step to avoid the ring-opening reaction of the thiazole ring of vitamin B_1.

[3] Benzaldehyde always contains some benzoic acid, since it readily oxidizes in presence of air. Therefore, freshly distilled benzaldehyde should be utilized for this reaction.

[4] After the starting materials have been added, the reaction system should be kept alkalescent by adjusting its pH at 9~10 using short range pH papers, because the carbanion can be formed easily under alkaline conditions.

[5] 95% ethanol can be used as the solvent for the recrystallization of benzoin. Benzoin has a solubility of 12~14 g/100 mL in boiled hydrous ethanol.

Questions

(1) Explain the role of vitamin B_1 in the benzoin condensation (the catalytic mechanism).

(2) What are the differences between benzoin condensation, aldol condensation and Cannizarro reaction?

(3) Why should the benzaldehyde be freshly distilled before using in this experiment? And why should we keep the pH value of the mixture at 9~10 after benzaldehyde is added? What will be the result if the pH value is too high or too low?

Exp.21   Preparation of Benzoic Acid

Objectives

(1) To learn and understand the principles and methods for the preparation of benzoic acid via the photohalogenation of toluene.

(2) To develop the skills in the basic operations such as refluxing, suction filtration and drying.

Principles

In this experiment, benzoic acid is prepared by the bromination of toluene initiated by light, followed by the hydrolysis and acidification.

$$\text{PhCH}_3 + 3\text{Br}_2 \xrightarrow{h\nu} \text{PhCBr}_3 + 3\text{HBr}$$

$$\text{PhCBr}_3 + \text{H}_2\text{O} \xrightarrow{\text{NaOH}} \text{PhCOONa} \xrightarrow[\text{H}_2\text{O}]{\text{HCl}} \text{PhCOOH}$$

Materials

Equipment: UV lamp, magnetic stirrer, infrared lamp, constant pressure dropping funnel, three-necked flask, plastic basin, rubber tube, glass stopper, receiving flask.

Reagents: toluene, bromine, boiling chips, 20% NaOH, 10% HCl.

Procedures

1. Synthesis of Halogenated Intermediate Via Photobromination

As shown in Fig. 3.11, set up the apparatus with the flask placed in a black box or wrapped in metal foil. Put a magnetic stir bar in the 500 mL three-necked flask. Then add 2 mL (0.0188 mol) of toluene into the flask. Turn on the magnetic stirrer and the water pump[1].

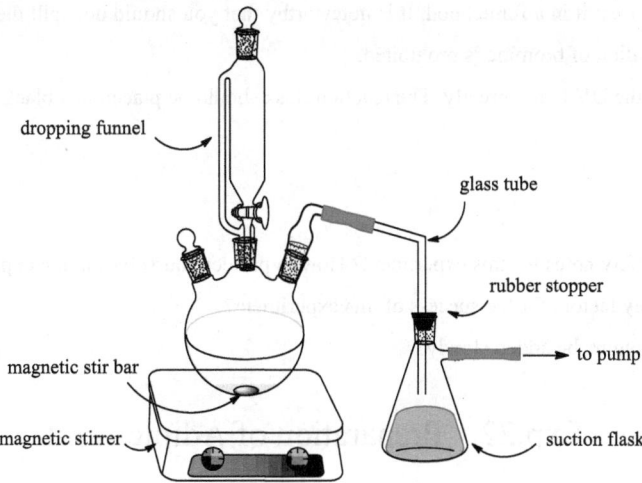

Fig.3.11 Apparatus for the photo promoted synthesis of benzoic acid

Then carefully add 3 mL (0.0585 mol) of bromine dropwise from the dropping funnel[2], keeping the adding speed at one drop every four to five seconds. Set the UV lampas close as possible to the flask to let the light shine on the mixture in the flask[3]. After bromine has been added completely, continue to stir the mixture under UV light until the solvent becomes light yellow. Finally, turn the UV lamp off.

2. Base Hydrolysis

Disconnect the dropping funnel and the water bump from the flask. Place 3 pieces of boiling chips and 20 mL of 20% NaOH solution into the flask. Sake the flask to mix the materials well. Heat the flask on a heating mantle and keep the mixture under refluxing for about 2 h with constant shaking constantly. After that, cool the flask to room temperature and add 20 mL of water. Filter the mixture under reduced pressure and collect the filtrate for further steps.

3. Acidification

To the above filtrate, add 10% HCl solution dropwise under stirring until the mixture becomes acidic indicated by a litmus paper. Filter the mixture under reduced pressure, dry and collect the crystals. Then put the crystals in an 80℃ oven for a deep drying.

Benzoic acid can be further purified by sublimation on a micro flame burner. The product is put into a preweighed bottle and the yield of the product could be calculated.

Notes

[1] In order to facilitate the reaction, it is necessary to use water pump to remove the hydrogen bromide formed in the reaction.

[2] Bromine is highly corrosive and irritant, therefore you should be very careful when using it. You should wear rubber gloves and treat it in a fume hood. It is noteworthy that you should not spill the liquid bromine out, and direct contact and inhalation of bromine is prohibited.

[3] **NEVER** view the UV lamp directly. The reaction flask should be placed in a black box or wrapped in metal foil.

Questions

(1) What are the safety notes for this experiment? How to protect yourselves in the experiment?

(2) What are the key factors for the success of this experiment?

(3) Why should bromine be added slowly?

Exp.22 Preparation of Adipic Acid

Objectives

(1) To learn the principles and method for the preparation of adipic acid by the oxidation of cyclohexanol or cyclohexene.

(2) To learn how to make a mechanical stirring.

(3) To be more skilled in the basic operations such as concentration, filtration and recrystallization.

Principles

Adipic acid or hexane dioic acid is an important dicarboxylic acid. It is a white solid powder with a slightly acidic taste. And it is slightly soluble in water, highly soluble in ethanol, methanol, isopropanol, ethers and etc. In industrial, the condensation polymerization of apidic acid with hexamethylenediamine can lead to the preparation of nylon-66. It is also an intermediate for the preparation of polyurethanes and hot melt adhesives. Adipic acid can also react with alcohols to generate adipates, which could be used as the plasticizers for the preparation of PVC and PVC copolymers, polystyrene (PS), ethylcellulose and nitrocellulose polymer.

At present, the common approaches for the preparation of adipic acid in the laboratory textbooks are the oxidation reactions of cyclohexanol with 50% HNO_3 or $KMnO_4$. However, these reactions can release toxic and harmful byproducts, and suffer from some other drawbacks such as long reaction times and/or tedious work-up procedures. In order to overcome these drawbacks and reduce pollution, in this experiment, H_2O_2 (30%) is used as the oxidant with Na_2WO_4/H_3PO_4 as the catalysts and cyclohexene as the reactant. The method is green with some advantages such as mild conditions, easily controllable process, shorter time, easy separation of the products and with no harmful substances produced. The reaction equation is shown as follows.

$$\text{cyclohexene} \xrightarrow[\Delta]{\substack{30\% \text{ } H_2O_2 \\ Na_2WO_4/H_3PO_4}} HO_2C\text{-}(CH_2)_4\text{-}CO_2H$$

Materials

Equipment: mechanical stirrer, three-necked flask, beaker, condenser, constant pressure dropping funnel, exhaust gas absorber, water bath, the equipment for suction filtration.

Reagents: cyclohexanol, 50% HNO_3, NH_4VO_3, cyclohexene, 30% H_2O_2, Na_2WO_4, H_3PO_4.

Procedures

1. The NH_4VO_3-catalyzed Oxidation of Cyclohexanol by HNO_3

To a 100 mL three-necked flask, add 8 mL (0.085 mol) of 50% HNO_3 and 1 small particle of NH_4VO_3[1]. Connect a mechanical stirrer, a constant pressure dropping funnel and a condenser to the flask. Equip an exhaust gas absorber (using a basic solution to absorb NO_2 gas produced in the reaction) on the top of the condenser. Place 2.7 mL (0.025 mol) of cyclohexanol in the funnel. The flask is preheated at about 50℃ in a water bath. Then remove the water bath and add 5～6 drops of cyclohexanol into the flask under stirring. After the reaction has been initiated, the reaction mixture would become warm and some reddish brown gases would be released[2]. Then add the rest of cyclohexanol slowly. Note that cyclohexanol should be added carefully so that the reaction mixture is kept gently boiling[3]. Use cold or hot water to control the temperature. After the addition of the starting material (it may need 15 min) is complete, heat the flask in a boiling-water bath for about 10 min till there are almost no observable reddish brown gases released from the system. After that, pour the reaction mixture fully into a beaker that has been cooled in cold water. After the crystals have precipitated completely, collected the crystals by filtration under reduced pressure. Wash the crystals with a small amount of cold water[4]. Dry and then weigh the obtained crystals (2～2.5 g, with m.p. 149～155℃). After recrystallization in water, the product is obtained as a solid (about 2 g) with m.p. 151～152℃. Pure adipic acid presents as white prismatic crystals with m. p. 153℃. This experiment would take 3～4 h.

2. The Na$_2$WO$_4$/H$_3$PO$_4$-catalyzed Oxidation of Cyclohexene by H$_2$O$_2$

The preparation of the peroxide: to a 100 mL three-necked flask, add a magnetic stir bar, Na$_2$WO$_4$ (0.88 g), H$_3$PO$_4$ (0.6 g) and 22.3 mL of 30% H$_2$O$_2$. Then connect a condenser, a thermometer and a constant pressure dropping funnel onto the flask. Stir the mixture at 50~60℃ for about 0.5 h. Note that the temperature should not exceed 70℃ and the mixture should not be stirred for a long time[5].

The preparation of the crude adipic acid: add 5 mL (0.049 mol) of cyclohexene slowly from the constant pressure funnel into the flask. After the addition is complete, stir the reaction mixture under reflux (102℃) for about 1 h. Stop heating once the solution becomes deep green[6].

The purification of the product: cool the above reaction mixture to room temperature, and poured it into a 100 mL beaker. Acidify the solution with 10% HCl till the pH value of the solution is between 1 and 2. Concentrate the solution to about 15 mL, and cool the condensed solution in an ice-water bath. Collect the solid by filtration under reduced pressure. Dry and weigh the obtained crude product. Then its melting point is measured to test its purity.

Notes

[1] Be cautious when you transfer the nitric acid, since it's highly corrosive and irritant.

[2] The rest of cyclohexanol should be added after the success initiation of the reaction.

[3] Cyclohexanol should be added slowly so that the reaction mixture is kept gently boiling.

[4] A minimum amount of water should be used, otherwise more products would be dissolved in water and lost.

[5] H$_2$O$_2$ would be decomposed at a higher temperature.

[6] During the early stage, a cyclohexene-water-H$_2$O$_2$ azeotrope is formed. Therefore the reaction temperature would be around 70℃ for 20~60 min. When the reaction temperature reached 102℃, the solution become light pale, the reaction should not be stopped at this moment. During this stage, Mo^{6+} species is reduced into Mo^{5+} or Mo^{4+} species by cyclohexene. However, when the reaction become deep green, the activity of the catalyst has been suppressed and the reaction can be stopped.

Questions

(1) Please list the advantages and disadvantages of the two methods for the preparation of adipic acid.

(2) In the first reaction, why should a basic solution be used to absorb the exhaust gas?

(3) In the second reaction, why can the reaction be stopped when the reaction mixture becomes deep green?

References

Ren D M, Wang J, Zhao Y, et al. 2016. The teaching improvement for the green synthesis of adipic acid. Technology Outlook, 01(01):185.

Exp.23 Preparation of Cinnamic Acid

Objectives

(1) To master the principle and method of preparing cinnamic acid (Perkin reaction).

(2) To understand the principles of steam distillation and to be familiar with the operations.

Principles

The Perkin reaction refers to the condensation reaction of an aromatic aldehyde (such as benzaldehyde) containing no α-H with an organic acid anhydride under the action of a basic catalyst.

The commonly used catalysts in Perkin reaction are sodium and potassium salts of the carboxylic acids corresponding to the anhydrides used in the reactions or K_2CO_3 as well as tertiary amines. The main role of the catalyst is to promote enolization of the anhydride for the generation of anhydride anion.

Perkin reaction of benzaldehyde with acetic anhydride in the presence of anhydrous potassium acetate yields cinnamic acid.

Reaction:

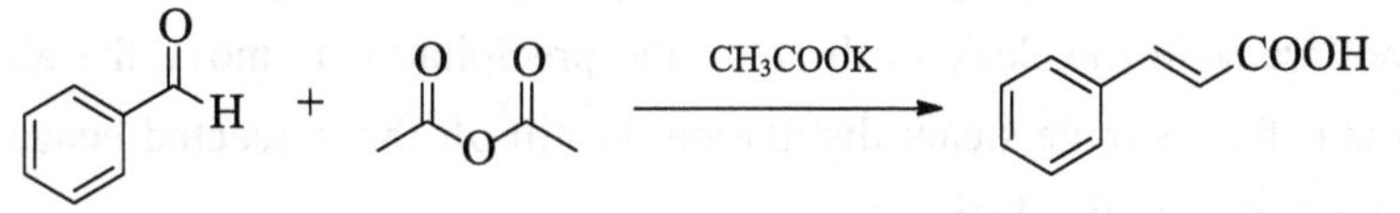

Mechanism:

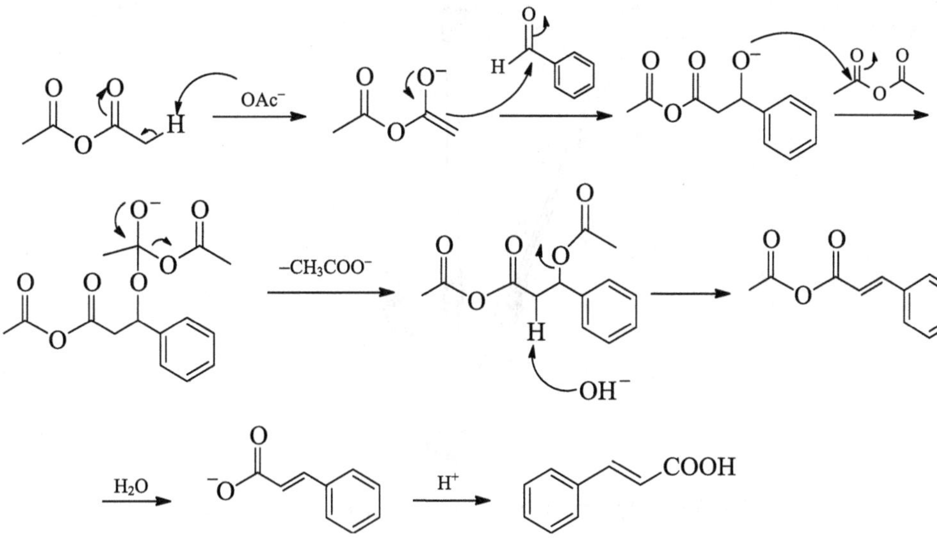

Materials

Equipment: three-necked flask, air condenser, beaker, thermometer (250 ℃), distillation adapter, Erlenmeyer flask, steam distillation unit, suction flask, Büchner funnel, heating mantle.

Reagents: freshly distilled benzaldehyde, acetic anhydride, anhydrous potassium carbonate, 10% NaOH solution, activated charcoal, conc. hydrochloric acid, ethanol, Congo red test paper. Physical properties of the reactants and product are shown in Table 3.1.

Table 3.1 Physical properties of the reactants and product

Compounds	M/(g/mol)	ρ/(g/mL)	m.p./°C	b.p./°C
PhCHO	106.12	1.04	−26	179
K_2CO_3	138.21	2.428	891	333.6
PhCH=CH—COOH	148.17	1.245	135~136	300

Procedures

1. Preparation

In a 100 mL three-necked flask, introduce 1.5 mL (0.015 mol) of freshly distilled benzaldehyde, 4 mL (0.036 mol) of acetic anhydride and 2.2 g of anhydrous potassium carbonate, Shake the mixture gently. Install an apparatus as shown in Fig. 3.12[1]. The reaction mixture is heated to boiling and maintained under reflux for 40 min with the temperature kept at 150~170°C[2]. After the reaction mixture is slightly cooled down, precipitation maybe occur, add approximately 5~10 mL of hot water in the flask to dissolve the precipitates. Remove the air condenser, and rebuilt an apparatus for a simple steam distillation. Distill off the unreacted benzaldehyde until no oily droplets are observed in the distillate.

Fig. 3.12 Apparatus for the preparation of cinnamic acid

2. Purification

Cool down the reaction mixture a little. Then add 10% sodium hydroxide solution into the flask dropwise under stirring to alkalinize the solution to reach a pH value of 9~10, so as to change cinnamic acid into water-soluble sodium salt. If there are still some solids undissolved, add some warm water in the solution under stirring until no more solids dissolve. A little activated carbon is added in the alkalinized solution when the solution has a deep color. After that, heat the solution to boiling for 2~3 min and make a hot filtration. The filtrate is transferred into a 250 mL beaker and acidified with 1 : 1 hydrochloric acid under stirring until the solution turns Congo red test strip into blue[3]. Cool the mixture thoroughly. Apply a suction filtration to get the crystals, wash the crystals with a small amount of water while filtering. Thereafter, the crude product can be purified by recrystallization in 30% ethanol. Dry the crystals under 85℃, weigh them and calculate the yield.

Pure cinnamic acid forms white flake crystals with m. p. 135~136℃ (*trans*).

Notes

[1] Glassware must be completely dry when running the reaction, because acetic anhydride is easily hydrolyzed to acetic acid and anhydrous K_2CO_3 is also highly susceptible to moisture.

[2] The temperature of the reaction must be set so that the reflux ring should only be one-third to half way up the condenser. And the suitable reaction time is about 40 min.

[3] Make sure the test paper show obvious color change when making the acidification procedures.

Questions

(1) Can the acidification be made by using concentrated sulphuric acid? Is it possible to acidify with concentrated sulphuric acid?

(2) Please write down the reaction equations for the reaction of propionic anhydride with benzaldehyde in the presence of potassium propionate.

(3) What is the meaning of using a steam distillation in this experiment? Could the steam distillation be replaced by a simple distillation?

Exp.24 Preparation of Coumarin-3-carboxylic Acid

Objectives

(1) To understand the existence of coumarin compounds in nature and their biological significance.

(2) To master the principles and methods of preparing coumarin-3-carboxylic acid by using Knoevenagel reaction.

Principles

1. Introduction

Coumarin, with its chemical name of 1, 2-benzopyrone or *o*-hydroxy cinnamic acid lactone, is a white rhombic crystal or crystalline powder found in many natural plants. It was first discovered from the seeds of coumarin in 1820 and also in essential oils of lavender and cinnamon. Coumarin has the aroma of lemongrass; therefore it is an important spice, commonly used as a flavoring agent and can be used for the preparation of perfume, toilet water, flavor, etc. It is also used in some rubber and plastic products. Coumarin derivatives can also be used as pesticides, rodenticides, medicine and so on. Due to the low content of coumarins in natural plants, most of them are obtained synthetically. In 1868, Perkin obtained coumarin with o-hydroxybenzaldehyde (salicylaldehyde) heated with acetic anhydride and potassium acetate. This method is called Perkin synthesis. The reaction process is as follows:

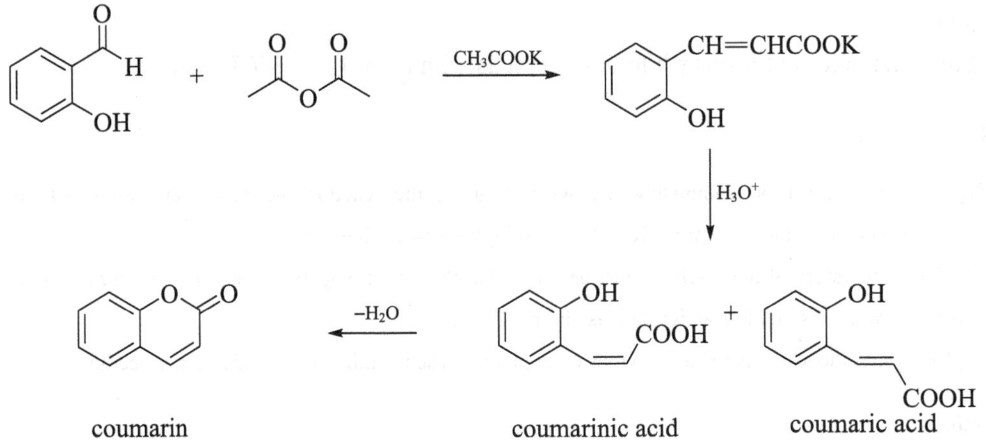

However, due to long reaction time, high reaction temperature and the unstable yield, Perkin method will not be used in this experiment.

2. The Method by Using Knoevenagel Reaction

Instead, coumarin-3-carboxylic acid is synthesized by the reaction of salicylaldehyde and diethyl malonate under the catalysis of organic base at low temperature. This modified aldol reaction under the action of an organic base is called the Knoevenagel reaction. The Knoevenagel reaction of salicylaldehyde and diethyl malonate under the catalysis of piperidine affords coumarin-3-carboxylic acid ethyl ester, which is used to form coumarin-3-carboxylic acid by a series of followed reactions including the hydrolysis in sodium hydroxide acidification with hydrogen chloride and lactonization.

In this method, an active methylene compound with one or two electron-withdrawing groups to increase the activity of methylene hydrogen, is used to take the place of the anhydride in the Perkin method. Meanwhile, the using of weaker organic bases can avoid self-condensation of aldehydes, expanding the scope of condensation reaction in the raw materials.

Reaction:

Materials

Equipment: water bath, round-bottom flask, condenser, drying tube, Erlenmeyer flask, vacuum filtration unit, melting point apparatus.

Reagents: salicylaldehyde, diethyl malonate, anhydrous ethanol, 95% ethanol, piperidine, glacial acetic acid, NaOH, conc. HCl, anhydrous $CaCl_2$.

Physical properties of the reagents are shown in Table 3.2.

Table 3.2 Physical properties of the reagents

Compounds	M/(g/mol)	ρ/(g/mL)	m.p./℃	b.p./℃
salicylaldehyde	122.12	1.17	−7	197
diethyl malonate	160.17	1.0551	−50	199.3
piperidine	85.15	3.0	−7	106
coumarin-3-carboxylic acid ethyl ester	218.2054	1.289	92~93	378
coumarin-3-carboxylic acid	190.15	—	190	—

Procedures

1. Synthesis of Coumarin-3-carboxylic Acid Ethyl Ester

To a 50 mL dry round-bottom flask, add 1.3 mL (1.8 g, 0.015 mol) of salicylaldehyde, 2.6 g (0.016 mol) of diethyl malonate, 10 mL of absolute ethanol, and 10 drops of piperidine and 1 drop of glacial acetic acid[1]. Then put a few boiling chips into the mixture. The reaction system was fitted with reflux condenser. The upper end of the condenser needs to be connected with a calcium chloride drying tube. The reaction mixture is heated under reflux on a water bath for 0.5 h.

After the reaction mixture is cooled slightly, the liquid is transferred to an Erlenmeyer flask with 8 mL of water and is then cooled down in a cooling bath to crystallize the product completely. Filter the mixture under reduced pressure to obtain the crystals. The crystals are washed with cold 50% ethanol twice (each 3~5 mL) and the crystals are compressed to drain the liquor[2]. The crude coumarin-3-carboxylic acid ethyl ester is obtained as white crystals.

After the product has dried, weigh it and calculate the percentage yield. If necessary, the crude product can be recrystallized from 25% ethanol. By measuring the melting point of crude product, its purity can be tested. The melting point of pure coumarin-3-carboxylic acid ethyl ester is 92~93℃.

2. Synthesis of Coumarin-3-carboxylic Acid

In a 50 mL round-bottom flask, 2.0 g of coumarin-3-carboxylic acid ethyl ester, 1.5 g of sodium hydroxide, 10 mL of 95% ethanol and 5 mL of water are added. A few boiling chips are added in the flask. Install the reflux condenser and heat the mixture to reflux. When the ester and sodium hydroxide are completely dissolved, continue to heat under reflux for 15 min.

The hot reaction solution is poured into dilute hydrochloric acid (7.5 mL of conc. HCl in 25 mL of water) for acidification. The reaction mixture is cooled in an ice-water bath for complete precipitation of the crystals; the crystals are filtered out under reduced pressure and washed with a small amount of ice water twice. The crystals can be dried by pressing them gently to afford the crude product, coumarin-3-carboxylic acid. Dry and weigh the crude product. The crude product can be purified by a further recrystallization with water to give a white powdery solid. The melting point of pure coumarin-3-carboxylic acid is 190℃ (decomposed).

Notes

[1] Adding a small amount of glacial acetic acid may cause an acid-catalyzed reaction of o-hydroxybenzaldehyde and piperidine first to form an imino compound, which then reacts with diethyl malonate carbanion.

[2] Reducing the solubility of coumarin-3-carboxylic acid ethyl ester in ethanol can reduce product loss.

Questions

(1) What is the purpose of using glacial acetic acid in the reaction?

(2) How to prepare coumarin by using coumarin-3-carboxylic acid?

Exp.25　Preparation of Furfuryl Alcohol and Furoic Acid

Objectives

(1) To understand the basic principles and methods for the preparation of furfuryl alcohol and furoic acid from furfural via the Cannizzaro reaction.

(2) To further consolidate the skills needed for operations of washing, extraction, simple distillation, vacuum filtration and recrystallization.

Principles

Cannizzaro reaction is a disproportionation reaction. It refers to a redox reaction in which two molecules of a non-enolizable aldehyde are reacted to produce an equimolar mixture of a primary alcohol and a carboxylic acid using a strong alkaline. Furoic acid and furfuryl alcohol can be prepared by the Cannizzaro reaction of furfural.

Reaction:

$$2 \text{ furfural-CHO} \xrightarrow{\text{NaOH}} \text{furfural-CH}_2\text{OH} + \text{furfural-COONa}$$

$$\text{furfural-COONa} \xrightarrow{\text{HCl}} \text{furfural-COOH} + \text{NaCl}$$

Mechanism:

$$\text{Ph}-\overset{\overset{H}{|}}{C}=O + OH^- \xrightleftharpoons{\text{fast}} \text{Ph}-\overset{\overset{H}{|}}{\underset{\overset{|}{OH}}{C}}-O^- \xrightarrow{\text{slow}} \text{Ph}-C=O \cdots H$$

$$\underset{\text{acid}}{\overset{\text{Ph}}{\underset{HO}{>}}C=O} + \underset{\text{alkali}}{\overset{\text{Ph}}{\underset{H}{>}}\overset{H}{\underset{|}{C}}-O^-} \xrightarrow{\text{proton exchange}} \overset{\text{Ph}}{\underset{-O}{>}}C=O + \text{Ph}-\overset{H}{\underset{H}{\overset{|}{C}}}-OH$$

$$\xrightarrow{H^+} \overset{\text{Ph}}{\underset{HO}{>}}C=O + \text{Ph}-\overset{H}{\underset{H}{\overset{|}{C}}}-OH$$

Materials

Equipment: beaker, magnetic stirrer, separatory funnel, round-bottomed flask, water condenser, thermometer (250 ℃), heating mantle.

Reagents: furfural, 40% NaOH solution, ether, HCl, anhydrous $MgSO_4$.

Physical properties of the reagents are shown in Table 3.3.

Table 3.3 Physical properties of the reagents

Compounds	M/(g/mol)	ρ/(g/mL)	b p /℃	Solubility
furoic acid	112.0835	—	—	—
furfuryl alcohol	98.10	1.1296	171	soluble in ether
furfural	96.09	2.428	161.7	soluble in ether

Procedures

1. Preparation

In a small beaker, place 4 g sodium hydroxide and 6 mL water. Then cool the mixture to 5 ℃ in an ice bath. Add 6.6 mL furfural dropwise to the sodium hydroxide solution (in about 10 min) while stirring with a magnetic stirrer[1]. Maintain the reaction temperature at 8~12 ℃. After the addition is completed, allow the reaction mixture to stand at room temperature under stirring for

20 min[2] to give yellow syrup. Add appropriate amount of water (about 5 mL) to the beaker with stirring to dissolve the syrup[3], and the solution turns dark red at this point.

2. Separation

Extract the reaction mixture with ether (3×10 mL). Combine all the ether fractions and keep the aqueous fraction.

3. Purification of Furfuryl Alcohol

Dry the combined extracts by using 2 g of anhydrous magnesium sulfate. Filter the ether solution into a distillation flask and heat the solution over a hot-water bath to remove ether. Heat the residue in a heating mantle, and collect the fraction at 169~172 ℃ to afford furfuryl alcohol.

4. Purification of Furoic Acid

Add concentrated hydrochloric acid in the aqueous fraction to acidify it to pH 3[4]. Cool down the mixture completely to precipitate crystals. Filter the mixture to obtain the crystals of furoic acid, wash the crystals with a little water, dry and weigh the crystals.

Pure furfuryl alcohol is a colorless transparent liquid with b. p. 171 ℃ and n_D^{20} 1.4868.

Pure furoic acid forms white needle crystals with m. p. 133~134 ℃.

Notes

[1] It is necessary to use the freshly distillated furan formaldehyde.

[2] The reaction mixture must be stirred vigorously to ensure efficient mixing because it is heterogeneous.

[3] Add minimum amount of water to wash the resulting precipitate, otherwise the dissolvation of furoic acid in water will result in loss of the product.

[4] Add enough acid when acidifying the aqueous fraction so as to make complete precipitation of furoic acid.

Questions

(1) Why furfural must be redistilled before being used? What impurities can be generated by furfural after a long-term storage? If the impurities are not removed beforehand, what impact can be made on this experiment?

(2) In this experiment, you should add furfural dropwise into sodium hydroxide solution. If the feeding sequence is reversed, what changes can be found in the reaction? Is there any impact on the yield?

(3) What are the key factors affecting the product yield?

Exp.26 Preparation of Ethyl Acetate

Objectives

(1) To understand the principle and methods for the preparation of ethyl acetate.

(2) To further consolidate the skills needed for the operations of reflux, distillation, extraction, etc.

Principles

Main reaction:

$$CH_3COOH + C_2H_5OH \xrightleftharpoons[120\,°C]{conc.\ H_2SO_4} CH_3COOC_2H_5 + H_2O$$

Side reaction:

$$2C_2H_5OH \xrightarrow[140\,°C]{conc.\ H_2SO_4} CH_3CH_2OCH_2CH_3 + H_2O$$

$$C_2H_5OH \xrightarrow[170\,°C]{conc.\ H_2SO_4} H_2C{=}CH_2 + H_2O$$

Materials

Equipment: round-bottom flask, thermometer (100 °C), thermometer adapter, separatory funnel, distillation head, Erlenmeyer flask, water condenser, heating mantle.

Reagents: anhydrous ethanol, glacial acetic acid, conc. H_2SO_4, 10% Na_2CO_3 solution, saturated NaCl solution, saturated $CaCl_2$ solution, anhydrous $MgSO_4$ (or anhydrous K_2CO_3).

Physical properties of the reagents are shown in Table 3.4.

Table 3.4 Physical properties of the reagents

Compounds	M/(g/mol)	ρ/(g/mL)	m.p./°C	b.p./°C
ethanol	46.07	0.789	−114.1	78.3
ethyl acetate	88.11	0.902	−84	77
ethylene	28.06	—	−169	−103.7

Procedures

1. Preparation of Ethyl Acetate

In a dry 50 mL round-bottomed flask, introduce 12 mL (0.2 mol) of anhydrous ethanol and 6 mL (0.1 mol) of glacial acetic acid[1]. Add 0.5 mL of conc. H_2SO_4[2] slowly in the flask under shaking[3]. Place a few boiling chips in the mixture and heat the reaction mixture under reflux for 0.5 h. After the reaction flask is cooled down slightly, change the process from reflux to distillation with the changing of apparatus. Distill out ethyl acetate in the flask until the volume of the distillate has reached about 1/2 the volume of the reactants.

2. Purification

Add 10 mL of water to the distillate, and then transfer the solution into a separatory funnel. The organic phase is separated from the mixed liquor and the aqueous phase is discarded. The organic phase is washed with 10 mL of 10% Na_2CO_3 solution[4] and 5 mL of saturated NaCl solution, 5 mL of saturated $CaCl_2$ solution and 5 mL of water in turn[5]. The organic phase is transferred to a dry Erlenmeyer flask with a stopper and dried over an appropriate amount of anhydrous magnesium sulfate[6].

Filter the dried solution to remove magnesium sulfate. Transfer the filtrate into a 25 mL distillation flask for distillation[7], collecting 73~78℃ fractions. Weigh the product and calculate the yield.

Pure ethyl acetate has a fruity flavor with b. p. 77℃ and n_D^{20} 1.3723.

The IR and ^{1}H NMR spectra of ethyl acetate are shown in Fig. 3.13 and Fig. 3.14, respectively.

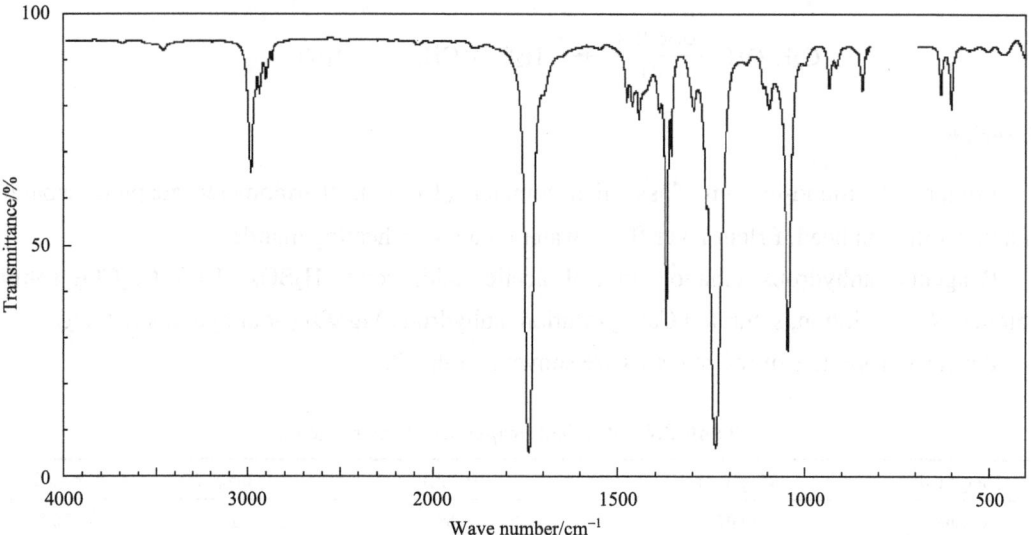

Fig. 3.13　The IR spectrum of ethyl acetate

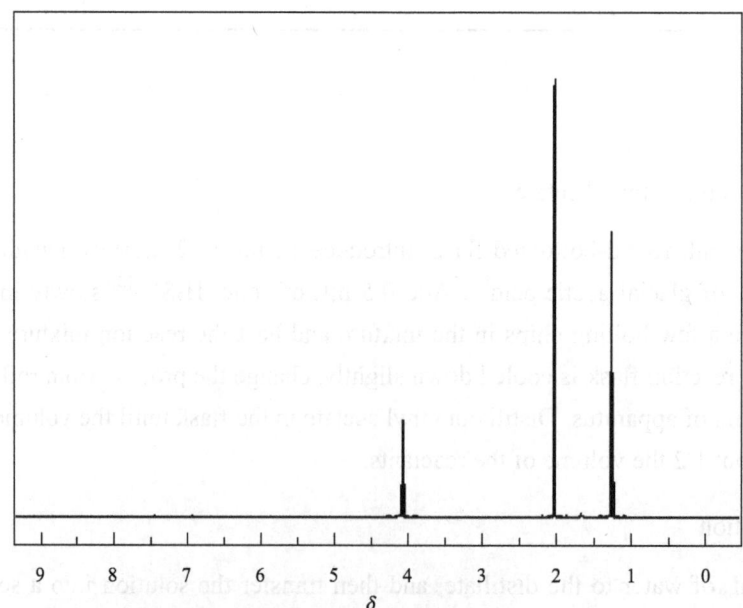

Fig. 3.14　The ^{1}H NMR spectrum of ethyl acetate

Notes

[1] The equipment used for the esterification reaction must be dry to prevent the water from affecting the balance of the reaction.

[2] The concentrated sulfuric acid used is very hazardous. Wash spills off yourself immediately with large amounts of water. Neutralize spills on the laboratory bench with baking soda.

[3] The reaction mixture must be mixed well before being heated; otherwise it will be easy to carbonize.

[4] Carbon dioxide is generated while the organic phase is washed with 10% Na_2CO_3 solution. Therefore, attention should be paid to the timely release of gas when washing the product.

[5] The distillate should be washed with saturated sodium chloride solution before being washed with saturated calcium chloride solution. Otherwise, it is difficult to make the following liquid separation because precipitation of calcium carbonate would occur.

[6] The organic phase should be thoroughly dried before distillation and the desiccant should not be transferred to the distillation flask. Ethyl acetate and water can form a binary azeotrope having a boiling point of 70.4℃ (containing 8.1% water). Ethyl acetate, ethanol and water can form a ternary, azeotrope-like mixture with a boiling point of 70.2℃ (containing 8.4% ethanol and 9% water).

[7] Do not forget to add boiling chips before reaction and distillation. The receiver used for distillation should be cooled with ice water to reduce product loss.

Questions

(1) What are the characteristics of esterification reaction? How to create the conditions so that the esterification reaction is driven toward product formation as complete as possible in this experiment?

(2) What are the possible side reactions of this experiment, and how can we avoid them?

(3) Whether can we use excessive acetic acid or not? Why?

Exp.27　Synthesis of Dibutyl Phthalate

Objectives

(1) To learn the principle and methods of preparing dibutyl phthalate.

(2) To learn to use the Dean-Stark trap and master the skills for the operations of vacuum distillation.

Principles

Dibutyl phthalate is an organic compound commonly used plasticizer. It is produced by the reaction of n-butanol with phthalic anhydride catalyzed by a strong acid (such as conc. H_2SO_4). The reaction equation is as follows:

$$\text{phthalic anhydride} + n\text{-}C_4H_9OH \xrightarrow{H_2SO_4} \text{o-}C_6H_4(COOC_4H_9\text{-}n)(COOH)$$

$$\text{o-}C_6H_4(COOC_4H_9\text{-}n)(COOH) + n\text{-}C_4H_9OH \xrightarrow{H_2SO_4} \text{o-}C_6H_4(COOC_4H_9\text{-}n)_2$$

Materials

Equipment: two-necked flask, round-bottom flask, condenser, Dean-Stark trap, Erlenmeyer flask, thermometer, Büchner funnel, suction flask, separatory funnel.

Reagents: phthalic anhydride, n-butanol, conc. H_2SO_4, saturated NaCl solution, anhydrous Na_2SO_4, 5% Na_2CO_3 solution.

Procedures

In a 100 mL two-necked flask, introduce 1.5 g (0.010 mol) of phthalic anhydride, 3.3 mL (0.036 mol) of n-butanol and 2 drops of conc. H_2SO_4, and stir the mixture well. Set up the apparatus as shown in Fig. 3.10 and fill the Dean-stark trap with n-butanol until the liquid lever is just below the level of the side-arm of the trap.

Heat the reaction mixture under gentle reflux until the temperature of the mixture reaches 140℃; this may cost about 2 h[1].

Cool the reaction solution to 70℃ or lower, then transfer it into a separatory funnel and neutralize it with 20~30 mL of 5% Na_2CO_3 solution. Wash the mixture twice with 20 mL of warm saturated NaCl solution each[2]. Then, the organic layer is dried over anhydrous Na_2SO_4.

The crude product is heated to distil off excessive n-butanol first on a water pump and finally on an oil pump to collect the product under vacuum. Weigh the weight of the pure product obtained.

The pure dibutyl phthalate is colorless oily liquid with b.p. 340℃ (175~176℃/ 666.5 Pa) and n_D^{20} 1.4911.

Notes

[1] The reaction should be stopped when the temperature reaches 140℃.

[2] To prevent the saponification reaction, the concentration of the alkaline used in the neutralization process should not be too high; the neutralization temperature should be 70℃ or lower.

Questions

(1) What are the reactions of n-butanol under heating with concentrated sulfuric acid as catalyst?

(2) What is the negative effect of heating too fast in the reaction process?

Exp.28 Synthesis of Ethyl Acetoacetate

Objectives

(1) To learn the principle and methods of preparing ethyl acetoacetate by Claisen condensation of ethyl acetate.

(2) To master the technique for the operations of reflux and vacuum distillation.

Principles

Ethyl acetoacetate (EAA) is mainly used as a chemical intermediate in the production of a wide variety of compounds. It is prepared via the Claisen condensation of ethyl acetate. The Claisen condensation is a carbon-carbon bond forming reaction that occurs between two esters or one ester and another carbonyl compound in the presence of a strong base, resulting in a β-keto ester or a β-diketone.In this experiment, ethyl acetoacetate is prepared by the condensation of two moles of ethyl acetate in the presence of sodium ethoxide (which is first generated by the reaction between sodium and trace of ethanol contained in ethyl acetate).

Reaction:

$$2CH_3COOC_2H_5 \xrightarrow{C_2H_5ONa} CH_3COCH_2COOC_2H_5 + C_2H_5OH$$

Materials

Equipment: round-bottom flask, water condenser, drying tube, separatory funnel, Erlenmeyer flask, thermometer, heating mantle, Büchner funnel, suction flask, sodium press.

Reagents: ethyl acetate, sodium, 50% acetic acid, $CaCl_2$, saturated NaCl solution, anhydrous Na_2SO_4.

Procedures

In a dry 50 mL round-bottom flask, add 10 mL of dry ethyl acetate[1], and then introduce 0.9 g of freshly extruded sodium wire into the flask rapidly[2]. Next, fit a water condenser into the flask with the joints lubricated with grease, and place a calcium chloride drying tube on the top of the condenser right away.

Heat the mixture under gentle reflux over a heating mantle until all the sodium beads disappear. Cool down the mixture slightly and add 50% acetic acid solution until the mixture turns acidic (pH=5~6).

Pour the resulting solution into a separatory funnel and add in saturated NaCl solution in the same amount of the resulting solution. Separate the phases, collecting the upper organic phase. Dry the obtained organic liquid over anhydrous Na_2SO_4, filter and remove ethyl acetate with a simple distillation over a warm water bath to obtain the crude product. Finally, distill the residue in the flask under reduced pressure[3], collecting the fraction boiling at a certain temperature. Weigh the mass of the pure product obtained.

Pure ethyl acetoacetate is a colorless liquid with b.p. 180.4℃ (decomposed) and n_D^{20} 1.4194.

Notes

[1] All the parts of the apparatus and ethyl acetoacetate used must be dried thoroughly before being used in the vacuum distillation.

[2] Sodium easily reacts with oxygen and water accompanying heat release. Sodium combustion can occur in the atmospheric air at a temperature high enough. Therefore, sodium metal should be handled carefully at all times.

[3] Ethyl acetoacetate will decompose when being collected by a simple distillation. So it is purified by vacuum distillation.

Questions

(1) What will happen if the equipment or ethyl acetoacetate in the vacuum distillation are not dried?

(2) What are the purposes of adding 50% acetic acid solution and saturated sodium chloride solution?

(3) Why should we use vacuum distillation for the purification of ethyl acetoacetate even though its boiling point is not very high?

Exp.29 Preparation of Acetylsalicylic Acid (Aspirin)

Objectives

(1) To synthesize aspirin from salicylic acid and acetic anhydride.

(2) To be acquainted with the general procedures of recrystallization and vacuum filtration.

Principles

Aspirin, also known as acetylsalicylic acid, is a medication used to treat pain, fever, or inflammation. The synthesis of aspirin is classified as an esterification reaction. Salicylic acid is treated with acetic anhydride, an acid derivative, causing a chemical reaction that turns salicylic acid's hydroxyl group into an ester group. Small amounts of sulfuric acid are almost always used as a catalyst.

salicylic acid + acetic anhydride $\xrightarrow[80\sim90℃, 5min]{H_2SO_4}$ acetylsalicyclic acid (aspirin) + acetic acid

Materials

Equipment: Erlenmeyer flask, thermometer, Büchner funnel, suction flask, beaker, watch glass.

Reagents: salicylic acid, acetic anhydride, ethyl acetate, conc. H_2SO_4, conc. HCl, saturated $NaHCO_3$ solution, 1% $FeCl_3$ solution.

Procedures

In a 150 mL Erlenmeyer flask, introduce 2.0 g (0.014 mol) of salicylic acid and 5 mL (0.05 mol) of acetic anhydride[1,2]. Then, add 5 drops of conc. H_2SO_4 and stir the mixture. Heat it in a hot water bath (80~90℃) for 5 min while stirring continually with a glass rod[3]. The solid will dissolve completely.

Take out the flask from the water bath and add 1 mL of cold water into the mixture while swirling the flask. Add 50 mL of cold water while swirling the mixture and let the solution cool to room temperature. As the solution cools, crystals of aspirin will appear. Cool the solution further by placing the reaction flask in an ice bath. If no crystals appear, scratch the inside wall of the flask with a glass rod to induce crystallization. Carry out a vacuum filtration to collect the crystals and wash the crystals with 5 mL of the cold water. Weigh and record the mass of the crude product (aspirin) obtained.

Purify the crude acetylsalicylic acid by recrystallization. Pour the resulting solution into a 100 mL beaker, and then add in 25 mL of saturated aqueous sodium bicarbonate solution while stirring continually with a glass rod. Stir the mixture until no more bubbles of carbon dioxide are observed. Filter the solution to remove any insoluble impurities. Carefully pour the filtrate with stirring into an ice cold HCl solution (5 mL of conc. HCl in 10 mL of water) in a 150 mL beaker and cool the mixture in an ice bath. Make sure that the resulting solution is acidic and that the aspirin has completely precipitated out. Filter the solid by suction and wash the crystals 2 times with 5 mL of cold water each. Remove all the liquid from the crystals by pressing with a clean stopper or cork.

Air dry the crystals and transfer them to a watch glass. Dissolve the final product in a minimum amount (no more than 2~3 mL) of hot ethyl acetate. Make sure that the product is completely dissolved while gently and continuously heating on a steam bath. Cool the solution to room temperature and then in an ice-bath. Collect the product by vacuum filtration and rinse the residue out of the flask with a few milliliters of cold petroleum ether. When the product is completely dry[4], weigh its weight, determine its melting point (lit. m. p. 135~136℃) and calculate the percentage yield.

Ferric chloride test: dissolve a few crystals in 5 mL of water in a test tube. Add 1~2 drops of 1% $FeCl_3$ solution to the solution and observe the color change in it to predict the purity of the product[5]. The test tube containing the product should only chang to a very slight purple color after this test, indicating a high purity.

Pure acetylsalicylic acid is a white needle crystal with b. p. 180.4℃ and n_D^{20} 1.4194.

Notes

[1] All the instrument must be completely dry during the reaction, the reaction reagents also must be dried before use, and the acetic anhydride used in this experiment must be freshly distilled, being collected at 139~140℃.

[2] Pay attention to the order when adding the reactants. If salicylic acid and conc. H_2SO_4 are added first, salicylic acid will be oxidized.

[3] Because acetic anhydride is a **HIGHLY CORROSIVE**, work with acetic anhydride in the fume hood only.

[4] The aspirin is prone to decompose when being heated to the decomposition temperature of 126~135℃. Therefore, do not heat it for a long time when making the recrystallization. The crude product can be purified by recrystallization from ethanol-water or benzene-petroleum ether (60~90℃).

[5] If the crude product contains unreacted salicylic acid, the ferric chloride solution reacts with unreacted salicylic acid, forming a purple color.

Questions

(1) Why must the apparatus used be dry when preparing aspirin?

(2) Give an alternative method of synthesis of aspirin, using salicylic acid as a starting material. Give the mechanism.

(3) What is the purpose of using the concentrated sulfuric acid in the acetylation reaction? Which substance can be used in the place of sulfuric acid?

(4) What side reactions will occur in this experiment?

(5) Why is the polymeric by-product insoluble in sodium bicarbonate solution, while salicylic acid itself is soluble?

(6) What product can be formed when salicylic acid reacts with excess ethanol in the presence of concentrated sulfuric acid?

(7) Why can we induce the crystallization in stubborn cases by scratching the inside wall of the flask with a glass rod? What other methods can be used to accelerate the crystallization?

Exp.30 Preparation of Acetanilide

Objectives

(1) To synthesize acetanilide by the reaction of aniline and acetic anhydride.

(2) To learn to purify acetanilide by crystallization.

(3) To learn how to make a fractional distillation.

Principles

Acetanilide is the first analgesic-antipyretic drug (relieving pain and fever) with another name of "antipyretic ice". Acetanilide itself is an important drug as well as an important intermediate in the synthesis of sulfa drugs (drug compounds that contain a sulfonamide functional group).

Acetanilide can be prepared by the acetylating reaction of aniline with acetylating reagents including acetyl chloride, acetic anhydride or acetic acid. The order of the reactivities is acetyl chloride > acetic anhydride > acetic acid. Due to the high price and corrosive properties of acetyl chloride and acetic anhydride, we use acetic acid as acetylating reagent in this experiment.

Reaction:

$$\text{C}_6\text{H}_5\text{NH}_2 + \text{CH}_3\text{COOH} \xrightarrow{\text{Zn}} \text{C}_6\text{H}_5\text{NHC(O)CH}_3 + \text{H}_2\text{O}$$

This reaction plays an important role in organic synthesis as well as in the protection of the amino group on an aromatic ring. The amino group on an aromatic ring is easily oxidized, therefore the amino group is first protected by acetylation before the introducing the desired groups onto the aromatic ring. The amino group can be thereafter deprotected by the hydrolysis of amide to amine.

Materials

Equipment: round-bottom flask, fractionating column, distillation adapter, heating mantle, thermometer (150 ℃), suction flask, Büchner funnel, measuring cylinder, filter paper, beaker, water vacuum pump.

Reagents: freshly distilled aniline, glacial acetic acid, zinc powder, activated charcoal.

Procedures

In a 50 mL round-bottom flask, place 5 mL of freshly distilled aniline[1], 7.5 mL of glacial acetic acid and a little zinc powder (about 0.1 g) [2]. Install the apparatus as shown in Fig. 3.15. Heat the reaction mixture to boiling and continue heating the mixture under reflux for 45 min, maintaining the reading of the thermometer at about 105 ℃[3]. About 4 mL of distillate could be got. Heating can be stopped when the reading of the thermometer drops continuously.

Decant the reaction solution into a beaker containing 100 mL of cold water with stirring while the solution is still hot (leaving the unreacted zinc behind) [4]. At this time, fine-grained solids will appear. After the mixture is cooled down completely, isolate the crystals from the solvent by vacuum filtration and wash the crystals with a little cold water. Remove all the liquid from the crystals by pressing with a clean stopper or cork, giving the crude product as a white or pale yellow solid.

Transfer the crude product to a 250 mL beaker, add 100 mL of water and heat the mixture to boiling with stirring. Continue to add hot water into the beaker with stirring and heating until all the crystals have dissolved. If the solution has a deep color or contains oily substance, cool

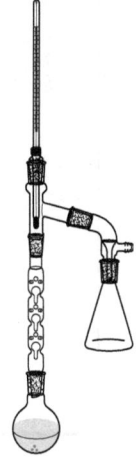

Fig. 3.15 Apparatus for the preparation of acetanilide by using acetic acid

down the mixture slightly, add 0.5 g of activated charcoal[5], continue to heat it to boiling for 2 min, and then remove the insoluble impurities by a hot filtration under reduced pressure[6]. Cool down the filtrate naturally. After the solution has come to room temperature, place it in an ice bath to complete the crystallization process[7]. Collect the crystals by suction filtration, washing the filtered crystals twice with a little cold water. Dry the crystals by pressing with a clean stopper or cork. Transfer the crystals onto a preweighed watch glass for air drying. Weigh and calculate the yield of the product.

Pure acetanilide forms white flake crystals with m. p. 114.3 ℃.

The IR and ^{1}H NMR spectra of acetanilide are shown in Fig. 3.16 and Fig. 3.17, respectively.

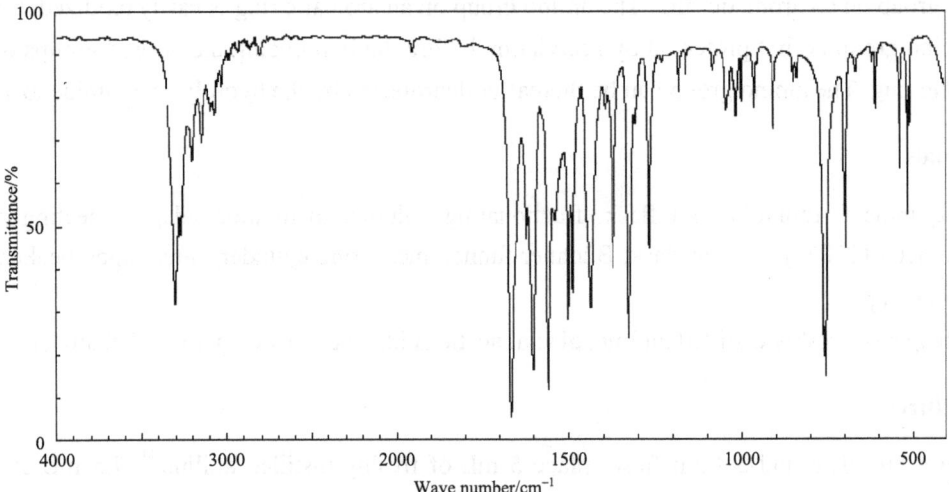

Fig. 3.16 The IR spectrum of acetanilide

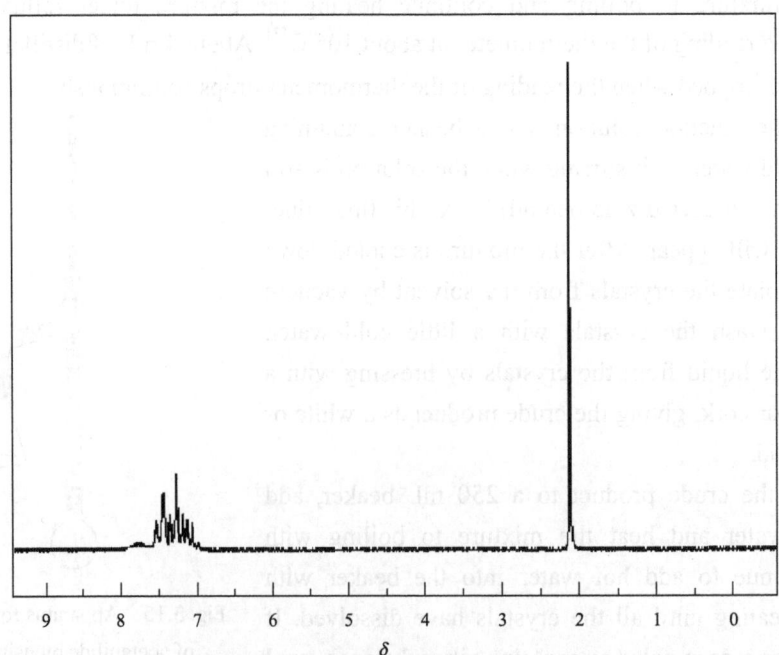

Fig. 3.17 The ^{1}H NMR spectrum of acetanilide

Chapter 3   Preparation Experiments of Organic Compounds

Notes

[1] Aniline becomes dark after a long-term storage because of being oxidized, so it needs to be redistilled before being used. Because of the high boiling point of aniline, simple distillation using an air condenser or vacuum distillation should be used for distilling it.

[2] The role of zinc powder is to prevent the oxidation of aniline. Only a small amount of zinc is needed, if being added too much, it will form water-insoluble zinc hydroxide.

[3] Do not allow the fractionation temperature rising too high, otherwise a large amount of acetic acid will be distilled out of the reaction system and the product yield will be reduced.

[4] The acetanilide precipitates when the solution cools, getting stuck to the inside wall of the flask. This will cause trouble in the transferring of the solid. Therefore, the reaction liquid should be poured out while it is still hot.

[5] Activated charcoal cannot be added to the boiling solution, otherwise it will cause bumping.

[6] Perform the hot filtration as fast as possible with a preheated Buchner funnel; otherwise the product would precipitate, clogging up the pores of the filter paper and getting stuck on the insides of the funnel.

[7] If the solution is supersaturated, you can use a glass rod rubbing the inner wall of the beaker or adding crystals to precipitate crystals when making the recrystallization.

Questions

(1) How to increase the yield of acetanilide when using acetic acid as the acetylating reagent?

(2) Why is it needed to control the reaction temperature at about 105℃? What kind of effect would cause if reaction temperature is too high or too low in the experiment?

(3) According to the reaction equation, how much water will be produced in theory? Why is the actual volume of the liquid collected greater than the theoretical one?

(4) Why does the reading of the thermometer drop when the reaction ends?

Exp.31   Preparation of *p*-Acetamidobenzenesulfonyl Chloride

Objectives

(1) To learn the principle of chlorosulfonation reaction and the general steps of the reaction.

(2) To be proficient in the operations for recrystallization.

Principles

Sulfonyl chloride is the basic raw material for preparing a series of sulfonamides. The reaction of *p*-acetaminobenzenesulfonyl chloride and ammonia or ammonia derivative is a key step in the preparation of sulfa drugs.

The chlorosulfonyl group can be introduced to *para*-position of the acetamide group of acetanilide in one step via the electrophilic aromatic substitution reaction known as chlorosulfonation. The electrophile that initially adds to the ring is probably SO_3, forming the sulfonic acid.

Substitution is essentially all *para* due to combined electronic and steric effects. The sulfonic acid is then converted to *p*-acetamidobenzenesulfonyl chloride by the reaction with excess chlorosulfonic acid, generating sulfuric acid as the co-product.

Reaction:

Materials

Equipment: round-bottom flask, constant pressure dropping funnel, angled cone adapter, Claisen distillation head, magnetic stirrer, thermometer adapter, ice-water bath, heating mantle, thermometer (150℃), suction flask, Büchner funnel, angled glass tube, filter paper, beaker, water vacuum pump.

Reagents: acetanilide, chlorosulfonic acid, 5% NaOH solution.

Procedures

In a 100 mL dry round-bottom flask equipped with a magnetic stirring bar, add 5 g of dry acetanilide. Place the flask in a heating mantle and heat it to melt the solid. Remove and swirl the flask, then stopper it and place it in an ice-water bath. A thin film is formed on the bottom of the flask [1]. Add 13 mL of chlorosulfonic acid into a constant pressure dropping funnel, stopper the dropping funnel and check it carefully to ensure that there is no leaking (caution: wear gloves while in fume hood!) [2]. After the reaction flask is cooled down, fit the flask with a Claisen distillation head, which is in turn fitted with the dropping funnel on the straight arm and an angled cone adapter on the side arm, connect the cone adapter to one end of an angled glass tube. The other end of the tube is inserted into a gas absorbing flask containing 5% NaOH solution through a thermometer adapter. The apparatus is shown in Fig. 3.18.

Place the reaction flask in a cold water bath maintained between 10 and 15℃. Then add chlorosulfonic acid into the flask rapidly all at once from the dropping funnel. Stir the solution rapidly, keeping it in the cooling bath with the water temperature below 20℃ [3].

After the acetanilide is mostly dissolved and the initial exothermic reaction has subsided, remove the cooling bath and allow the solution to warm to room temperature with continuous stirring. Then heat the flask in a hot water bath at 70~80℃ for 20 min to complete the reaction.

Cool the reaction mixture in an ice bath to room temperature or below. Pour the reaction mixture carefully into a 500 mL beaker containing 100 g of crushed ice (caution: wear gloves and pour the reaction solution over the ice slowly and carefully to avoid splattering). Stir the ice slurry using a stir rod to prevent formation of large lumps as the product precipitates. Any large lumps that form should be carefully broken up using the stir rod.

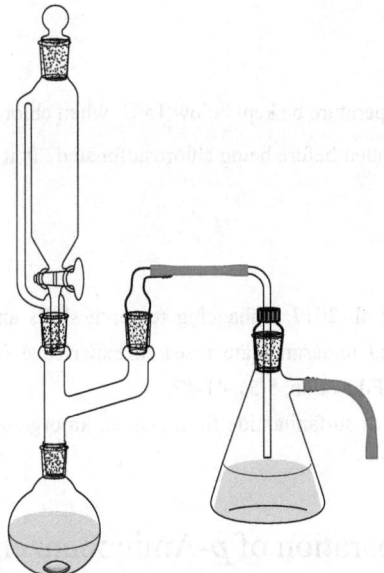

Fig. 3.18   Apparatus for the preparation of *p*-acetamidobenzenesulfonyl chloride

Filter the mixture to collect the precipitate under reduced pressure, wash the precipitate with 15 mL portions of cold water until the filtrate tests neutral to pH paper. Air dry the product on the Büchner funnel by pulling air through it for at least 10 min. The obtained crude product is used immediately for the synthesis of *p*-acetamidobenzenesulfonamide.

Pure *p*-acetamidobenzenesulfonyl chloride can be obtained by crystallization as follows: in a 250 mL round-bottom flask, place the crude product and a little chloroform. Heat the mixture to boiling. Gradually add some more chloroform to the flask until the solids dissolve completely. Transfer the solution quickly to a 250 mL Erlenmeyer flask and cool it in an ice water bath for precipitation. Filter to collect the crystals under reduced pressure, wash the crystals with a small amount of chloroform, dry and weigh the product. The melting point of pure *p*-acetamino-benzenesulfonyl chloride is 149 ℃.

Notes

[1] Acetanilide should be condensed before reacting because the chlorosulfonation reaction is intense.

[2] Chlorosulfonic acid is strong corrosive and extremely hazardous. Wear lab jackets, gloves and safety goggles when handling this chemical, and only do so inside a fume hood. Chlorosulfonic acid reacts violently with water, or even moisture, and generating gaseous HCl in the process, which will make people suffocate when being inhaled. The equipment and reagents used in the reaction should be very dry. The waste liquid containing chlorosulfonate should not be poured into the sink, it must be collected into a waste acid tank.

[3] The key to this experiment is to prevent local overheating; otherwise it will cause the decomposition of chlorosulfonic acid.

Questions

(1) Why should the reaction temperature be kept below 15℃ when chlorosulfonic acid is added in the flask?

(2) Why should aniline be acetylated before being chlorosulfonated? Is it plausible to make a chlorosulfonation starting from aniline directly?

References

Coppock P, Park S H, Paredes J, et al. 2017. Enhancing research skills and attitudes in undergraduate organic chemistry with course-embedded undergraduate research experience (CURE) via green organic synthesis. Journal of Laboratory Chemical Education, 5(3): 41-47.

Hurdis E, Wang J. 1969. Preparation of sulfanilamide from aniline: an organic chemistry experiment. J Chem Edu, 46: 697-698.

Exp.32 Preparation of *p*-Aminobenzenesulfonamide

Objectives

(1) To understand the principle and methods for the preparation of *p*-aminobenzenesulfonamide, as well as the principle of the ammonolysis of acyl chloride and the hydrolysis of acetamido derivatives.

(2) To be more skilled in the operations of reflux, decolorization, recrystallization, and so on.

Principles

Aminobenzenesulfonamide, also known as sulfonamide, presents as white granules or powdery crystals. It has a molecular formula of $C_6H_8N_2O_2S$, molecular weight of 172.22, and melting point of 164.5~166.5℃. Aminobenzenesulfonamide is odorless with slightly bitter taste. It is slightly soluble in cold water, ethanol, methanol and acetone; soluble in boiling water, glycerol, hydrochloric acid, potassium hydroxide and sodium hydroxide solution; insoluble in benzene, chloroform, ether and petroleum ether. The first sulfonamide was a red azo dye, which was discovered by chance to have a strong inhibitory effect on bacteria and can be used as drugs. Sulfonamide is the basis of the structures and medicinal properties of sulfa drugs. Itself is a drug for the prevention and treatment of bacterial infections, and is an anti-inflammatory drug commonly used in modern medicines.

Reaction:

Materials

Equipment: Erlenmeyer flask, suction flask, round-bottom flask, Büchner funnel.

Reagents: acetanilide, chlorosulfonic acid, conc. aqueous ammonia, conc. HCl, Na_2CO_3.

Procedures

1. Preparation of p-acetamidobenzenesulfonamide

In a fume hood, place the crude product of acetamidobenzenesulfonyl chloride obtained in Exp.31 in a 125 mL Erlenmeyer flask, add 23 mL of conc. aqueous ammonia solution gradually in the Erlenmeyer flask while stirring[1]. The mixture makes a very rapid exothermic reaction, affording a white paste immediately. After finishing adding ammonia solution, continue to stir the mixture for 15 min to complete the reaction. Heat the resulting mixture in a warm water bath (70℃) under continuous stirring for 10 min to remove excess ammonia. Cool down the mixture, filter it under reduced pressure to obtain the crystals, wash the crystals with a little cold water, and dry the crystals to get the crude product. The crude product will be used directly in the following process without a further refinement.

2. Preparation of p-aminobenzenesulfonamide

Transfer the crude product of p-acetamidobenzenesulfonamide to a 50 mL round-bottom flask. Add 20 mL of 10% HCl solution into the flask and heated the mixture under reflux until the solids dissolve completely (about 0.5 h needed). When the mixture is cooled down to the room temperature, a nearly clear solution will be obtained. If there are some solids re-precipitating, test the acidity or alkalinity of the solution. If the solution is not acidic, add appropriate amount of hydrochloric acid in it, and continue to boil the mixture for 15 min. Repeat these operations until no solids have precipitated after the solution is cooled down to room temperature. Transfer the resulting solution to a beaker. Add Na_2CO_3 solids slowly with stirring until a pH of 7~9 is obtained[2]. In the process of neutralization, the solid will precipitate. Cool the mixture in an ice-water bath, filter to obtain the crystals and wash the crystals with a small amount of ice water. Recrystallize the crude product from water (about 12 mL of water for each gram of the product). The resulting p-aminobenzenesulfonamide forms white leaf crystals with m. p. 165~166℃.

Notes

[1] Because ammonia is toxic and has a very strong, unpleasant odor, this entire procedure must be performed inside a fume hood.

[2] When sodium carbonate is used to neutralize hydrochloric acid, a large amount of carbon dioxide gas is generated, so it is necessary to make continuous stirring.

Questions

How to understand the amphotericity of *p*-aminobenzenesulfonamide? Show the reaction equations for the reactions of sulfonamide with a dilute acid and a dilute alkali.

Exp.33 Green Synthesis of *N*-bromosuccinimide

Objectives

(1) To master the principle, technics and operations for the preparation of *N*-bromosuccinimide.

(2) To understand the concept, process and standards of green synthesis.

Principles

In this experiment, *N*-bromosuccinimide (NBS) is prepared by the reaction of succinimide with bromine which is formed by in situ redox reaction of sodium bromate and sodium bromide under acidic conditions. Because without direct use of bromine, this method is green with simple and safe performance, high yield and low pollution.

Reaction:

$$6\,\text{(succinimide-NH)} + 2NaBrO_3 + 4NaBr + 3H_2SO_4 \longrightarrow 6\,\text{(succinimide-N-Br)} + 6H_2O + 3Na_2SO_4$$

Materials

Equipment: three-necked flask, constant pressure dropping funnel, funnel, U-shaped glass tube, condenser, beaker.

Reagents: succinimide, sodium bromide, sodium bromate, 5% NaOH solution, 50% H_2SO_4 solution.

Procedures

As shown in Fig. 3.19, in a 100 mL three-neck flask, place 1.8 g (18 mmol) of succinimide, 1.0 g (7 mmol) of sodium bromate, 1.4 g (13.8 mmol) of sodium bromide and 10 mL of water. Separately equip a constant pressure dropping funnel containing 2 mL of 50% solution of sulfuric acid and a water condenser on each of the two necks of the flask, and stopper the rest neck of the flask. Connect a one-hole stopper on top of the condenser, insert one end of a U-shaped glass tube in the stopper and connect another end of the tube with an inverted funnel. Suspend the edge of the funnel on top of the surface of 5% NaOH solution contained in a beaker. Dissolve the mixture under stirring at 25 ℃ by using a stir bar. Add the sulfuric acid solution dropwise to the flask from the dropping funnel[1]. The solution turns brown red in this process.

After all the sulfuric acid solution has been added, stir the reaction mixture until the brown red color fades, this may need about 0.5 h. Filter the solution under reduced pressure[2], dry the obtained crystals to give the crude product. Weigh the product and record the yield. The melting point of pure *N*-bromosuccinimide is 178 ℃.

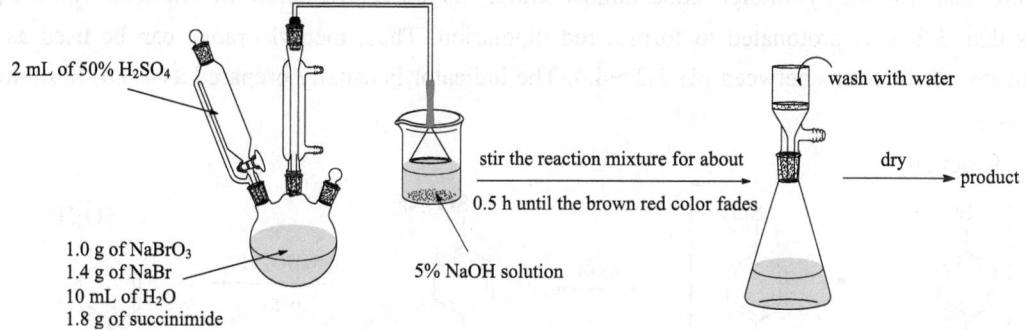

Fig. 3.19 Apparatus and flew chart for the preparation of NBS

Notes

[1] The dropping rate of sulfuric acid should be strictly controlled, so that no more bromine vapor is generated on the top surface of the reaction solution.

[2] The waste liquid should not be poured into the sink; it must be collected into a waste acid tank.

Questions

Why should we add the sulfuric acid solution into the reaction flask dropwise rather than all at once?

References

Fujisaki S, Hamura S, Eguchi H, et al. 1993. Organic synthesis using sodium bromate. II. A facile synthesis of *N*-bromoimides and amides using sodium bromate and hydrobromic acid (or sodium bromide) in the presence of sulfuric acid. Bull Chem Soc Jan, 66: 2426-2428.

Kajigaeshi S, Nakagawa K, Fujisaki S, et al. 1985. A practical synthesis of *N*-bromo imides by use of sodium bromite. Bull Chem Soc Jan, 58: 769-770.

Wang C, Wang L M, Wang F, et al. 2011. Green synthetic conditions of *N*-bromosuccinimide. Fine Chemical Intermediates, 41(2): 63-65.

Exp.34   Synthesis of Methyl Orange

Objectives

(1) To learn the principle and method for synthesizing a common diazo dye via the diazo-reaction and coupling reaction.

(2) To prepare methyl orange.

Principles

Methyl orange is an azo dye, which can be synthesized by the coupling reaction of p-aminobenzenesulfonic acid diazonium salt with an acetic acid salt of N, N-dimethylaniline in a neutral or weakly acidic medium. Methyl orange is often used as acid-base indicator. In solutions with the pH greater than 4.4, methyl orange exists almost entirely as the yellow anion. In solutions with the pH less than 3.2, it is protonated to form a red dipolarion. Thus, methyl orange can be used as an indicator for titrations between pH 3.2~4.4. The indicator is usually prepared as a 0.01% solution in water.

Reaction:

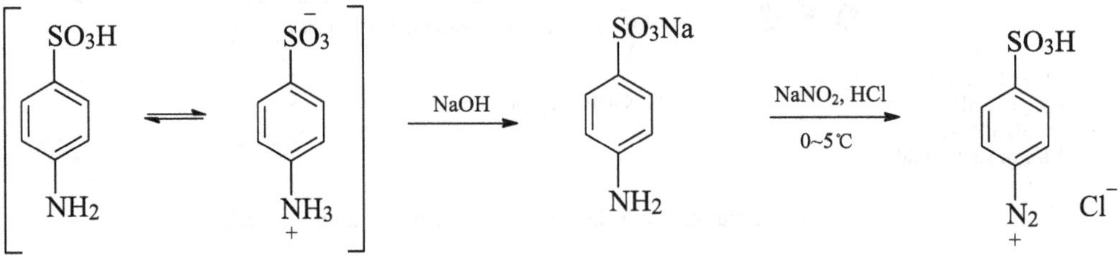

sulfanilic acid(zwitterion)   sodium sulfanilate

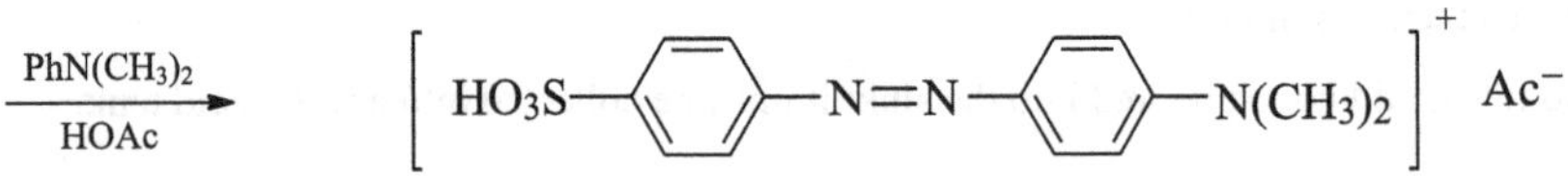

acidic methyl orang(red)

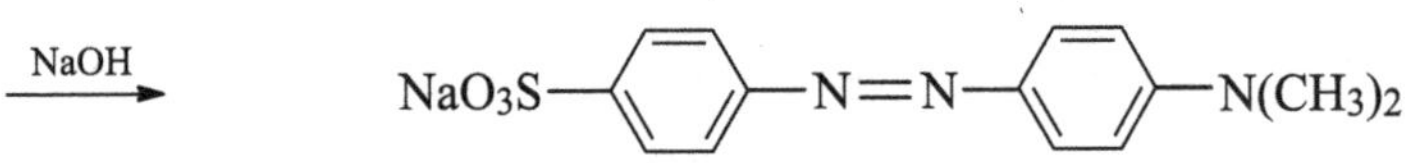

methyl orange(orange)

Materials

Equipment: beaker, test tube, thermometer, heating mantle, watch glass, glass rod, Büchner funnel, suction flask, drying oven.

Reagents: sulfanilic acid, sodium nitrite, ice, NaOH (1%, 5%, 10%), glacial acetic acid, conc. HCl, N, N-dimethylaniline, saturated NaCl solution, ethanol, ether, KI-starch test paper.

Procedures

1. Diazotization of Sulfanilic Acid

In a 100 mL beaker, dissolve 2.0 g (0.012 mol) of sulfanilic acid in 10 mL of 5% NaOH solution (with gentle heating if necessary)[1]. Cool the solution to room temperature and add a sodium nitrite solution (0.8 g of $NaNO_2$ in 3 mL of water) with stirring until dissolution completes.

Slowly pour the solution into a 250 mL beaker containing about 13 mL of ice water and 2.5 mL of conc. HCl. During the reaction, the temperature should be controlled bellow 5℃[2].

The aqueous layer is monitored periodically using KI-starch test paper. If a positive test is not obtained, more sodium nitrite solution should be added[3]. Keep the mixture in an ice-water bath for about 15 min to complete the reaction. The white suspension is used directly for the synthesis of methyl orange without isolation.

2. Coupling Reaction

Add a cooled solution of dimethylaniline (1.3 mL) in glacial acetic acid (1 mL) to the diazonium salt solution with constant stirring. Stir and mix the mixture thoroughly for 10 min and a stiff paste of red dye will be formed.

Add 15 mL of 10% NaOH solution to the mixture and heat the mixture to boiling to dissolve the solid completely. When most of the methyl orange is dissolved, cool the mixture with an ice-water bath to precepitate the product. When the precipitation completes, wash the precipitates with a small amount of saturated NaCl solution, ethanol and ether, successively. Then collect the crude product by vacuum filtration.

3. Recrystallization

Recrystallize the crude product with 1% NaOH solution[4]. Collect the product by vacuum filtration and wash it with saturated NaCl solution, ethanol and ether, successively. Dry and weigh the product (about 2.5 g). Methyl orange presents as orange slice-shaped crystals. Dissolve a little methyl orange in water, add a few drops of diluted hydrochloric acid, then neutralize it with basic solution, observe the color changes and make a record.

Notes

[1] Sulfanilic acid is an amphoteric compound, which is insoluble in acid solution and soluble in basic solution. Its acidity is slightly stronger than its basicity. It is nevertheless necessary to carry out diazotization reaction in an acid (HNO_2) solution. To resolve this problem, sulfanilic acid is first dissolved in basic solution, rather than in acidic solution.

[2] Temperature is important in this reaction. If the temperature of the reaction is above 5℃, the diazonium salt will be hydrolyzed to a phenol, thus reducing the yield of the desired product.

[3] If the KI-starch test paper unchanged, a few sodium nitrite solutions should be added.

[4] The color of wet methyl orange can deepen quickly when being exposed in the natural light. The ordinary crude methyl orange is a red-purplish (mauve) crystal. It will dry quickly if being washed with alcohol and ether in turn.

Questions

(1) Why should sodium hydroxide be added before the diazotization reaction in this experiment? Is it plausible

for performing the diazotization by mixing sulfanilic acid with hydrochloric acid followed by adding sodium nitrite solution?

(2) Why should the temperature for the diazotization reaction of sulfanilic acid be controlled below 5 ℃?

(3) Why does the diazonium coupling not occur in either strongly acidic or strongly basic media?

(4) Why is a *para-* product formed by the coupling reaction of *N*, *N*-dimethylaniline with diazonium salt?

(5) What will happen if cuprous chloride is used as a catalyst for the preparation of diazonium salt?

Exp.35 The Preparation and Beckmann Rearrangement of Benzophenone Oxime

Objectives

(1) To learn the methods for the synthesis of oximes.

(2) To learn the principle of Beckmann rearrangement.

Principles

Like all aldehydes and ketones, benzophenone condenses with hydroxylamine to give an oxime, a commonly used characteristic solid derivative. When being treated with various acids, an oxime undergoes a fascinating rearrangement to an acid amide, which had been discovered by Beckmann in 1886. Since then, this reaction has been called after his name, the Beckmann rearrangement.

Reaction:

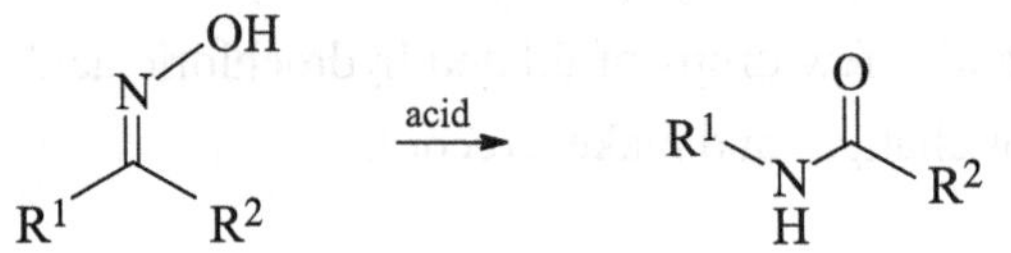

Mechanism:

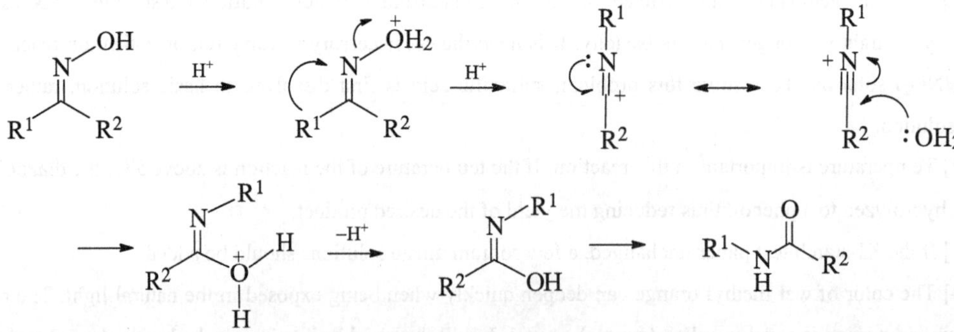

The Beckmann rearrangement is often catalyzed by acids. In the Beckmann rearrangement, the oxime nitrogen atom is inserted into the C_{oxime}—C_α bond of aldehydes and ketones. In most cases, the Beckmann rearrangement of ketoximes is stereospecific, N—O bond cleavage occurs with simultaneous migration, and the migrating group is anti-periplanar to the leaving group on the nitrogen. The dominant application of the Beckmann rearrangement is the conversion of cyclohexanone to caprolactam via the oxime. Caprolactam is the feedstock in the production of Nylon-6.

In this experiment, benzophenone oxime is synthesized by the reaction of hydroxylamine and benzophenone. The resulting benzophenone oxime is subjected to make a rearrangement catalyzed by polyphosphoric acid (PPA), and benzoylaniline is then obtained by pouring the reaction mixture into ice water. Polyphosphoric acid, a syrupy liquid that reacts rapidly with water to give a solution of phosphoric acid, is the best medium in which to carry out the reaction because when the reaction is finished, a simple dilution with water forces out the product. A recrystallization completes the experiment.

Reaction:

Materials

Equipment: Erlenmeyer flask, beaker, Büchner funnel, suction flask, melting point apparatus, heating mantle.

Reagents: benzophenone, hydroxylamine hydrochloride, sodium hydroxide, polyphosphoric acid, ethanol, conc. HCl.

Procedures

1. Preparation of Benzophenone Oxime

In a 125 mL Erlenmeyer flask, place 2.5 g (0.0137 mol) of benzophenone, 1.5 g (0.0216 mol) of hydroxylamine hydrochloride, 5 mL of ethanol and 1 mL of water. Then add in 18~20 pellets of NaOH and swirl the flask well to dissolve the solids.

Boil the resulting solution gently on a steam bath for about 5 min. With continuous stirring, pour the solution into a beaker containing a solution composed of 8 mL of conc. HCl and 50 mL of water. Soon afterwards, white solids will be observed.

Cool mixture down completely, collect the precipitate by vacuum filtration, wash the solids with a little cold water, press the filter cake for drying and absorb the excess liquid of the filter cake with a filter paper to afford the crude product. The crude product can be purified by recrystallization from about 20 mL methanol. Pure benzophenone oxime forms colorless needle crystals with m. p. 142~143 ℃.

2. Beckmann Rearrangement of Benzophenone Oxime

Take 100 mL beaker, add the obtained benzophenone oxime in this experiment and 25 mL of polyphosphoric acid in the beaker. Place a thermometer (200℃) in the mixture and stir the mixture continuously with a glass rod.

Heat the mixture gently in a heating mantle until its temperature reaches 100℃. At this moment, the exothermic rearrangement can be expected to take place. Keep the temperature maintained at about 100℃ for 20 min. After that, heat the mixture under vigorous stirring until its temperature reaches at 125~130℃, and stop heating. Wait at least 10 min before cautiously pouring the warm syrup into a beaker with 350 mL of cold water in it, stirring well[1].

The precipitate is suction filtered and recrystallized from ethanol (about 20 mL will be needed). Dry and weigh the product[2]. Determine the melting point.

Pure benzoyl aniline forms silvery white needle crystals with m. p. 163~164℃.

Notes

[1] Much heat is released when you pour the PPA into water, and this step is to be carried out with appropriate care.

[2] The product is best to be squeezed to drying on the filter paper first, and then dried in the air.

Questions

(1) What type of catalysts can be used in Beckmann rearrangement?

(2) Describe the reaction mechanism for preparing caprolactam from cyclohexanone.

Exp.36 Preparation of Aniline

Objectives

(1) To understand the principle and method for the preparation of aniline by the reduction of nitrobenzene.

(2) To be skilled in some basic operations, such as steam distillation and simple distillation.

Principles

The preparation of aromatic amines cannot be accomplished by the electrophilic substitution reaction of aromatics because the amino groups cannot be introduced directly onto the aromatic ring by the electrophilic substitution reaction. The aromatic amines can be prepared by the reduction of aromatic nitro compounds in acidic medium. Typical reducing agents include Fe/HCl, Fe/CH_3COOH, Sn/HCl, and so on. In industry, nitrobenzene can be reduced to aniline by using iron powder and hydrochloric acid. However, this method is not enviromental-friendly because there is much aniline-containing iron waste generated. Nowadays, the catalytic hydrogenation of nitrobenzene is used as a replacement for this reaction, and nickel, platinum, and palladium based catalysts are often used. Iron powder is still used for this reaction in laboratory.

Reaction:

$$4 \text{C}_6\text{H}_5\text{NO}_2 + 9\text{Fe} + 4\text{H}_2\text{O} \xrightarrow{\text{H}^+} 4 \text{C}_6\text{H}_5\text{NH}_2 + 3\text{Fe}_3\text{O}_4$$

Materials

Equipment: two-necked flask, condenser, steam generator, distillation adapter, Erlenmeyer flask.

Reagents: nitrobenzene, ether, Na₂CO₃, NaOH, glacial acetic acid, iron powder, saturated NaCl solution.

Procedures

Place 9 g (0.16 mol) of iron powder, 17 mL of H₂O and 1 mL of glacial acetic acid in a 150 mL two-necked flask, which is fitted with a reflux condenser and a dropping funnel containing 7 mL of nitrobenzene. Add nitrobenzene into the flask dropwise from the dropping funnel. It is advisable to shake the flask after each addition[1]. When all the nitrobenzene has been added, heat the reaction mixture under reflux on a hot water bath for 0.5 h with occasional shaking of the flask[2]. Cool the flask and add Na₂CO₃ solution until the solution is strongly alkaline[3]. Set up an apparatus for a steam distillation[4], and continue the distillation until oily droplets of aniline are no longer visible in the distillate coming over. Add sodium chloride until the solution is saturated[5].

Transfer the solution to the separatory funnel. Retain the organic layer and extract the aqueous phase three times with 7 mL of ether for each. Combine all organic layers in a dry Erlenmeyer flask and dry the solution by standing it over sodium hydroxide pellets[6]. Decant the ethereal solution into a 50 mL round-bottomed flask fitted with a water condenser, heat the solution over a 70℃ water bath until no more ether distils. Replace the water condenser by an air condenser[7]. Fit a thermometer adapter and a 250℃ thermometer onto the flask, and heat the flask over a heating mantle. Collect the fraction which boils at 182～185℃[8].

The boiling point of pure aniline is 184.4℃, n_D^{20} is 1.8563.

The IR and ¹H NMR spectra of aniline are shown in Fig. 3.20 and Fig. 3.21, respectively.

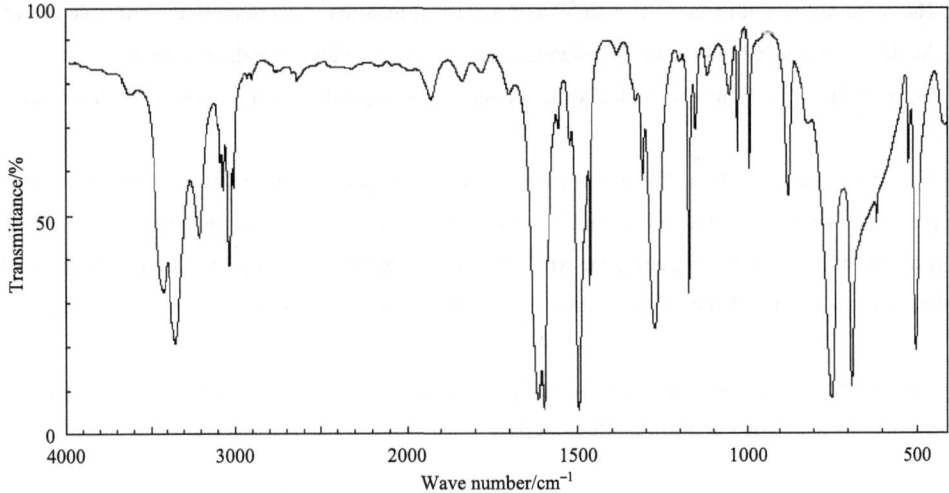

Fig. 3.20 The IR spectrum of aniline

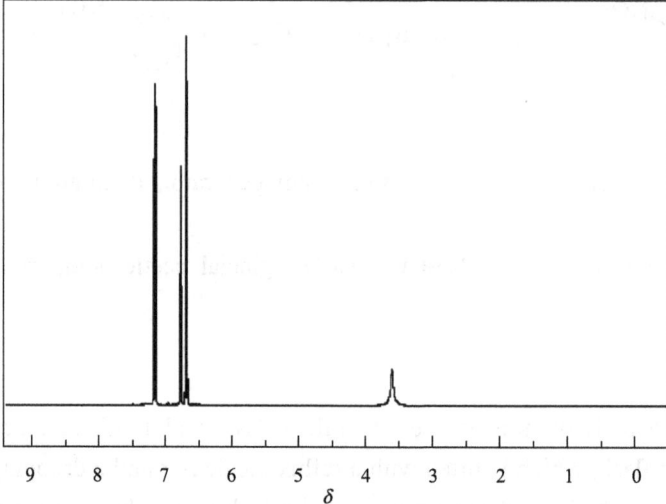

Fig. 3.21 The ^{1}H NMR spectrum of aniline

Notes

[1] This experiment is not only an exothermic reaction, but also a heterogeneous reaction (oil/water/solid). A sharp reaction will take place each time nitrobenzene is added, so it is necessary to control the reaction temperature carefully so as to avoid bumping caused by overheating or heat accumulating. Shake adequately to mix the reagents well in the bottle.

[2] Nitrobenzene is a yellow oily liquid. If the yellow oil disappears in the reflux solution and turns into milky white oil beads, it indicates that the reaction has been completed.

[3] After the reaction is completed, part of the aniline forms aniline hydrochloride salt. Decompose aniline hydrochloride completely to set aniline free by modulating the solution to strong basic.

[4] After the steam distillation, phenol salt and benzidine remain in the residue, while aniline and nitrobenzene contain in the distillate.

[5] The solubility of aniline in water at 20℃ is 3.4 g/100 g of water. According to the principle of salting out, sodium chloride can be added to saturate the solution so as to reduce the loss of aniline in water.

[6] Since aniline can form a complex with calcium chloride, the resulting solution is dried over sodium hydroxide.

[7] In addition to distillation, the separation of aniline from nitrobenzene can also be carried out by the following method: acidify the mixture obtained by steam distillation with hydrochloric acid, forming water-soluble aniline hydrochloride, and extract nitrobenzene with ether. Then add base in the aqueous layer to set aniline free and extract the aniline with ether. Note: avoid contacting aniline with the skin or inhalation of its vapor because aniline is toxic.

[8] After the reaction is completed, add hydrochloric acid solution (1 : 1, V/V) with slight heating to remove the dark brown substance adhering to the round-bottom flask.

Questions

(1) What kind of organic substances can be separated by the steam distillation? In this experiment, why can the steam distillation be chosen for separating aniline from the reaction mixture?

(2) When refining the aniline, why should we use granular sodium hydroxide as the desiccant rather than magnesium sulphate or calcium chloride?

Exp.37 Preparation and Resolution of Racemic α-Phenylethylamine

Objectives

(1) To master the principle and method of preparing (±)-α-phenylethylamine by Leuckart reaction.

(2) To separate enantiomers of α-phenylethylamine by forming diastereomeric complexes with (+)-tartaric acid.

(3) To be skilled in some basic operations, such as extraction and steam distillation.

Principles

In this experiment, racemic α-phenylethylamine is prepared by Leuckart reaction of acetophenone and ammonium formate. The reaction equation is as follows:

$$\text{PhCOCH}_3 + \text{HCOONH}_4 \xrightarrow{\Delta} \text{PhCH(NH}_2\text{)CH}_3 + H_2O + CO_2$$

The product obtained in this reaction is a mixture of the enantiomers of α-phenylethylamine. If an optically pure enantiomer is wanted, the further separation of the enantiomeric mixture is needed.

In this experiment, (+)-tartaric acid is used as the resolving agent for the separation of racemic α-phenylethylamine. When racemic α-phenylethylamine reacts with (+)-tartaric acid, (+)-amine-(+)-tartaric acid salt and (−)-amine-(+)-tartaric acid salt are formed. These two diastereomeric complexes have a significant solubility difference in methanol. The (+)-amine-(+)-tartaric acid salt is more soluble in methanol than the (−)-amine-(+)-tartaric acid salt. The (−)-amine-(+)-tartaric acid salt can be isolated by selectively crystallizing out from a methanolic solution. The isolated (−)-amine-(+)-tartaric acid complex reacts with excess aqueous sodium hydroxide, converting the salt to the free amine (organic soluble) and disodium tartrate (aqueous soluble). The isolated salt will be converted to the optically active free base by treatment with sodium hydroxide followed by extraction. A flowchart for the chemical resolution of α-phenylethylamine by using (+)-tartaric acid is shown in Fig. 3.22.

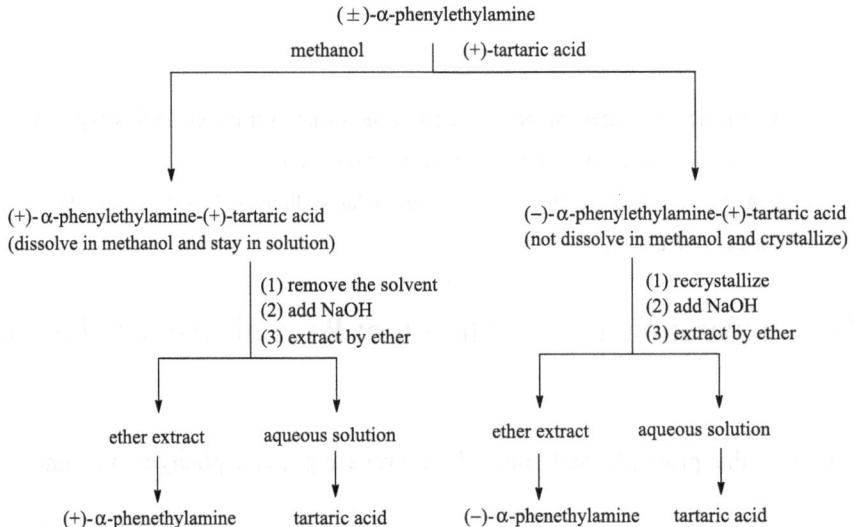

Fig. 3.22 A flowchart for the chemical resolution of α-phenylethylamine by using (+)-tartaric acid

A polarimeter will be used to determine the optical rotation of the product, from which we can determine enantiomeric excess and evaluate the efficiency of the resolution.

Materials

Equipment: water vacuum pump, polarimeter, refractometer, distillation adapter, separatory funnel, three-neck flask, distillation head, round-bottom flask, volumetric flask, Allihn condenser, suction flask, Erlenmeyer flask, Claisen distillation head.

Reagents: acetophenone, ammonium formate, sodium chloride, methanol, hydrochloric acid, benzene, sodium hydroxide, (+)-tartaric acid, ether, anhydrous sodium sulfate.

Procedures

1. Preparation of (±)-α-phenylethylamine

In a 100 mL round-bottom flask, add 12.0 g (0.1 mol) of acetophenone, 20.0 g (0.32 mol) of ammonium formate, place 2~3 boiling chips, and set up a simple apparatus with the bulb of a thermometer inserted below the surface of the reaction mixture.

Heat the mixture slowly in a heating mantle. When the temperature of the reaction mixture is above 150℃, the mixture separates into two phases, and there is a sharp demarcation line between the two layers. Continue to heat the mixture and the mixture will be changed into a single phase. Raise the temperature until it reaches 185℃ and then stop heating. Some water and acetophenone will be distilled out of the flask in this process.

Transfer the distillate to a separatory funnel, separate the layers and transfer the upper organic layer back into the reaction flask. Heat the reaction mixture again and keep the reaction temperature at 185~190℃ for about 1.5 h. Cool the reaction mixture to room temperature, transfer the distillate to a separatory funnel, separate the layers and wash the organic layer with saturated salt solution (2×10 mL). Transfer the organic layer into the original reaction flask, add in 10 mL of conc. HCl and a couple of boiling chips. Heat the reaction mixture until it begins to boil gently. Keep the mixture under reflux for 50 min. After that, cool the mixture to room temperature, transfer the mixture to a separatory funnel, and extract it with toluene (2×10 mL). Retain the organic layer and drain the bottom aqueous layer into a beaker.

Cool down the aqueous layer in an ice-water bath, and slowly add in 20 mL of 50% sodium hydroxide solution under continuous stirring. Transfer the solution to a separatory funnel, separate the layers and retain the organic layer, extract the aqueous layer three times with 10 mL of ether each. Combine all the organic layers and the ether extract, dry the combined liquid over granular sodium hydroxide. Decant the clear liquid into a dry distillation flask.

Distill and collect the fraction that boils between 189~190℃. Weigh the product and calculate the percentage yield. Refractive index of (±)-α-phenylethylamine is 1.5238.

2. Separation of (±)-α-phenylethylamine

Add 50 mL of methanol and 3.8 g (0.119 mol) of (+)-tartaric acid into a 100 mL round-bottom flask, heat the mixture in a warm water bath under stirring until all the crystals have dissolved.

Heat the solution to nearly boiling, add 3.0 g (0.025 mol) of (±)-α-phenylethylamine in the solution[1]. After the addition is completed, cool the solution to room temperature naturally under shaking. Plug the flask and keep the resulting solution stand still at room temperature for 24 h. White rhombic crystals will be obtained after that (if the crystals are not rhombic but needle-like or shapeless, reheat the mixture for another crystallization until white rhombic crystals have formed) [2]. Filter the mixture by suction filtration and wash the collected crystals three times with an appropriate amount of cold methanol. Keep the aspirator running until the crystals become dry. The crystals of (−)-amine-(+)-tartaric acid salt can be isolated. Weigh the crystals and calculate the percentage yield.

Add the obtained crystals, 10 mL of water and 2 mL of 50% sodium hydroxide solution into an Erlenmeyer flask. Mix to dissolve the solid completely by swirling the Erlenmeyer flask. Transfer the solution into a separatory funnel, extract the solution with ether (3×10 mL), combine the ether extracts and dry them over anhydrous sodium sulfate.

Decant the clear liquid into a round-bottom flask. Remove ether from the system by a constant pressure distillation and collect the fraction that boils between 81~81.5℃ at a pressure of 2.4 kPa using a vacuum distillation. (−)-α-phenylethylamine will be obtained. Weigh the product and calculate the percentage yield. Determine the refractive index of (−)-α-phenylethylamine. If (+)-α-phenylethylamine is to be isolated, you can use the filtrate in the filtration for the purification of the crystals of (−)-amine-(+)-tartaric acid salt. Remove the solvent in the filtrate to obtain the crystals of (+)-amine-(+)-tartaric acid salt. Thereafter, treat the crystals in the same way as that for (+)-amine-(+)-tartaric acid salt, and (+)-α-phenylethylamine can be isolated.

3. Calculate the Specific Optical Rotation

Take a clean, dry, 50 mL volumetric flask. Weigh the volumetric flask accurately by an electronic balance. Transfer the prepared sample into the volumetric flask. Weigh the volumetric flask accurately for a second time, and calculate the weight of the sample. Then make up a solution in the volumetric flask by adding methanol up to the level just below the graduation mark.

Next, take a clean 1 dm of polarimeter sample holder, rinse the sample holder with the sample solution you have prepared. Then, fill the sample holder with the sample solution. Ensure no bubbles are present; if there are bubbles present, tilt the sample holder to ensure the bubbles are in the curved part of the holder and not in the top or bottom of the sample holder. Next, open the hatch of the polarimeter and place the sample holder into the tube opening and then close the hatch again[3]. And begin to take the reading. Measure the angle of optical rotation of the sample. The difference between the reading and the zero point is the angle of optical rotation. Then calculate the specific optical rotation.

Notes

[1] Be careful when adding (±)-α-phenylethylamine into the hot solution, avoiding bumping of the liquid.

[2] It's a key factor that rhombic crystals should be formed.

[3] The polarimeter should be used carefully.

Questions

(1) What are the advantages of the method in the synthesis of α-phenylethylamine by using acetophenone as a reactant?

(2) What is the key factor in this experiment?

(3) What is a racemate? How many ways can be used to separate optical enantiomers?

Exp.38 Optical Resolution of *Trans*-1, 2-diaminocyclohexane

Objectives

(1) To learn the principle of the optical resolution of isomers.

(2) To learn the method of separating enantiomers of *trans*-1,2-diaminocyclohexane by forming diastereomeric complexes with L-tartaric acid.

(3) To learn how to calculate the specific optical rotation.

Principles

Vicinal diamines (or 1, 2-diamines) are one of the most widely used building blocks for making catalysts and drugs. Among them, 1, 2-diaminocyclohexane (DACH) is the most popular chiral diamine due to its commercial availability.

Optical resolution is the method of separating enantiomers from racemates. Optical resolution is widely used as it is an easy and practical method. One of the most common optical resolution methods utilizes diastereomers which are obtained from the reaction of racemic compounds and an optical resolving agent. Diastereomers will be separated based on the differences of physical properties between diastereomers.

In this experiment, L-tartaric acid is used as the resolving agent for the separation of *trans*-1, 2-diaminocyclohexane. L-tartaric acid reacts with racemic 1, 2-diaminocyclohexane, forming DACH-L-tartaric acid complexes as the diastereomeric mixture. The complex of (R, R)-DACH and L-tartaric acid is much less soluble in water than (S, S)-DACH-L-tartaric acid complex. Therefore, only (R, R)-DACH-L-tartaric acid complex comes out from the aqueous solution as a white crystal leaving the other complex in the solution. And (R, R)-DACH-L-tartrate will be isolated. Finally, (R, R)-DACH can be recovered from the salt by alkalinization. A flow chart for this experiment is shown in Fig. 3.23.

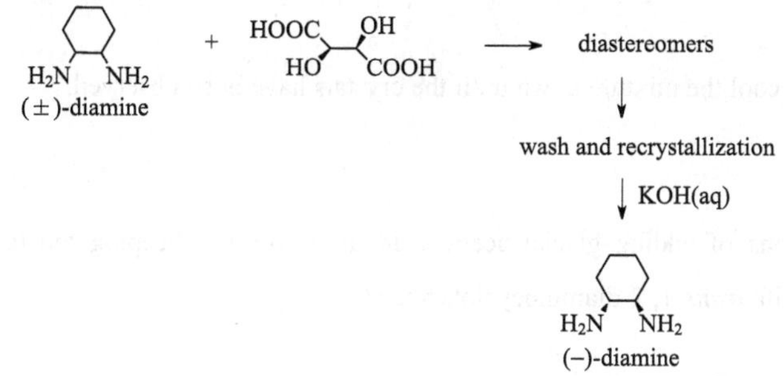

Fig. 3.23   Chemical resolution of *trans*-1,2-diaminocyclohexane by using L-tartaric acid

Materials

Equipment: heating mantle, round-bottom flask, thermometer, measuring cylinder, magnetic stirrer, separatory funnel, Büchner funnel, polarimeter.

Reagents: *trans*-1, 2-diaminocyclohexane, glacial acetic acid, L-tartaric acid, dichloromethane, potassium hydroxide, anhydrous sodium sulfate, ethanol.

Procedures

1. Resolution of *Trans*-1, 2-diaminocyclohexane

In a 50 mL round-bottom flask, dissolve 5.0 g (0.0438 mol) of *trans*-1, 2-diaminocyclohexane in 8.5 mL of water, add 3.3 g (0.0220 mol) of L-tartaric acid slowly with stirring, heat the mixture and control its temperature at around 70℃.

Add 2.2 mL of glacial acetic acid dropwise into the mixture with stirring until the tartaric acid crystals have dissolved completely. The temperature should be controlled at around 90℃. Cool down the mixture with continuous stirring until the precipitation has occurred[1].

Stop stirring and cool the solution in an ice water bath for 30 min. Filter the mixture to obtain the crystals, and wash the crystals twice (one with 2 mL of ice-water and the other with 4 mL of ethanol). Collect the white solids by vacuum filtration and wash them with water.

Transfer the white solids obtained to a beaker, add 3 mL of 4 mol/L potassium hydroxide solution to the beaker to dissolve the crystals. Transfer the solution to a separatory funnel, extract it with ethyl acetate (2×15 mL).

Combine the organic layers and dry the solution over anhydrous sodium sulfate in a dry Erlenmeyer flask. Remove the solvent with a rotator evaporator and collect the product of (−)-*trans*-1, 2-diaminocyclohexane. Weigh the product and calculate the percentage yield.

2. Calculate the Specific Rotatory Power

Accurately weigh 500 mg of the product and dissolve it in ethanol to afford 5 mL of solution. Determine the optical rotation degree with a polarimeter. Record the optical rotation degree and calculate the specific optical rotation.

Notes

[1] Keep stirring to cool the mixture down until the crystals have been observed.

Questions

What are the reasons of adding glacial acetic acid dropwise and keeping the temperature at 70℃ when (+)-tartaric acid reacts with *trans*-1, 2-diaminocyclohexane?

Exp.39　Preparation of 2, 3; 5, 6-Di-*O*-isopropylidene-α-D-mannofuranose

Objectives

(1) To learn the principle and methods for the protection of hydroxyl group in sugars.

(2) To learn the operations for the protection of D-mannose by using isopropylidene as the protecting group.

Principles

The hydroxyl groups in D-mannose can be protected by ketal formation with acetone catalyzed by *p*-toluene sulfonic acid. The reaction equation is as follows:

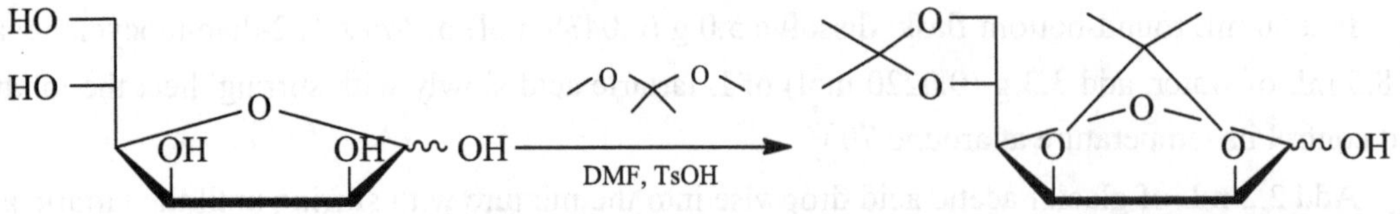

Materials

Equipment: round-bottom flask, magnetic stirrer, condenser, heating mantle, separatory funnel, Erlenmeyer flask, thermometer, measuring cylinder, Büchner funnel.

Reagents: D-mannose, DMF, 2, 2-dimethoxypropane, *p*-toluenesulfonic acid, sodium hydroxide, ethyl acetate, NaCl, anhydrous sodium sulfate, ether, *n*-hexane.

Procedures

Add 1 g (0.0056 mol) of D-mannose and 10 mL of DMF into a 50 mL round-bottom flask, stir the mixture to dissolve the solid with a magnetic stirrer[1], and then add 5 mL of 2, 2-dimethoxypropane and 10 mg of *p*-toluenesulfonic acid into the mixture. Set up a reflux apparatus with a drying tube equipped on the top of the condenser.

Heat the mixture until its temperature reaches 80 ℃, keep the mixture maintained at this temperature under stirring for about 1 h. Let the reaction mixture cool down, add an appropriate amount of 10% NaOH solution to neutralize the solution[2].

Transfer the neutral solution to a separatory funnel, dilute it with 20 mL of water and extract the diluted solution with ethyl acetate (2×15 mL), combine the organic layers and wash the combined organic liquid three times with saturated NaCl solution. Dry the solution over anhydrous sodium sulfate, filter and remove the solvent from the filtrate by a rotary evaporator.

Pure 2, 3; 5, 6-di-*O*-isopropylidene-α-D-mannofuranose is obtained as colorless crystals with m.p. 122~123 ℃. The procedure can be shown as Fig. 3.24.

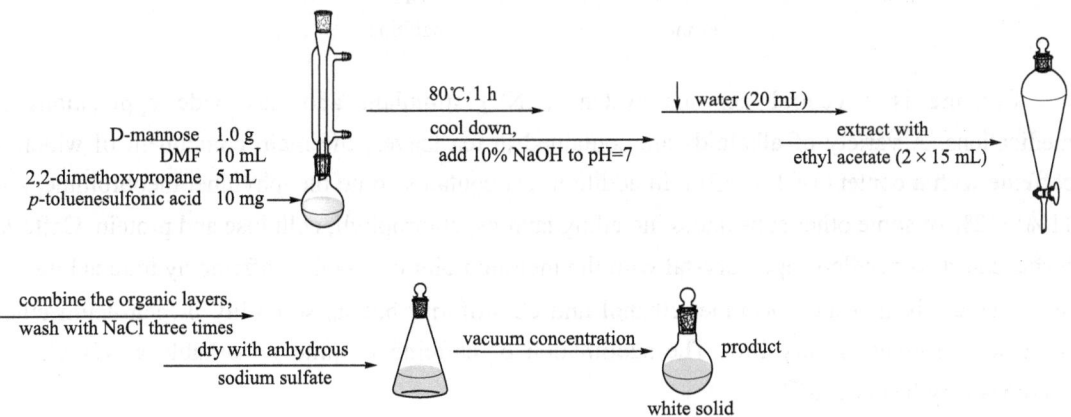

Fig. 3.24　The flowchart for the preparation of 2, 3; 5, 6-di-*O*-isopropylidene-α-D-mannofuranose

Notes

[1] When stirring the reaction mixture with a magnetic stirrer, the speed should be controlled to avoid liquid splashing.

[2] Do not add overdosed 10% NaOH solution when making the neutralization.

Questions

(1) What is the purpose of connecting a drying tube on the condenser?

(2) What measures can be taken if no crystals of the product precipitate after a rotary evaporation?

Exp.40 Isolation of Caffeine from Tea Leaves

Objectives

(1) To learn the principle and methods for the isolation of caffeine via Soxhlet extractor.

(2) To further consolidate the skills for some basic operations, such as extraction, distillation and sublimation.

Principles

Caffeine is a methylxanthine, with its IUPAC name of 1,3,7-trimethyl-3,7-dihydro-1H-indole-2,6-dione.

purine   caffeine

Caffeine is a central nervous system (CNS) stimulant and has wide applications in medications. A variety of alkaloids are contained in tea leaves, the main component of which is caffeine with a content of 1%~5%. In addition, tea contains some theophylline, theobromine, and 11%~12% of some other substances including tannins, chlorophyll, cellulose and protein. Caffeine forms colorless needle-shaped crystal with the melting point of 234℃. Caffeine hydrate is bitter; it dissolves easily in water, acetone, ethanol and chloroform, but its solubility in petroleum ether, benzene and ether is very low. The sublimation of caffeine is quite remarkable at 120℃, and becomes very fast at 178℃.

In this experiment, a Soxhlet extractor is used to isolate caffeine from tea leaves. The crude caffeine solution (using an organic solvent such as ethanol or chloroform) would first be isolated via continuous extractions in the Soxhlet extractor and a further sublimation will lead to pure caffeine.

A Soxhlet extractor is a piece of laboratory apparatus invented in 1879 by Franz von Soxhlet. It was originally designed for the extraction of a lipid from a solid material. As shown in Fig. 3.25, a Soxhlet extraction apparatus is composed of a condenser, a Soxhlet extractor, and a round-bottom flask.

When making a Soxhlet extraction, the solid sample is loaded into a thimble made by a piece of thick filter paper, which is placed inside the Soxhlet extractor. When the solvent is heated to reflux, the solvent vapour travels up a distillation arm, and floods into the chamber housing the thimble of a solid substance to be extracted. The chamber containing the solid material slowly fills with warm solvent. Some of the desired compound dissolves in the warm solvent. When the chamber is almost full, the chamber is emptied by the siphon. The solvent is returned to the distillation flask. This gives one cycle of Soxhlet extraction and the cycle may be allowed to repeat many times, over hours or days.

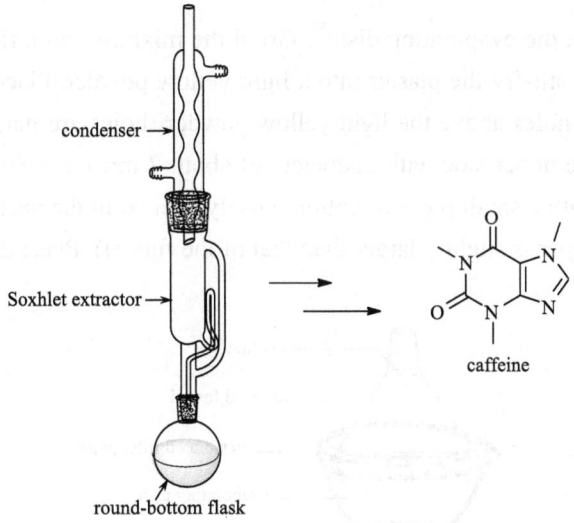

Fig. 3.25 Soxhlet extraction apparatus

During each cycle, a portion of the non-volatile compound dissolves in the solvent. After many cycles, the desired compound is concentrated in the distillation flask. The advantage of this system is that instead of many portions of warm solvent being passed through the sample, just one batch of solvent is recycled.

After extraction the solvent is removed typically by means of a rotary evaporator, yielding the extracted compound. The non-soluble portion of the extracted solid remains in the thimble, and is usually discarded.

Materials

Equipment: Soxhlet extractor, condenser, round-bottom flask, heating mantle, evaporating dish, funnel.

Reagents: 95% ethanol, CaO, tea leaves.

Procedures

1. Isolation of Caffeine

Set up an apparatus as shown in Fig. 3.25. Weigh 10 g of tea leaves, load the crushed tea leaves into a thimble made by a filter paper[1], and place this inside the Soxhlet extractor[2]. Add 100 mL of 95% ethanol and several boiling chips into the round-bottom flask. Heat the ethanol solution in the flask under reflux for 2~3 cycles of Soxlet extraction[3]. After that, cool down the flask slightly, change the apparatus for a simple distillation, and distill out most of the solvent. When there is only about 5 mL of liquid left in the flask[4], stop heating and transfer the residual liquid to an evaporating dish while it is still hot.

2. Purification of Caffeine by Sublimation

Add 4 g of CaO in the evaporating dish[5]. Grind the mixture into a fine-mixed plaster with a pestle, and furthermore stir-fry the plaster into a light yellow powder. Place a prepared filter paper with dozens of pierced holes above the light yellow powder (holes are made from the bottom side of the filter paper to the upper side with diameters of about 2 mm). As shown in Fig. 3.26, put an inverted glass funnel with a small piece of cotton loosely packed in the neck on the filter paper (the diameter of the filter paper is slightly larger than that of the funnel). Place the evaporating dish on a heating mantle.

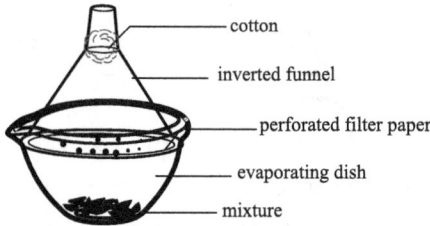

Fig. 3.26 Set-up for the sublimation of caffeine

Turn on the heat and carefully control the temperature. Once there are white vapors of caffeine observed obviously, turn off the heat, remove the evaporating dishes from the heating mantle, and keep the sublimation continuing for another 5 min[6]. After the system has cooled to room temperature, carefully pour out the residue from the evaporating dish into the trash and scrape the purified caffeine from the filter paper onto a piece of pre-weighed weighing paper. Weigh the purified caffeine and determine the recovery of the sublimation process. Record the color and shape of crude and sublimed caffeine crystals.

Notes

[1] Fit the paper thimble well to the chamber of the extractor and make its upper edge of the thimble lower than the top point of the siphon tube. The filter paper should be carefully and tightly sealed to prevent the leakage of the solid.

[2] Handle the Soxhlet extractor carefully because the siphon tube is fragile.

[3] Stop the extraction when the extract is light-colored.

[4] Difficult transferring will be resulted in if the residual liquid is very sticky, therefore the extracted liquid should not be distilled to dry.

[5] Calcium oxide is used to absorb water and remove some acidic impurities.

[6] Control the temperature during the sublimation process. If the temperature is too low, the sublimation will be very slow; if the temperature is too high, the product will decompose.

Questions

(1) What is the purpose of using calcium oxide for the sublimation of caffeine?
(2) Why can caffeine be purified by sublimation?

Chapter 4　Reactivity Experiments of Organic Compounds

Exp.41　The Preparation and Reactivitics of Ethylene and Ethyne

Objectives

(1) To understand the principle and method for the generation of ethylene from dehydration of ethanol.

(2) To learn the principle and method of producing ethyne from calcium carbide.

(3) To study the general properties of ethylene and ethyne, and their identification methods.

Principles

Ethanol and acid (sulfuric acid, phosphoric acid, etc.) are commonly used in the laboratory to produce ethylene, through a dehydration process. Concentrated sulfuric acid acts as a catalyst and dehydrating agent. A side reaction producing diethyl ether also occurs. The overall reaction is described as blow.

$$CH_3CH_2OH \xrightarrow[170\,°C]{H_2SO_4} H_2C=CH_2 + H_2O$$

$$2CH_3CH_2OH \xrightarrow[140\,°C]{H_2SO_4} CH_3CH_2OCH_2CH_3 + H_2O$$

The generation of ethyne in the laboratory is usually achieved by a reaction between calcium carbide and water. Besides ethyne, one molecule of calcium hydroxide is also produced. The reaction equation is shown as follows:

$$CaC_2 + 2H_2O \longrightarrow HC\equiv CH + Ca(OH)_2$$

Materials

Equipment: round-bottom flask, gas washing bottle, distillation head, test tube, thermometer, constant pressure dropping funnel, adapter with T-connection, alcohol burner.

Reagents: conc. H_2SO_4, 95% ethanol, 10% sulfuric acid, 10% NaOH solution, 2% Br_2/CCl_4 solution, 1% $KMnO_4$ solution, 2% ammonia solution, saturated NaCl solution, 10% $CuSO_4$ solution, ammoniacal solution of CuCl, 2% $AgNO_3$ solution, calcium carbide.

Procedures

1. Preparation and Reactivities of Ethylene

The apparatus used for producing ethylene is shown in Fig. 4.1. In a 250 mL round-bottom flask, add 4 mL of 95% ethanol followed by careful addition of 12 mL of conc. H_2SO_4 and a few boiling chips. Introduce the ethylene gas through 6 mL of 10% sodium hydroxide solution in a gas washing bottle to wash it. Turn on the heater, and control the temperature well at 160~170℃[1]. Ethylene gas is thereby generated. When the experiment is ended, take out the gas inlet tube from water and remove the alcohol burner, avoiding back-flowing of water.

(1) Addition reaction: take a test tube and add 2 mL of 2% of Br_2/CCl_4 solution. The resulting ethylene gas is introduced and the color change of the solution in the test tube will be observed.

(2) Oxidation reaction: take a test tube, add 6 drops of 1% potassium permanganate solution and 2 mL of 10% sulfuric acid, introduce ethylene into the solution and observe the phenomenon.

(3) Igniting experiment: ignite the ethylene gas at the tip of the tube. Observe the burning flame and extinguish the flame with cold water in a test tube at the end of the experiment.

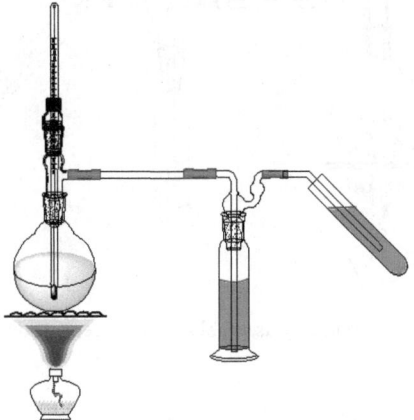

Fig. 4.1　A laboratory assembly for the preparation of ethylene

2. Preparation and Reactivites of Ethyne

The apparatus used for producing ethyne is shown in Fig. 4.2. In a 250 mL round-bottom flask, add 5 g of calcium carbide, place a constant pressure dropping funnel containing 40 mL of saturated NaCl solution on top of the flask via an adapter with T-connection[2]. Connect the branch of the adapter to a gas washing bottle containing 10% copper sulfate solution. Adjust the constant pressure dropping funnel, so that the saturated NaCl solution drops into the round-bottom flask slowly. The generated ethyne could be collected by displacement of water from a container.

(1) Reaction with halogen: take a test tube and add 2 mL of 2% of Br_2/CCl_4 solution. Introduce the resulting ethyne gas into the test tube and observe the color change of the solution in the test tube.

(2) Oxidation reaction: take a test tube, add 6 drops of 1% potassium permanganate solution and 2 mL of 10% sulfuric acid, introduce the ethyne into the solution and observe the phenomenon.

(3) Formation of silver acetylide[3]: take a test tube, add 2 mL of 2% silver nitrate solution, 1 drop of 10% NaOH solution and a few drops of 2% ammonia water, shake to dissolve the precipitate formed, introduce acetylene gas into the solution and observe the reaction phenomenon.

(4) Formation of copper acetylide[3]: take a test tube, add 3 mL of ammoniacal solution of CuCl, introduce acetylene gas into the solution and observe the reaction phenomenon.

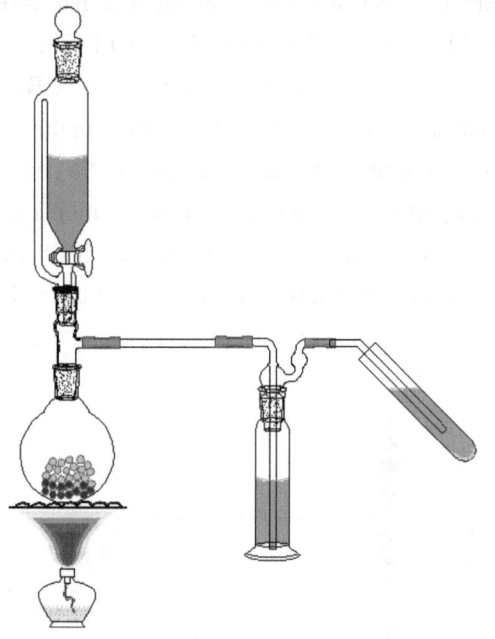

Fig. 4.2 A laboratory assembly for the preparation of ethyne

Notes

[1] Ethyl hydrogen sulfate decomposes at 170℃ to form ethylene, but reacts with ethanol to form ether at 140℃. Therefore, to reduce the formation of ether, it is necessary to raise the reaction temperature rapidly to above 160℃.

[2] Water reacts with calcium carbide vigorously; therefore saturated NaCl solution is used instead for the preparation of ethyne, which can react with calcium carbide smoothly and uniformly.

[3] Dry silver acetylide and copper acetylide can detonate when exposed to heat or vibration. In order to prevent the explosion from happening, the silver or cuprous acetylide formed in this experiment should be destroyed while still wet by warming with dilute nitric acid or dilute hydrochloric acid.

Questions

(1) When concentrated sulfuric acid is used as the catalyst for the preparation of ethylene by dehydration of ethanol, what impurities may be generated in the product and what effects may be resulted on the experiment by the impurities?

(2) What effects will it have if the temperature is raised slowly in the preparation of ethylene?

(3) How to remove smelly odor in the acetylene gas prepared with calcium carbide?

(4) Analyze and explain the different reaction rates of ethylene and ethyne in addition and oxidation reactions.

Exp.42 The Reactivities of Aromatic Compounds

Objectives

(1) To understand the chemical properties of aromatics compounds.
(2) To identify aromatics by applying experimental techniques.

Principles

Although benzene has unsaturated bonds, its chemical properties are more stable than olefins and alkynes due to the delocalized (conjugated) system. It is less likely to take addition reactions and oxidation reactions. However, it is more likely to make aromatic substitution reactions, which are characteristic reactions of benzene and other aromatic hydrocarbons. On the other hand, alkyl benzene, such as toluene, can be oxidized to benzoic acid by potassium permanganate, thereby discoloring the potassium permanganate solution.

Materials

Equipment: light bulb of 60 W or above, glass rod, test tube, alcohol burner, test tube holder, beaker.

Reagents: benzene, toluene, cyclohexene, potassium permanganate solution (0.5%), sulfuric acid solution (10%), conc. H_2SO_4, conc. HNO_3, naphthalene, acetanilide acetic acid-water (9 : 1) solution (0.2 mol/L), chlorobenzene acetic acid-water (9 : 1) solution (0.2 mol/L), p-nitrophenol acetic acid-water (9 : 1) solution (0.2 mol/L), phenol acetic acid-water (9 : 1) solution (0.2 mol/L), benzene acetic acid-water (9 : 1) solution (0.2 mol/L), bromine acetic acid-water (9 : 1) solution (0.05 mol/L), Br_2/CCl_4 solution (3%), benzyl chloride, chloroform, anhydrous aluminum trichloride, $AgNO_3$ solution (50 g/L).

Procedures

1. Reactions with Potassium Permanganate Solution

In three separate test tubes, place 0.5 mL of each of the following: benzene, toluene and cyclohexene. Add 0.2 mL of potassium permanganate solution (0.5%) and 0.5 mL of sulfuric acid solution (10%) to each tube, shake vigorously (if necessary, heat in a 60~70℃ water bath for a few minutes), observe and compare the reaction phenomena of the three compounds with the oxidant and give the explanation.

2. Sulfonation

1) Sulfonation of Monocyclic Aromatic Hydrocarbons

Take two test tubes, place 1 mL of benzene to one tube and the same amount of toluene to the other test tube. Add 3 mL of conc. H_2SO_4 to each tube. Stopper each test tube with a one-hole rubber stopper and insert a glass tube in the hole. Heat the test tubes in a water bath (no more than 30℃) and shake the test tubes occasionally[1]. Carefully observe and report in which test tube the delamination disappears first. When the delamination disappears, the reaction mixture in each of the test tubes is poured into a small beaker containing 10 mL of water, respectively. After stirring lightly, record what you observed and give the explanation.

2) Sulfonation of Naphthalene

Obtain one dry test tube and add 1 g of naphthalene into the test tube. Heat the test tube in a water bath until naphthalene is melted. Cool the test tube a bit and add 1 mL of conc. H_2SO_4 to the tube. Heat the test tube carefully again for 1~2 min. Shake the tube as it is being heated until the mixture becomes a homogeneous liquid. Cool the mixture to get a dark viscous liquid, add 2 mL of water to this prepared liquid, heat the solution slightly to dissolve the liquid, and the solution is then cooled to 15~20℃. Record and explain what you observed.

Compare the results of the three sulfonation reactions in the above two experiments, determine which one is the easiest and explain why.

3. Nitration

Add 1.5 mL of conc. HNO_3 and 2 mL of conc. H_2SO_4 to each of two dry test tubes. After thorough mixing these two acids, cool the mixture to room temperature with cold water bath. Slowly add 1 mL of benzene in one of the test tubes and 1 mL of toluene in another one. Continue to shake and mix the solutions. If there is too much heat released, it should be cooled in cold water to keep the temperature lower than 50℃. After 3~5 min, pour the mixture in each of the tubes into a beaker with 20 mL of cold water. After the mixtures stay still, observe what happens.

4. Halogenation

1) Aromatic Halogenation

In five labeled test tubes, place 2 mL of each of the following: acetaniline, benzene, chlorobenzene, p-nitrophenol, and phenol acetic acid-water (9 : 1) solutions (0.2 mol/L). Place these tubes in a 400 mL beaker filled with warm water [(35±2)℃]. Add 10 mL of bromine acetic acid-water (9 : 1) solution (0.05 mol/L) into another test tube, and place this tube into the same beaker holding the labeled test tubes. Keep these test tubes in warm water for 5 min until they reach the temperature of the water. Graduate one portion of 2 mL of bromine acetic acid-water (9 : 1) solution (0.05 mol/L) into each of the labeled test tubes, rapidly mix each portion of the mixtures with a glass rod, and observe the time (in seconds) for discoloration. Record the time (the solution becomes colorless or light yellow to reach the end point). Compare their reaction rates.

2) Side-chain Halogenation

Obtain two dry test tubes and add 1 mL of toluene and 10 drops of Br_2/CCl_4 solution (3%) into each tube. Shake the mixtures well and put one of them in the sunlight (if there is no sunlight, use magnesium light or light produced by a bulb with the voltage higher than 60 W). Wrap the other test tube with black paper and place it in a dark place (such as a cabinet). Wait for a while, take it out, and blow air into it. Observe what happens.

Obtain two strips of filter paper and insert them into the above-mentioned two test tubes. Soak one end of the filter paper strips with the solution, and then take them out and make them dry slightly in the air. Carefully smell the odor of the filter paper strips, and describe what the difference is (caution: benzyl bromide causes irritation of nose and throat with a very sharp, pungent gas odor).

3) Comparison of Aromatic Halogenation and Side-chain Halogenation

Add 2 mL of distilled water in each of two clean test tubes, add 0.5 mL of chlorobenzene in one of the test tubes and 0.5 mL of benzyl chloride into the other one. Heat the mixture to boiling and then add 0.3 mL of $AgNO_3$ solution (50 g/L) dropwise into each of the test tubes. Observe and compare the results.

5. Friedel-Crafts Reaction

In a dry and clean test tube, add 2 mL of chloroform and 3 drops of anhydrous benzene. Shake well and make sure the inside wall of the tube is wet. Then add in a little anhydrous aluminum trichloride with a strip along the wall of the tube, making the powder stick to the tube wall. Observe the color changes of the powder on the tube wall and the solution in the tube[2].

Notes

[1] The key to this experiment is to shake well, because the aromatics and the mixed acid are difficult to dissolve mutually in each other.

[2] Reactions of aromatic compounds (including aromatic halides) with chloroform using an aluminium trichloride catalyst give triarylmethanes, which are often brightly colored. For example, benzene and its homologs, aromatic halides give orange to red colors; naphthalene gives blue; anthracene forms green; and biphenyl and phenanthrene affords purplish red. You must observe the colors in time; otherwise it is easy to change.

Questions

(1) Why can alkylbenzenes and cyclohexene be oxidized by $KMnO_4$, but benzene can not?

(2) What method can be used to identify aromatic halogenations and side-chain halogenations of aromatics, and what method can be used to identify benzene from toluene?

(3) Rank the groups ($NHCOCH_3$, Cl, OH and other groups) in the decreasing order of their activating effects on bromination of substituted benzenes based on the experimental results.

Exp.43   The Reactivities of Some Alkyl Halides

Objectives

(1) To understand the mechanism of nucleophilic substitution reactions of alkyl halides and be familiar with the elimination reactions and the reaction conditions.

(2) To understand how to identify primary (1°), secondary (2°), tertiary (3°) alkyl halides according to their relativities in nucleophilic substitution reactions.

Principle

The most characteristic reaction of alkyl halides is nucleophilic substitution reaction. The C—X bonds of halogenated hydrocarbons are polar and much more active than C—H bonds of hydrocarbons. The reactivities of alkyl halides in nucleophilic substitution reactions are dominated by the halogen atoms and the structures of the hydrocarbyl groups. For the alkyl halides, the order of reactivity is: RI>RBr>RCl; for halogenated alkanes with different hydrocarbyl structures, the order of reactivity is different due to their different reaction mechanisms. In a monomolecular nucleophilic substitution (S_N1) reaction, such as the reaction of an alkyl halide with an ethanolic solution of silver nitrate, the typical order of reactivities of different alkyl halides is tertiary (3°) > secondary (2°) > primary (1°). However, in a bimolecular nucleophilic substitution (S_N2) reaction, such as the reaction of an alkyl bromide (or chloride) with sodium iodine in acetone, the order of reactivities of alkyl halides is primary (1°) > secondary (2°) >tertiary (3°). The chemical properties of halogenated olefins and halogenated aromatic hydrocarbons are quite different from those of halogenated alkanes. Their reactivities are lower comparing to those of alkyl halides, and the order of the reactivities is allylic or benzyl halides > primary and secondary alkyl halides > vinyl and phenyl halides. Alkyl halides undergo β-elimination reactions under strong basic condition to form alkenes. When polyhalogenated hydrocarbons (such as $CHCl_3$) are treated with strong bases, α-elimination reactions will occur and carbenes will be producted. Cabenes are important intermediates in organic synthesis.

Materials

Equipment: test tube, alcohol burner, test tube holder, beaker.

Reagents: saturated ethanolic $AgNO_3$ solution, 1-bromobutane, 2-bromobutane, 2-methyl-2-bromopropane, bromobenzene, HNO_3 (3 mmol/L), 1-chlorobutane, 1-iodobutane, sodium iodide-acetone solution, benzyl chloride, chlorobenzene, 2, 4-dinitrochlorobenzene, potassium hydroxide, 20% NaOH solution, chloroform, 2% $KMnO_4$ solution.

Procedures

1. Reactions with Silver Nitrate in Ethanol

(1) Label four dry test tubes and place 1 mL of saturated ethanolic $AgNO_3$ solution in each test tube[1]. Then add two drops of each of the following halides into each test tube: 1-bromobutane, 2-bromobutane, 2-methyl-2-bromopropane and bromobenzene. Shake the mixture and record the time at which the initial precipitation is observed in each test tube. If no precipitation occurs in one test tube, heat it in a 70℃ water bath for 3 min. Add a drop of 3 mmol/L HNO_3 into each test tube that has precipitate. If precipitate in the test tube dissolves, it indicates that there is no silver halide

formed. Based on your observations, determine the order of reactivities of the four brominated substrates and explain how the structure of an alkyl halide affects the rate of a S_N1 reaction.

(2) Label three dry test tubes and place 1 mL of saturated ethanolic AgNO₃ solution in each test tube. Add 2 drops of 1-chlorobutane, 1-bromobutane and 1-iodobutane in each separate tube. Shake the mixture and record the time at which the initial precipitation is observed for each reaction. If no precipitation occurs in some of the test tubes, place only those test tubes into a 70 ℃ water bath. Based on your observations, determine the order of reactivities of the three halogenated butanes, and explain how the halogen atom affects the rate of a S_N1 reaction.

2. Reactions with Sodium Iodide in Acetone

Place 1 mL of sodium iodide-acetone solution in each of four separate test tubes. Then add three drops of each of the samples into each test tube. Samples are 1-bromobutane, benzyl chloride, chlorobenzene and 2, 4-dinitrochlorobenzene. Shake the mixture and record if there is any precipitation observed for each reaction. If no precipitation occurs in some of the test tubes, place only those test tubes into a warm water bath and check for precipitation again. If precipitation occurs, it indicates that sodium bromide or sodium chloride is formed[2].

3. Elimination Reaction

1) Alkene-forming Elimination: β-Elimination Reaction

Place 1 g of solid KOH and 4~5 mL of ethanol into a test tube. Heat the test tube a bit to dissolve all the KOH. Then add 1 mL of 1-bromobutane. Mix the mixture well and plug the test tube with a one-hole stopper. Insert one end of U-bend glass tubing into the stopper and insert the other end of the tubing into a test tube filled with bromine water (or KMnO₄ solution). If bubbles are formed and the color of the bromine water (or KMnO₄ solution) fades, it indicates that an alkene is formed.

2) Carbene-forming Elimination: α-Elimination Reaction

Place 3 mL of 20% NaOH solution and 8 drops of chloroform into a test tube. Heat and shake the test tube for 1~2 min. When the solution is boiled, cool it to room temperature by placing the test tube into cold water. Add two or three drops of 2% KMnO₄ solution into the test tube. Record your observation and write down the chemical reactions[3].

Notes

[1] At 18~20 ℃, the solubility of silver nitrate in anhydrous ethanol is 2.1 g/100 g. Since alkyl halides are soluble in ethanol but not in water, ethanol is better choice as the solvent. It not only makes the reaction homogeneous, but also helps to smooth the reaction process.

[2] Sodium iodide dissolves in acetone, while sodium bromide or sodium chloride does not.

[3] Under strong basic condition, chloroform produces dichlorocarbene, which is easily hydrolyzed into formate in alkaline solution. Potassium permanganate is reduced to manganate, so that the solution turns green.

Questions

What is the difference between the reactivity order of primary, secondary and tertiary haloalkanes in silver nitrate ethanolic solution and in sodium iodide-acetone solution? Explain the reasons.

Exp.44 The Reactivities of Alcohols and Phenols

Objectives

(1) To understand the reactivities of alcohols and phenols.
(2) To learn the experimental methods of distinguishing alcohols from phenols.

Materials

Equipment: test tube.

Reagents: *n*-butanol, 2-butanol, *tert*-butanol, benzoyl chloride, benzyl alcohol, cyclohexanol, ethanol, glycerin, phenol, resorcinol, hydroquinone, ammonium cerium (IV) nitrate reagent, glacial acetic acid, hydrochloric acid, zinc chloride, 5% potassium dichromate solution, 10% sodium hydroxide solution, bromine water, ferric trichloride.

Procedures

1. The Reactivities of Alcohols

1) Reactions with Benzoyl Chloride

Obtain three test tubes with stoppers; add 0.5 mL of each of the following into each tube: *n*-butanol, 2-butanol and *tert*-butanol. Add 1 mL of water and 10 drops of benzoyl chloride, and then add 2 mL of 10% sodium hydroxide solution. Stopper each test tube and shake it vigorously to make sure the solution in the test tube is basic. Observe whether the solution separates into layers and sniff each test tube and identify each scent.

2) Reactions with Ammonium Cerium (IV) Nitrate Reagent

Add 5 drops of each of ethanol and glycerol in each of two test tubes and dilute them with 1 mL of water. Then add 0.5 mL of ammonium cerium (IV) nitrate reagent into each tube and shake. Observe the reaction phenomenon[1].

3) Reactions with Lucas Reagent

Add 5 drops of each of the following in each of three dry test tubes: *n*-butanol, 2-butanol, and *tert*-butanol. Add 1 mL of Lucas reagent to each test tube. Stopper the test tubes, shake, and cool at room temperature (preferably at 26~27 ℃). Let them stand for 5 min and observe the phenomenon[2].

4) Reactions with Chromic Acid Reagent

Add 5 drops of each of the following in each of three dry test tubes: *n*-butanol, 2-butanol, and *tert*-butanol. Dissolve the samples by adding 1 mL of acetone first, then 5 drops of chromic acid reagent in each of the test tubes and shaking. Observe and record the phenomenon.

2. The Reactivities of Phenols

1) The Acidities of Phenols

Add 6 mL of saturated phenol aqueous solution to a clean test tube. Test its acidity on a wide range pH test strip by dipping a drop of solution on the test strip with a glass rod. Then the solution is separated into two parts, one part is kept as a blank control; the other is used by adding in 5% sodium hydroxide solution. While adding, shake the mixture and make sure the solution is clear and colorless. Then add 1 mol/L hydrochloric acid in the solution till it is acidic, observe the phenomenon and compare it with that in the blank reagent.

2) Reactions with Bromine Water

Obtain a clean test tube and add 2 drops of saturated phenol aqueous solution. Dilute it to 2 mL with water, and then add bromine water dropwise in the solution. Observe whether precipitation occurs and whether the color of bromine water fades during dropping.

3) Reactions with Ferric Trichloride

Add a few drops of phenol and 2 mL of water to a test tube and then add 1~2 drops of 1% ferric trichloride solution. Obtain another test tube, add distilled water and ferric trichloride reagent into it and use it as a blank control. Compare the color of the two solutions.

Perform the above experiment using resorcinol and hydroquinone instead of phenol. Observe the phenomenon.

Notes

[1] Ammonium cerium (IV) nitrate reagent is usually only applicable in tests for alcohols of ten or fewer carbon atoms. The color of the reaction solution turns red.

[2] Lucas reagent is only suitable for the identification of primary, secondary and tertiary alcohols of 3~6 carbon atoms. The alcohols of six or more carbon atoms are insoluble in Lucas reagent. The products obtained from methanol and ethanol are volatile, resulting unobvious phenomenon.

Questions

(1) There are 6 bottles of unlabeled reagents, known as *tert*-butyl chloride, cyclohexanol, ethanol, *tert*-butyl alcohol, 2-chlorobutane and *n*-chlorobutane, respectively. Identify them with appropriate reagents.

(2) What happens if you add excess bromine water to the phenol solution? Explain why.

Exp.45　The Reactivities of Aldehydes and Ketones

Objectives

(1) To understand the characteristic reactivities of aldehydes and ketones.

(2) To learn the experimental methods of determing aldehydes and ketones.

Materials

Equipment: test tube.

Reagents: 95% ethanol, acetone, butyraldehyde, benzaldehyde, acetophenone, Tollen reagent, Fehling reagent, I_2-KI solution, NaOH solution.

Procedures

1. Reactions with 2, 4-Dinitrophenylhydrazine

Obtain three test tubes and add 1 mL of 2, 4-dinitrophenylhydrazine into each tube. Then add 1~2 drops of each of the following carbonyl compounds to each test tube: acetone, butyraldehyde and benzaldehyde, observe and record the phenomenon[1].

2. Reactions with Tollen Reagent

Add 1 mL of Tollen reagent to each of three test tubes and then add 2 drops of each of the following compounds to each test tube: acetone, butyraldehyde, and benzaldehyde. Leave them at room temperature for a few minutes. If there is no silver mirror on the tube, heat it in a water bath at 50~60 ℃ for a few minutes[2] and observe whether there is a silver mirror.

3. Reactions with Fehling Reagent

Take three test tubes and add 1 mL of Fehling reagent A and 1 mL of Fehling reagent B into each tube. Mix them into a solution. Add 2 drops of each of acetone, butyraldehyde, and benzaldehyde into each tube. Shake and place them into a boiling water bath for 3~5 min. Observe the reaction.

4. Reactions with Iodoform

Obtain three test tubes and add 4 drops of each of acetone, ethanol and acetophenone into each test tube, respectively. Add 1 mL of iodine-potassium iodide solution and slowly drop 3 mol/L sodium hydroxide solution into each tube. Observe the color change of the reaction solution and whether there are yellow crystals precipitated[3].

Notes

[1] The color of the precipitate is related to the conjugate system within the aldehyde and ketone molecules: non-conjugated ketones form yellow precipitates, and the conjugated ketones form orange to red precipitates.

[2] The heating time should not be too long; otherwise explosive silver fulminate will be generated.

[3] In addition to acetaldehyde and α-methyl ketone, α-methyl alcohol can also make iodoform reaction because it can be oxidized by sodium hypoiodate to form α-methyl ketone.

Questions

Base on the experiments above, identify each of the following groups of compounds: ①benzyl alcohol, benzaldehyde and cyclohexanone; ②2-butanone and 3-pentanone.

Exp.46 The Reactivities of Carboxylic Acids and Their Derivatives

Objectives

To understand the reactivities of carboxylic acids and their derivatives.

Materials

Equipment: test tube.

Reagents: formic acid, acetic acid, oxalic acid, 0.5% potassium permanganate solution, acetyl chloride, 2% silver nitrate solution, ethyl acetate, sulfuric acid, 30% sodium hydroxide solution, acetic anhydride, acetamide, hydroxylamine hydrochloride solution in ethanol, 95% ethanol.

Procedures

1. The Reactivities of Carboxylic Acids

1) Acidity

Obtain three test tubes, add 5 drops of each of formic acid and acetic acid in each of two tubes, and 0.5 g of oxalic acid into the other test tube. Then add 2 mL of distilled water into each tube. Shake each tube, dip a clean glass rod in each acid solution and draw a line on the same Congo red test paper with the rod[1]. Compare the color and the color intensity of each line. Explain the difference.

2) Oxidation Reactions

Place each of the following substances into each of three test tubes: 0.5 mL of formic acid, 0.5 mL of acetic acid and 0.2 g of oxalic acid. Add 1 mL of water into each tube, followed by 1 mL of diluted sulfuric acid (1 : 5) and 3 mL of 0.5% potassium permanganate solution. Heat to boiling and observe the phenomenon.

2. The Reactivities of Carboxylic Acid Derivatives—Hydrolysis

1) Hydrolysis of Acyl Chlorides

Add 1 mL of distilled water into a test tube and then add in 3 drops of acetyl chloride. Slightly shake the test tube and observe the phenomenon. After the reaction ends, add 3~4 drops of 2% silver nitrate solution. Observe what happens again.

2) Hydrolysis of Esters

Obtain three clean test tubes and add 1 mL of ethyl acetate and 1 mL of water into each tube. Add 2 drops of 15% sulfuric acid to the first tube and 2 drops of 30% sodium hydroxide solution to the second tube. Put these three test tubes into a 70~80℃ water bath at the same time. Shake the test tubes and compare the rates of disappearance of the ester layers in the three test tubes.

3) Hydrolysis of Anhydrides

In a test tube containing 1 mL of water, add 3 drops of acetic anhydride. Acetic anhydride is insoluble in water. Bead-shaped drops will form and sink to the bottom of the tube. Slightly heat the tube, observe the phenomenon and identify the odor.

4) Hydrolysis of Amide

Acid hydrolysis: add 5 g of acetamide and 3 mL of 6 mol/L sulfuric acid in a test tube and heat to boiling. Determine if there is acidic smell.

Base hydrolysis: add 5 g of acetamide and 3 mL of 6 mol/L sodium hydroxide solution in a test tube and heat to boiling. Use a wet red litmus paper to identify ammonia.

3. Hydroxamic Acid Reaction

In a test tube add 1 mL of ethanolic solution of hydroxylamine hydrochloride. Then add 1 drop of liquid sample or 50 mg of solid sample. Shake it well and add 0.2 mL of 6 mol/L sodium hydroxide solution. Heat the test tube to boiling and then slightly cool it down a bit. Add 2 mL of 1 mol/L hydrochloric acid into the test tube. If precipitation occurs, add 2 mL of 95% ethanol and then 1 drop of 5% ferric trichloride solution. Until the solution becomes purple and the color does not disappear any more[2].

Notes

[1] Congo red can be used as a pH indicator with a color change from blue to red at pH 3.0~5.2. It appears dark blue in weak acids, and blue in strong acids.

[2] This reaction cannot be used for the identification of a sample containing a functional group which gives a colored reaction with ferric trichloride. For this sample, it can be mixed with ferric trichloride solution first. If no color develops, the solution appears yellow.

Questions

Compare the reactivities of carboxylic acids and their derivatives in hydrolysis and alcoholysis reactions.

Exp.47　The Reactivities of Amines

Objectives

(1) To be familiar with the basicity of amines.

(2) To understand how to differentiate primary, secondary and tertiary amines by using simple chemical methods.

Materials

　　Equipment: test tube.

Reagents: aniline, diphenylamine, N-methylaniline, N, N-dimethylaniline, anhydrous ethanol, β-naphthol, benzenesulfonyl chloride (or p-toluenesulfonyl chloride), conc. HCl, sodium hydroxide, sodium nitrite.

Procedures

1. Basicity Test

Add 1~2 drops of aniline and 0.5 mL of water into a test tube. Shake the tube and observe the phenomenon. Then add 1~2 drops of conc. HCl and observe the changes. Dilute the solution with water and observe the phenomenon of dilution.

Add a few crystals of diphenylamine and 0.5~1 mL of anhydrous ethanol into a test tube. Shake the tube to completely dissolve the diphenylamine. Then add 0.5~1 mL of water and shake it again. Observe the reaction. Add a few drops of conc. HCl dropwise and shake to see if the solution turns transparent. Finally, dilute it with water and observe the result. .

2. Nitrosation of Amines

Add 10 drops of aniline and 4 mL of 20% hydrochloric acid to a test tube. Stir with a glass rod, and place the test tube in an ice bath to cool it to 0℃. Add 25% sodium nitrite solution dropwise, and stir the mixture until the mixture is clear and can turn a potassium iodide-starch paper into blue immediately[1]. Add a few drops of this solution to a β-naphthol solution[2], and observe if any orange-red substances form.

Add 5 drops of N-methylaniline and 2 mL of 20% hydrochloric acid to a test tube and stir to dissolve. Place the test tube in an ice bath and cool the mixture to 0℃. Add approximately 10 drops of 25% sodium nitrite solution dropwise to the test tube while stirring. Observe if any yellow oil forms.

Add 2 drops of N, N-dimethylaniline and 0.8 mL of 20% hydrochloric acid to a test tube. Stir the mixture, place the tube in an ice bath and cool it to 0℃. Add 4~5 drops of 25% sodium nitrite solution dropwise to the tube while stirring. And then add 10% sodium hydroxide solution dropwise until the solution is basic. Observe whether any green solid forms[3].

3. Hinsberg Test

Obtain three test tubes and add 3 drops of each of aniline, N-methylaniline and N, N-dimethylaniline into each test tube. Add 5 mL of 10% sodium hydroxide to each test tube and mix the solution well. Then add 6 drops of benzenesulfonyl chloride (or 0.2 g of p-toluenesulfonyl chloride)[4] to each tube. Stopper the test tubes and shake them for 3~5 min. Open the stopper and place the tube on a steam bath for 1 min[5]. Check if the test solution is basic. If it is not basic, add 10% sodium hydroxide solution dropwise. Observe whether there are solid or oily substances. If there are precipitates, separate them out (filter or use dropper, etc.). Then place the solid into 5 mL of water and 5 mL of 5% hydrochloric acid and test their solubility. If no precipitation occurs, add

20% hydrochloric acid to the solution until it reaches a pH value of 6, allow the test tube to cool while scraping the wall of the tube with a glass rod. Observe if there is any precipitation. Make a conclusion from the experimental results.

Notes

[1] Check the end point of the diazotization reaction by dipping a glass rod into the reaction solution and contact its end on a potassium iodide-starch test paper to observe the color change of the test paper.

[2] Preparation of β-naphthol solution: dissolve 0.1 g of β-naphthol in 1 mL of 5% sodium hydroxide solution.

[3] The nitrosation product of N, N-dimethylaniline forms a reddish-brown compound when reacting with hydrochloric acid, which should further react with a base to form a green substance.

[4] Aromatic sulfonyl chlorides are toxic and corrosive; avoid inhaling their vapors or allowing them to come into contact with your skin. This reaction should be carried out in a fume hood.

[5] Some N, N-dialkylanilines form purple-red dyes when they are heated with benzenesulfonyl chloride. Once this happens, carry out a new reaction in a 15~20℃ water bath.

Questions

Compare the basicity of aniline and diphenylamine.

第 1 章　有机化学实验的一般知识

　　有机化学实验是学习有机化学的重要部分。有机化学实验的目的是培养学生良好的实验工作方法和习惯、严谨的科学态度、实事求是的科学素养，并激发学生科研兴趣和科学钻研、创新精神。

　　有机化学实验教学的基本任务是：

　　(1) 使学生掌握有机化学实验的基本操作技能，能正确地进行有机物的合成、分离、表征及鉴定。

　　(2) 使学生了解开展有机化学研究的一般程序，培养学生查阅文献的能力，能够从文献中获取信息、分析数据，分析问题和解决问题，以及正确撰写科研报告。

　　有机化学实验室有某些潜在的危险，学生在进实验室之前必须正确地了解并学习如何处理这些危险，以免对学生本人及周围的人造成伤害。为此，下面先介绍有机化学实验的一般知识及规则。学生在进实验室之前，须确保已经认真学习并理解这部分内容，并且在实验过程中严格遵守相关规则。

1.1　有机化学实验室规则

　　为了保证有机化学实验正常、有效、安全地进行，学生必须严格遵守下列规则：

　　(1) 实验前，认真预习有关实验内容，写好预习报告。任何人没有预习报告，不得进入实验室开始进行实验。

　　(2) 清楚实验室水、电、安全出口、洗眼器、灭火器材、急救药箱等的位置和使用方法。如发生意外事故，应及时采取正确应急措施并立即报告指导教师。

　　(3) 进入实验室应穿实验服，衣服和鞋子应尽量盖住皮肤，不允许穿拖鞋、短裤、裙子等过多裸露皮肤的服装进入实验室。

　　(4) 严禁在实验室内饮食。

　　(5) 实验过程中需保持安静，不得擅自离开实验岗位。

　　(6) 实验过程中须遵从教师的指导，严格按照操作规程和实验步骤进行实验。若要对实验进行更改必须经指导教师同意。实验中认真操作、仔细观察、如实记录。

　　(7) 实验过程中保持实验室的整洁。废弃物应放入指定容器，不得随意丢弃或倒入水槽。试剂及仪器用完后放回原处。

　　(8) 实验结束后，将仪器洗净、归位、摆放整齐，做好实验室清洁工作，然后检查水、电等是否关好，洗干净手，请指导教师检查后方可离开实验室。

1.2　有机化学实验室的安全知识

　　有机化学实验所用药品很多都易燃、易爆或有毒，如使用不当，则可能发生着火、爆炸或中毒等事故。此外，所用仪器主要为玻璃仪器，如不小心易造成割伤。因此，进行有机化学实验必须特别注意安全。

各种实验室事故的发生通常是由不熟悉仪器、药品的使用方法，未严格按规程操作或不重视引起的。但只要认真准备、高度重视、规范操作，事故是可以避免的。因此，学生应在实验前了解、熟悉实验室安全知识，并在实验过程中严格遵守。

1.2.1 实验室安全守则

(1) 实验开始前，应检查仪器是否完好，装置是否正确、稳妥，经指导教师同意后方可开始实验。

(2) 实验过程中，不得擅自离开实验现场，须严密监测反应进行的情况，并不时检查装置有无漏气或破裂等问题。

(3) 当进行有可能发生危险的实验时，应根据实验情况采取必要的安全措施，如佩戴防护眼镜、面罩或橡皮手套等。

(4) 使用易燃、易爆药品时，应远离火源。

(5) 实验试剂不得入口。严禁在实验室吸烟、进食或饮水，实验完毕后要认真洗手。

(6) 熟悉安全用具如灭火器材、沙箱、安全出口、洗眼器和急救药箱的放置地点和使用方法。

(7) 实验过程中使用或产生的试剂不得随意丢弃。废液、废渣要投放至指定容器，统一处理。

1.2.2 事故预防

1. 防火

引起着火的原因很多，如用敞口容器加热低沸点的溶剂、加热方法不正确等均可引起着火。为了防止着火，实验中应注意以下基本原则：

(1) **不得**用敞口容器盛放或加热易燃、易挥发的化学药品。处理和使用易燃物时，应远离明火，同时注意室内通风，及时将蒸气排出。加热低沸点的液体应采用水浴。

(2) **不得**将易燃、易挥发的废弃物倒入废液缸和垃圾桶中，量大时应专门回收处理。

(3) 万一着火，应沉着、冷静，尽快采取正确措施控制火势。首先，立即切断电源，移走易燃物。然后，根据易燃物的性质和火势采取适当的方法进行扑救。有机物着火通常不得用水进行扑救。小火可用湿布盖灭。火势较大时，应用灭火器扑救。

实验室常用灭火器有二氧化碳灭火器、干粉灭火器及泡沫灭火器。

有机化学实验室中常用的是标准干粉灭火器，特殊干粉灭火器用于易燃金属。但注意不要在密闭空间使用干粉灭火器，因为其粉状物容易进入呼吸道而造成呼吸道受损，而且其使用后的残留物很难清理。使用干粉灭火器时，拔出销钉，将出口对准着火点（不要对着上升的火焰，否则干粉会被吹走），将上手柄压下，喷出干粉灭火。

二氧化碳灭火器也是有机化学实验室常用的灭火器，适用于电器及较贵重的仪器着火。

泡沫灭火器也具有较好的灭火性能。但泡沫灭火器使用含水的试剂，因此不适用于电器及流动的火势。

不管采用哪一种灭火器,都要从火的外围向中心扑灭。

地面或桌面着火时,还可用沙子扑灭。

衣物着火时,应就近在地上打滚(速率不要太快)将火焰扑灭。千万不要在实验室内乱跑,以免造成更大的火灾。

2. 防爆

为了防止爆炸事故的发生,应注意以下几点:

(1)**不能**在密闭体系内进行加热或反应。

(2)常压蒸馏或化学反应应保证体系与大气相通。减压蒸馏时,不能用平底烧瓶、锥形瓶、薄壁玻璃仪器等不耐压材料,实验前要先检查玻璃仪器是否有破损。无论是常压蒸馏还是减压蒸馏,均切记**不能**将液体蒸干。

(3)切勿使易燃、易爆物质接近火源。

(4)使用乙醚等醚类时,必须先检查有无过氧化物存在。如发现过氧化物,应先用硫酸亚铁等除去再使用。

(5)对于易爆炸的固体,如重金属乙炔化物、多硝基化合物(如苦味酸或苦味酸重金属盐)、三硝基甲苯等,**不得**重压或撞击,以免引起爆炸。含有此类物质的残渣,**必须**现场采用适当方法分解销毁。

3. 防中毒

(1)不得直接用手接触药品,接触有毒物质必须戴橡皮手套。操作后应立即洗手。任何药品不得用嘴尝。

(2)使用和处理挥发性的有毒或腐蚀性物质时,必须在通风橱中进行,并戴好防护用品。使用后的器皿应及时清洗。

(3)如果有毒物质溅在皮肤上或嘴上,应先尽量拭去污染物,然后用大量水冲洗被污染部位 10 min 以上。如已吞下,先尽可能吐出残留物,然后用大量水清洗口腔,并喝些饮用水。如有必要,送医院治疗。如果吸入有毒物发生中毒现象,应将中毒者及时移到空气新鲜的地方,严重者应及时送往医院。

4. 防灼伤

皮肤接触腐蚀性或高温、低温物质后均可能被灼伤。实验过程中必须穿实验服。为避免灼伤,在接触这些物质时,应戴橡皮手套和防护眼镜。当腐蚀性物质接触皮肤时,先迅速擦去,用大量水冲洗,并根据不同的灼伤,按下列要求处理:

(1)被碱灼伤时,用 1%乙酸或 3%硼酸溶液冲洗,然后用水冲洗,最后涂上烫伤膏。

(2)被酸灼伤时,用 3%碳酸氢钠溶液清洗,最后涂上烫伤膏。

(3)被溴灼伤时,处理方式与酸灼伤相同。

(4)由于接触热物体的小面积烫伤应将伤口置于冷水下冲洗冷却至少 20 min;严重烧伤时需要立即就医。

(5)如果酸溅入眼睛,应立即滴入 2%硼砂溶液清洗,然后用生理盐水冲洗。如果碱溅入眼中,应立即滴入 2%硼酸清洗,然后用生理盐水冲洗,并及时去医院治疗。

5. 防割伤

为了防止割伤，首先不能对玻璃仪器的任何部位施加过度的压力；使用玻璃仪器前要检查是否有破损；不要用手直接拿破损的玻璃仪器。

如果被玻璃碎片或尖锐仪器割伤，应按下列步骤进行：小伤口应彻底冲洗，并检查是否有异物。如伤口处有玻璃碎片，应将其取出，再用生理盐水将伤口洗净，涂上碘伏，用纱布包好伤口。若流血不止，应在伤口上方 5~10 cm 处用绷带扎紧或用双手掐住，压迫止血，然后送医院就诊。如伤口有可能接触到传染性物质，应立即就医。

为处理事故需要，实验室应备有急救箱，包括以下物品：

(1) 镊子、剪刀、纱布、药棉、绷带、橡皮膏等。

(2) 凡士林、创可贴、烫伤膏、玉树油、硼酸软膏等。

(3) 1%乙酸或硼酸溶液、1%碳酸氢钠溶液、2%硫代硫酸钠溶液、甘油、止血粉、医用酒精、红药水、龙胆紫等。

6. 用电安全

使用电器前，应检查线路连接是否正确，确保插头紧贴插座。连接松动的插头可能会因为过热而损坏电线和插头。**切勿**将不匹配的插头插入插座。

使用电器时，应防止人体与电器导电部分直接接触，**不能**用湿手或手握湿物接触插头。切勿将金属物体置于电器或插座的活动部件中。

实验结束后，应先切断电源，再将电源插头拔下。

1.3 有机化学实验常用仪器和装置

实验室玻璃仪器按其口塞及接口是否标准，可分为标准磨口玻璃仪器及普通玻璃仪器两类。因为标准磨口仪器使用更方便、有效，故一般情况下尽量使用标准磨口玻璃仪器。但一些普通玻璃也是不可缺少的。

1.3.1 标准磨口玻璃仪器

标准磨口玻璃仪器是按国际通用的技术标准制造，具有标准化接口或磨塞的玻璃仪器。属于同类规格的接口均可任意连接，各部件能组装成各种配套仪器，使用方便。标准磨口玻璃仪器的玻璃塞可避免因使用橡皮塞对反应物带来的污染，另外其口塞磨砂性能好，使设备内更易达到高真空度，对蒸馏尤其减压蒸馏更有利，对于毒物或挥发性液体的实验更安全。

标准磨口玻璃仪器口径的大小常用数字编号标明规格。通常标准磨口有 10、12、14、16、19、24、29、34、40 等多种型号。这些数字代表仪器接口部分最大端的直径(单位：mm)。

标准磨口玻璃仪器也常用两个数字表示磨口的大小。第一个数字表示接口大端的直径，第二个数字表示接口的长度。例如，14/20 接口表示接口大端的直径为 14 mm，接口的长度为 20 mm。有机化学实验室通常采用规格为 19/22 的标准磨口仪器。相同编号的接口、磨塞，无论接口长度如何都可以紧密连接。编号不同的仪器可借助不同编号的磨口接头(缩管变径接头或扩管变径接头)连接，通常用两个数字表示变径的大小，如接头 14×19，表示该接头的一端为 14 号磨口，另一端为 19 号磨口。

1. 有机化学实验常用标准磨口玻璃仪器

有机化学实验常用的标准磨口玻璃仪器参见图 1.1。

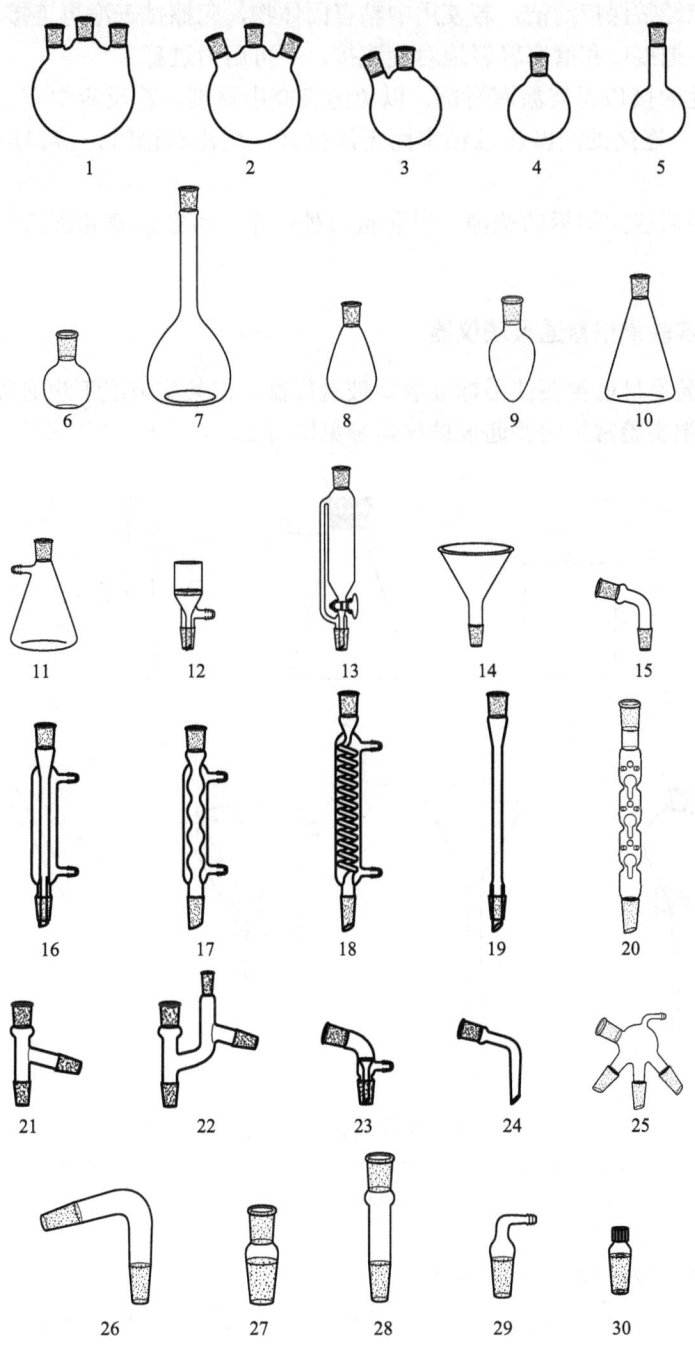

图 1.1 有机化学实验常用的标准磨口玻璃仪器

1. 直形三颈烧瓶；2. 斜形三颈烧瓶；3. 二颈烧瓶；4. 圆底烧瓶；5. 长颈圆底烧瓶；6. 平底烧瓶；7. 长颈平底烧瓶；8. 茄形烧瓶；9. 梨形烧瓶；10. 锥形瓶；11. 抽滤瓶；12. 砂芯漏斗；13. 恒压滴液漏斗；14. 漏斗；15. 干燥管；16. 直形冷凝管；17. 球形冷凝管；18. 蛇形冷凝管；19. 空气冷凝管；20. 刺形分馏柱；21. 蒸馏头；22. 克氏蒸馏头；23. 真空接液管；24. 接液管；25. 多尾接收管；26. 转弯接头；27. 缩管变径接头；28. 扩管变径接头；29. 弯形接头；30. 温度计套管

2. 使用标准磨口玻璃仪器注意事项

(1) 使用前应检查玻璃仪器是否有裂缝或划痕等，确保其无破损方可使用。

(2) 接口处应经常保持清洁，若使用中粘有固体物，应擦拭干净再连接。

(3) 装配时，把接口和磨塞轻轻地对旋连接，不可用力过猛。

(4) 一般用途的接口无需涂润滑剂，以免污染瓶中试剂。若反应中有强碱，则应涂少量润滑剂，以免接口连接处因碱腐蚀粘牢而无法拆开。减压蒸馏时，接口应涂真空脂，以免漏气。

(5) 实验结束后应立即拆卸洗净，以防接口处粘牢。可在活塞和磨口间垫一张纸片防止粘连。

1.3.2 有机化学实验常用普通玻璃仪器

有机化学实验室目前主要使用标准磨口玻璃仪器，但某些情况下也必须用到一些普通玻璃仪器。有机化学实验常用的普通玻璃仪器参见图 1.2。

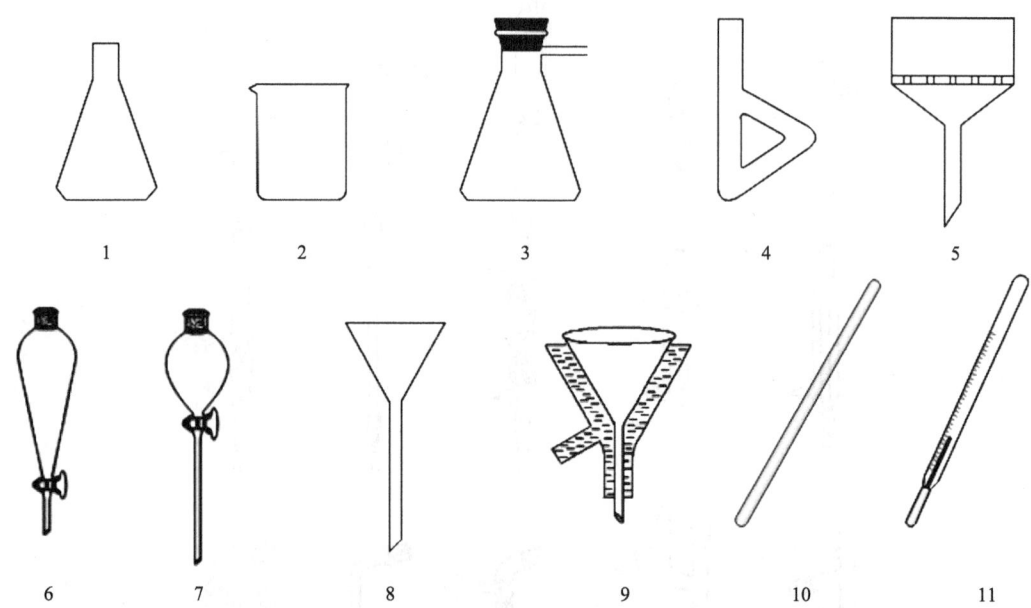

图 1.2 有机化学实验常用普通玻璃仪器

1. 锥形瓶；2. 烧杯；3. 抽滤瓶；4. 提勒（Thiele）管（b形管）；5. 布氏漏斗；6. 分液漏斗；7. 滴液漏斗；8. 漏斗；9. 热水漏斗；10. 玻璃棒；11. 温度计

使用普通玻璃仪器时要注意以下几点：

(1) 除少数玻璃仪器（如试管等）外，一般的普通玻璃仪器都**不能**直接用火加热。

(2) 厚壁玻璃仪器（如抽滤瓶等）不耐热，不能直接加热。锥形瓶不耐压，不能减压用。广口容器（如烧杯）不能存放有机溶剂。

(3) 带活塞的玻璃器皿用过洗涤后，在活塞与磨口间垫上纸片，防止粘住。

(4) 温度计**不能**用作搅拌棒，使用后要缓慢冷却，不可立即用冷水冷却。温度计的测量范围不可超出其刻度范围。温度计如不慎摔碎，应立即将洒出的水银用硫磺或锌粉覆盖。

1.3.3 有机化学实验常用装置

有机化学实验中常用的有回流、蒸馏、搅拌及气体吸收等基本实验装置。

1. 回流装置

回流装置可以使反应物或溶剂的蒸气不断地在冷凝管内冷凝而返回反应器中，防止反应瓶中的物质逸出。有机化学实验中的回流装置参见图 1.3。其中(a)为普通回流装置；(b)为带干燥管的回流装置；(c)为带气体吸收的回流装置；(d)为带分水器的回流装置；(e)为带滴液漏斗的回流装置。

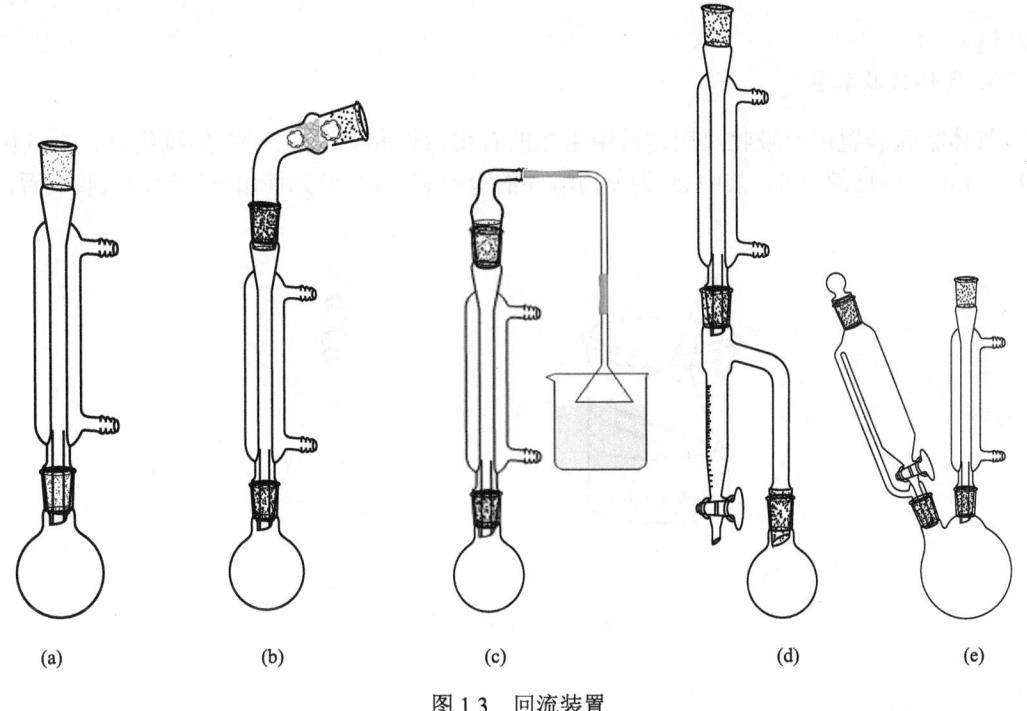

图1.3 回流装置

2. 蒸馏装置

蒸馏可以通过加热液体至沸腾变成蒸气后再冷却至液态来纯化液体。蒸馏是分离两种以上沸点相差较大的液体(30℃以上)的常用方法。常用的蒸馏装置参见图 1.4。其中(a)为使用水冷凝管的蒸馏装置；(b)为使用空气冷凝管的蒸馏装置；(c)为带滴液漏斗的蒸馏装置。

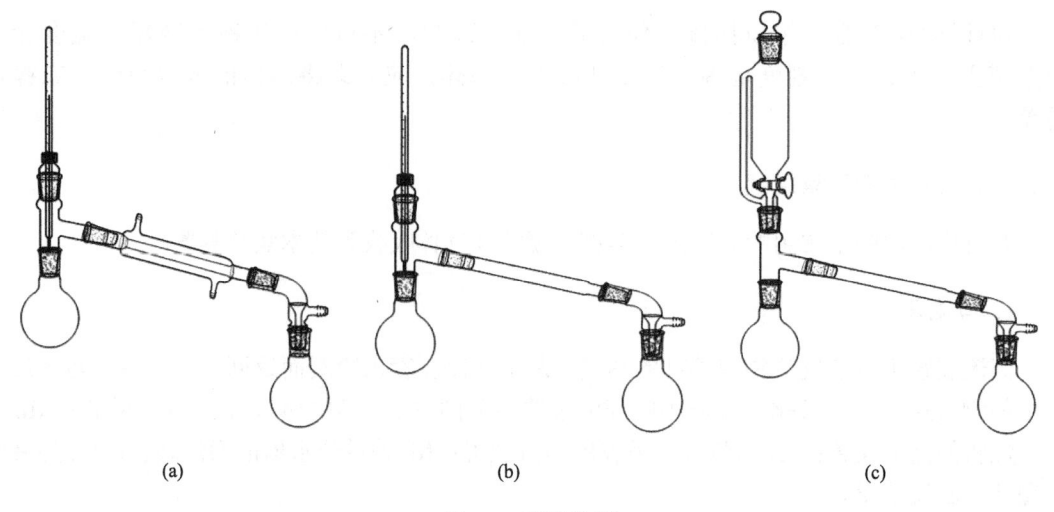

图 1.4 蒸馏装置

3. 气体吸收装置

气体吸收装置用于吸收反应过程中生成的有刺激性和有毒的气体（如氯化氢、二氧化硫等），装置图参见图 1.5。其中(a)为使用漏斗的吸收装置；(b)为使用导气管的吸收装置。

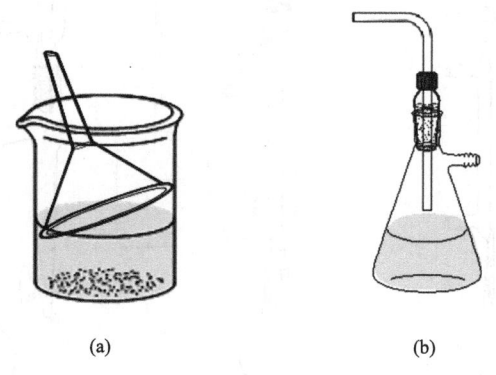

图 1.5 气体吸收装置

4. 搅拌装置

在非均相体系中进行的反应或进行需不断加入某些反应物的反应时，为了尽可能迅速均匀地混合，则需配备搅拌装置。

常用的机械搅拌、回流装置图参见图 1.6。图 1.6 中(a)是搅拌、回流和监测反应温度装置；(b)是搅拌、回流和滴加装置；(c)是搅拌、回流、滴加和监测反应温度的装置。磁力搅拌、回流装置参见图 1.7，其中(a)是搅拌、回流、滴加和监测反应温度的装置；(b)是搅拌、回流和分水装置。

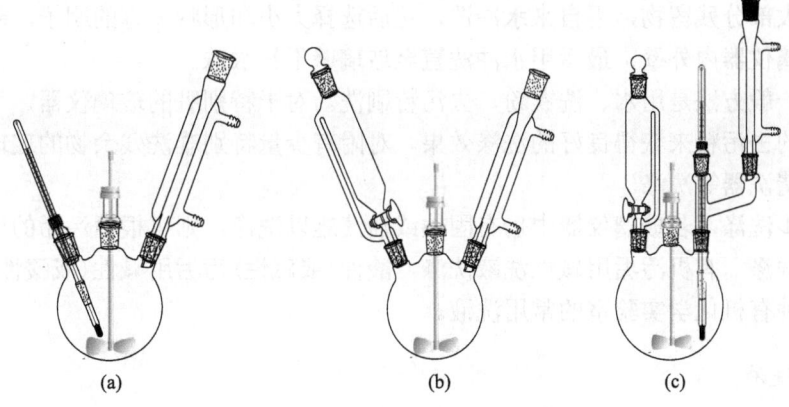

图 1.6　机械搅拌、回流装置

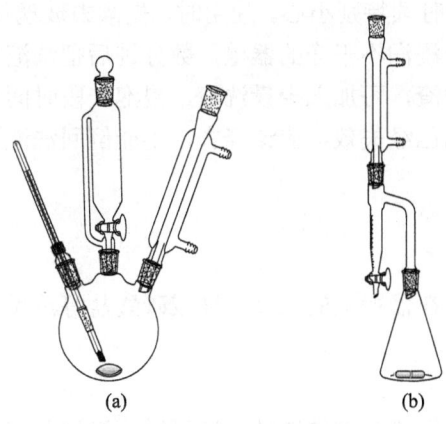

图 1.7　磁力搅拌、回流装置

1.3.4　仪器的安装

安装实验装置应注意以下事项：

(1) 所用的玻璃仪器和配件要干净。

(2) 所用的仪器大小要合适。例如，选用圆底烧瓶时，所盛反应物占其容积的 1/3～1/2 为宜，不要超过 2/3。

(3) 装配时应按照先下后上、从左到右的顺序逐个装配。拆卸时，则按相反的顺序逐个拆卸。

(4) 常压下进行的反应装置应与大气相通，**不得**密闭。

(5) 实验装置要求做到严密、正确、整齐和稳妥，便于操作。

1.4　玻璃仪器的洗涤和干燥

1.4.1　玻璃仪器的洗涤

每次实验前应检查玻璃仪器是否干净，实验后应立即清洗。对大多数玻璃仪器，先倒出或刮出瓶中大部分残留物，用自来水冲洗，然后选择大小和形状适宜的刷子，蘸取水和洗涤剂，刷洗玻璃仪器内外壁，最后用水冲洗直至玻璃壁不挂水珠。

洗涤的一般方法是用水、洗衣粉、去污粉刷洗。对于特别脏的玻璃仪器，可使用具有轻微研磨作用的去污粉来获得良好的洗涤效果。对附有少量特别难洗残余物的玻璃仪器，也可使用超声波清洗器等洗涤。

经过初步洗涤，若玻璃仪器中仍有固体或油渍难以洗净，则可根据污垢的性质选用适当的洗液进行洗涤。有机污垢用碱性洗液洗涤，酸性（或碱性）污垢用碱性（或酸性）洗液洗涤。下面介绍几种有机化学实验室的常用洗液。

1. 铬酸洗液

如果玻璃仪器过脏或含有凝结的有机物，则须使用铬酸洗液洗涤。**注意**：铬酸洗液腐蚀性很强且可致癌，因此使用时须特别小心。使用时，先倾去玻璃仪器内的水，慢慢倒入洗液，转动玻璃仪器，使洗液充分浸润不干净的器壁，数分钟后把洗液倒回洗液瓶中，用自来水冲洗。若壁上粘有少量炭化残渣，可加入少量洗液，浸泡一段时间后在小火上加热，直至冒出气泡。洗液颜色变绿时表示已经失效，应该弃去，不能倒回洗液瓶中，处理时必须使用正确方法，且需格外小心。

2. 浓盐酸

用浓盐酸可以洗去附着在器壁上的二氧化锰或碳酸盐等污垢。

3. 碱性洗液

含有氢氧化钠、氢氧化钾等强碱的碱性洗液可以洗涤油脂和一些有机物（如有机酸）。

4. 有机溶剂

当胶状或焦油状有机污垢用上述方法不能洗去时，可选用工业级的有机溶剂如丙酮等浸泡（**不要使用试剂级的溶剂洗涤**），要加盖以免溶剂挥发，使用后可回收重复使用。用于有机分析的玻璃仪器，除用上述方法处理外，还需用蒸馏水冲洗至玻璃壁不挂水珠。

但一般情况下，不建议使用强腐蚀性的酸、碱或有机溶剂作洗液，以免污染环境。

1.4.2 玻璃仪器的干燥

干燥玻璃仪器的方法通常有如下几种：

(1) 自然晾干：不急等用的仪器，可将仪器口倒置，放在仪器架上自然干燥。

(2) 吹干：急等用的仪器可用气流烘干器干燥（温度以 60~70℃为宜）。

(3) 烘干：将玻璃仪器沥干水滴后，口部朝上自上而下依次放入烘箱，烘箱温度保持在 100~105℃。带有磨砂口玻璃塞的仪器必须取出活塞后，才能烘干。待烘干的容器内不得有可燃性溶剂。

(4) 有机溶剂干燥：将水尽量沥干后，加入少量丙酮或乙醇摇洗并倾出，吹入冷风至干燥。

带有刻度的容器**不能用加热的方法进行干燥**，一般采用方法(1)或(4)干燥。

1.5 有机化学实验中试剂的加热、冷却和干燥

1.5.1 加热

某些化学反应在室温下难以进行或进行得很慢，为了加速反应，往往需要加热。实验室常用的热源有煤气、酒精和电能。在有机化学实验室中，为保证加热均匀和防止有机物因局部过热分解或燃烧，一般不使用直接加热，而采用更温和、更均匀的热浴间接加热。主要的加热方式有如下几种。

1. 空气浴

空气浴就是利用热空气加热。沸点在80℃以上的液体均可采用空气浴加热。直接利用热源隔着石棉网对容器加热，这是最简单的空气浴，但受热不均匀，因此不适合低沸点易燃液体或减压蒸馏。电热套是比较安全、简便的空气浴，具有不易着火、受热较均匀、热效高等优点，加热温度可达400℃，且其加热温度可通过调压变压器控制。

2. 水浴

当加热温度在100℃以下时，可使用水浴加热。水浴利用将反应容器浸入装有热水的容器中来间接加热。使用水浴时，勿使容器底触及水浴锅底或锅壁。若长时间加热，水浴中的水会不断蒸发，适当时要添加热水，使水浴液面经常保持略高于容器中的液面。水浴锅配有专门的由一组直径递减的同心圆环组成的盖子，可以有效地减少水分的蒸发。

3. 油浴

加热温度为100～250℃可用油浴。一般情况下，反应温度应比油浴温度低20℃左右。常用的油浴液有：

(1) 甘油：甘油可以加热到140～150℃，温度过高则会分解。甘油吸水性强，放置过久的甘油，使用前应首先加热蒸去所吸的水分，之后再用于油浴。

(2) 植物油：植物油如棉籽油、蓖麻油和花生油等，可以加热到220℃。若在植物油中加入1%对苯二酚等抗氧化剂，则可增加油的热稳定性。

(3) 液体石蜡：液体石蜡可加热到220℃左右，温度过高易燃烧。

(4) 固体石蜡：固体石蜡可加热到200℃左右，冷至室温时凝成固体，保存方便。

(5) 硅油：硅油在250℃时仍较稳定，透明度好，安全，是目前实验室中较为常用的油浴之一。

用油浴加热时，要在油浴中装置温度计（温度计感温头如水银球等，不应放到油浴锅底），以便随时观察和调节温度。加热完毕取出反应容器时，仍用铁夹夹住反应容器离开液面悬置片刻，待容器壁上附着的油滴完后，用纸或干布擦干。

使用油浴应防止水溅入油中，否则加热时会产生泡沫或暴溅。使用油浴时，要特别注意防止着火，当油受热冒烟时，应立即停止加热。

4. 沙浴

加热沸点在 80℃以上的液体时可以采用沙浴，特别适用于加热温度在 220℃以上者。使用时，将细沙或金属粒装在烧杯中，将反应容器半埋在沙中，加热烧杯。通过沙或金属粒的传热加热反应容器。由于沙浴温度上升较慢，且不易控制，因而使用不广泛。

1.5.2 冷却

有些有机化学反应在反应、分离或提纯等过程中需要使用冷却剂，进行冷却操作。例如：
(1) 某些反应需要在特定的低温条件下进行，如重氮化反应一般在 0~5℃下进行。
(2) 某些反应放出大量的热，需要降温来控制反应速率。
(3) 为了加速结晶的析出，需要冷却。
(4) 沸点很低的有机物，为了减少损失，使用时需要冷却。
(5) 高度真空蒸馏装置需要使用冷阱冷却。

冷却的方法很多，通常根据不同的要求选用合适的冷却方法和冷却剂。

1. 冷水

可用冷水在容器外壁流动，或把反应器浸在冷水中交换走热量。也可用水和碎冰的混合物作冷却剂，可冷却至 0~5℃。如果水对反应无影响，有时也可把冰块投入反应器中进行冷却。

2. 冰-盐

按不同比例将碎冰和无机盐混合可得到冷却剂。冰-盐混合物冷却温度参见表 1.1。

表 1.1 冰-盐混合物的质量分数和提供的冷却温度

盐	盐的质量分数	冰的质量分数	冷却温度/℃
$CaCl_2·6H_2O$	100	246	−9
$CaCl_2·6H_2O$	100	123	−21.5
$CaCl_2·6H_2O$	100	70	−55
$CaCl_2·6H_2O$	100	81	−40.3
NH_4NO_3	45	100	−16.8
NH_4NO_3	50	100	−17.8
NaBr	66	100	−28
NH_4Cl	25	100	−15
NaCl	33	100	−21

3. 干冰或干冰与有机溶剂混合

干冰(固体二氧化碳)可冷却到−60℃以下。如将干冰与甲醇、丙酮或氯仿等溶剂混合，可冷却至−78℃，混合时会猛烈起泡。

4. 液氮

液氮可冷却至 $-196\,^\circ\text{C}$（77 K），一般在科研中应用。

液氮和干冰是两种方便又廉价的冷却剂，应将其放在杜瓦瓶（广口保温瓶）或其他绝热效果好的容器中，以保持其冷却效果。

5. 低温浴槽

低温浴槽是一个小冰箱，冰室口向上，蒸发面用筒状不锈钢槽代替，内装乙醇。外设压缩机，循环氟利昂制冷。可装外循环泵，使冷乙醇与冷凝器连接循环。还可装温度计等指示器。反应瓶浸在乙醇液体中。适合 $-30\sim30\,^\circ\text{C}$ 的反应使用。

以上冷却方法供选用。**注意**：温度低于 $-38\,^\circ\text{C}$ 时，由于水银会凝固，因此不能用水银温度计。当温度低于 $-200\,^\circ\text{C}$ 时，应采用含有有机溶剂混合物（乙醇、甲苯和正戊烷）的温度计。

1.5.3 干燥

干燥是除去固体、液体或气体中少量水分或少量有机溶剂的常用方法。在有机化学实验中，试剂和产品的干燥具有重要的意义。

1. 原理

干燥方法可分为物理干燥和化学干燥两种。

物理干燥：物理干燥方法有烘干、晾干、吸附干燥和冷冻干燥等。近年来，还常用离子交换树脂和分子筛等方法进行干燥。

化学干燥：化学干燥方法采用干燥剂除水。根据除水作用原理又可分为两种：

(1) 干燥剂能与水可逆地结合，生成水合物。例如：

$$CaCl_2 + nH_2O \longrightarrow CaCl_2 \cdot nH_2O$$

(2) 干燥剂与水发生不可逆的化学变化，生成新的化合物。例如：

$$2Na + 2H_2O \longrightarrow 2NaOH + H_2 \uparrow$$

使用干燥剂时要注意以下几点：

(1) 干燥剂与水的反应为可逆反应时，反应达到平衡需要一定时间。因此，加入干燥剂后，一般最少要两个小时或更长一点的时间后才能收到较好的干燥效果。因反应可逆，不能将水完全除尽，故干燥剂的加入量要适当，一般为溶液体积的 5% 左右。当温度升高时，可逆反应的平衡向脱水方向移动，所以在蒸馏前必须将干燥剂滤除，否则被除去的水将返回液体中。

(2) 干燥剂与水发生不可逆反应时，使用这类干燥剂在蒸馏前不必滤除。

(3) 干燥剂只适用于干燥少量水分。若水分含量大，干燥效果不好。为此，萃取时应尽量将水层分净，这样干燥效果好，且产物损失少。

2. 液体有机化合物的干燥

1) 干燥剂的选择

干燥剂应与被干燥的液体有机化合物不发生化学反应，包括溶解、络合、缔合和催化等。

例如，酸性化合物不能用碱性干燥剂干燥。各类液体有机化合物的常用干燥剂列于表1.2。

表 1.2　各类液体有机化合物的常用干燥剂

液体有机化合物	适用的干燥剂
醚、烷烃、芳香烃	$CaCl_2$, Na, P_2O_5
醇	K_2CO_3, $MgSO_4$, Na_2SO_4, CaO
醛	$MgSO_4$, Na_2SO_4
酮	$MgSO_4$, Na_2SO_4, K_2CO_3
羧酸	$MgSO_4$, Na_2SO_4
酯	$MgSO_4$, Na_2SO_4, K_2CO_3
卤代烃	$CaCl_2$, $MgSO_4$, Na_2SO_4, P_2O_5
有机碱（胺）	NaOH, KOH

2) 干燥效能和吸水容量

吸水容量是指干燥剂能结合的水的最大摩尔数。干燥效能是指干燥剂在干燥过程结束后留在有机溶液中的水量。对于形成水合物的无机盐干燥剂，常用干燥后结晶水的蒸气压来表示干燥效能。例如，硫酸钠形成 10 个结晶水，其吸水容量为 1.27，水合盐的蒸气压为 260 Pa；氯化钙最多能形成 6 个水的水合物，其吸水容量为 0.97，在 25℃时水合盐的蒸气压为 39 Pa。因此，硫酸钠的吸水容量较大，但干燥效能弱；而氯化钙的吸水容量较小，但干燥效能强。在干燥含水量较大的化合物时，常先用吸水容量较大的干燥剂除去大部分水，再用干燥效能强的干燥剂进一步干燥。

3) 干燥剂的用量

根据水在液体中的溶解度和干燥剂的吸水容量，可算出干燥剂的最低用量。但是，干燥剂的实际用量是大大超过计算量的。一般干燥剂的用量为 10 mL 液体需 0.5～1 g 干燥剂。但在实际操作中，主要是通过现场观察判断干燥剂用量。

(1) 观察被干燥液体：干燥前液体呈浑浊状，经干燥后变成澄清，这可简单地作为水分基本除去的标志。

(2) 观察干燥剂：加入干燥剂后，因其吸水变黏，粘在器壁上，摇动不易旋转，表明干燥剂用量不够，应适量补加，直到新加的干燥剂不结块、不粘壁，干燥剂棱角分明、摇动时旋转，则表示所加干燥剂用量合适。

由于悬浮干燥剂(尤其 $MgSO_4$ 等小晶粒干燥剂)还能吸收一部分有机液体，影响产品的产率，故干燥剂用量应适中。应加入少量干燥剂后静置一段时间，观察用量不足时再补加。

4) 干燥时的温度

对于生成水合物的干燥剂，加热虽可加快干燥速度，但远远不如水合物放出水的速度快。因此，干燥通常在室温下进行。

5) 干燥操作步骤与要点

(1) 把液体中的水分尽可能除净，不应有任何可见的水层或悬浮水珠。

(2) 把待干燥的液体放入锥形瓶中，取颗粒大小合适(如无水氯化钙，应为黄豆粒大小并不夹带粉末)的干燥剂放入液体中，用塞子盖住瓶口，轻轻振摇，经常观察，判断干燥剂是否足量，静置(半小时，最好过夜)。

(3)把干燥好的液体滤入蒸馏烧瓶中，然后进行蒸馏。

3. 固体有机化合物的干燥

干燥固体有机化合物主要是除去残留在固体中的少量低沸点溶剂，如水、乙醚、乙醇、丙酮、苯等。由于固体有机化合物的挥发性比溶剂小，所以采取蒸发和吸附的方法达到干燥的目的。常用干燥方法如下：

(1)晾干。

(2)烘干：①用恒温烘箱烘干或恒温真空干燥箱烘干；②用红外灯烘干。

(3)冻干。

(4)吸附干燥：若遇难抽干溶剂时，把固体从布氏漏斗中转移到滤纸上，上下均放 2~3 层滤纸挤压，使溶剂被滤纸吸干。

(5)干燥器干燥：①普通干燥器；②真空干燥器；③真空恒温干燥器(干燥枪)。

4. 气体的干燥

有机化学实验中常用的气体有 N_2、O_2、H_2、Cl_2、NH_3、CO_2。有时要求气体中含很少或几乎不含 CO_2、H_2O 等，因此就需要对上述气体进行干燥。干燥气体的常用仪器有干燥管、干燥塔、U 形管、各种洗气瓶(常用来盛液体干燥剂)等。常用气体干燥剂列于表 1.3。

表 1.3 常用气体干燥剂

干燥剂	气体
CaO, 碱石灰, NaOH, KOH	NH_3
无水 $CaCl_2$	H_2, HCl, CO_2, CO, SO_2, N_2, O_2, 低级脂肪醚，烯烃，卤代烃
P_2O_5	H_2, N_2, O_2, CO_2, SO_2, 烷烃，烯烃
浓硫酸	H_2, N_2, HCl, CO_2, Cl_2, 烷烃
$CaBr_2$, $ZnBr_2$	HBr

1.6 实验预习、实验记录和实验报告的基本要求

学习撰写实验报告是整个实验教学中的重要环节，是对学生科研能力的一项基本训练。对实验报告及实验记录的最低要求是记录的实验结果能被准确复制。

1.6.1 预习

实验前要做好充分的预习和准备，这有利于学生了解实验目的以及实验中可能存在的危险。预习通常包括实验目的、实验材料、步骤、参考文献、实验前的问题等内容。学生在进行每个实验时，必须做好预习并完成预习报告，未完成预习的学生不得开始实验。

预习应包括以下内容：

(1)实验目的。

(2) 反应式或被分离物质的信息：对于制备实验，需列出主反应中反应物和产物的结构及名称，以及反应条件；对于分离实验，要求给出被分离物质的结构及名称。

(3) 试剂列表：原料、产物和副产物的物理常数(密度，分子量，熔、沸点等)；原料用量(单位：g，mL，mol)，计算理论产量；反应物及产物的危害性，这有利于学生了解实验中潜在的危险。

(4) 正确、清楚地画出装置图。

(5) 本次实验所涉及相关基本操作内容、实验的关键步骤、难点及实验过程的安全问题。

(6) 实验步骤：用图表形式表示实验步骤。例如，1-溴丁烷的制备(粗产物纯化过程)可用图 1.8 表示。

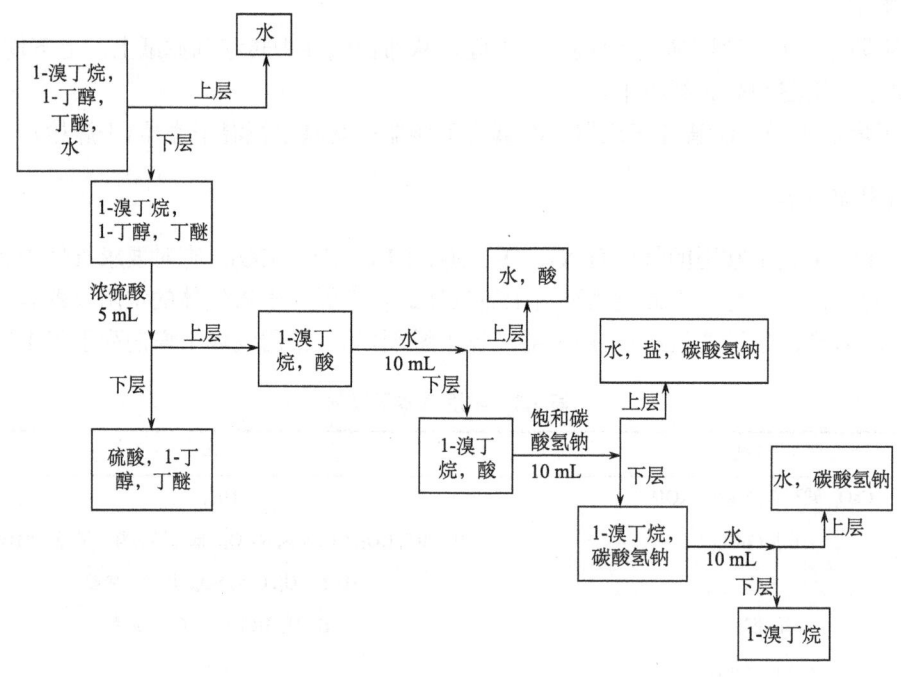

图 1.8　流程图示例：1-溴丁烷的提纯

(7) 参考文献：说明所有参考文献的来源。按下列顺序列出相关的参考文献：作者. 年. 期刊或书名，起始页码。

1.6.2　实验记录

实验记录对开展实验研究来说非常重要。实验记录本应是装订本，不得用活页纸或散页。在实验过程中需要记录日期及所有观察到的现象和结果，以及与实验计划有出入的地方。

实验记录本前面空出 4～5 页，留作编目录用。目录的格式多样，但无论何种格式，都应包括记录日期、记录的实验名称及对应页码。

每做一个实验，应从新的一页开始，并按照下列格式做实验记录：实验日期、实验名称、实验目的、实验背景(包括实验方法、实验数据、使用的试剂和仪器等)，以及所有的实验现象和结果。表 1.4 给出了实验记录的格式，可供参考。

表 1.4 实验记录的格式

实验日期		实验名称		
实验目的				
实验背景				
时间	步骤		现象	备注
分析				
签名				

在表中需详细记录观察到的实验现象或结果,如颜色的变化、晶体的外观、反应温度、测试结果、得到的物理性质,以及试剂和产物的质量等。

实验记录应直接记录在实验记录本上,而不是先记录在一张纸上再转抄到记录本上。记录应忠实、详尽,所提供的信息应能支撑实验目标的完成。记录需及时,如不及时,会导致实验讨论和结论部分难以准确给出。实验完毕,必须将实验记录交给指导教师签字后,才可离开实验室。

1.6.3 实验报告

在实验操作完成并在实验记录本上记录下所需的实验信息后,必须对实验结果进行总结,写出实验报告。撰写实验报告的目的是清楚、准确地表达在实验中得到的结果并对结果进行分析和解释。这是完成整个实验的重要环节。

实验报告内容大致分为如下几项,可根据实验实际情况进行删减:
(1) 实验名称。
(2) 实验目的和实验原理(反应方程式),包括可能的反应机理。
(3) 主要试剂及产物的物理性质(包括计算量)。
(4) 实验装置图。
(5) 实验步骤。
(6) 数据和现象(如 R_f 值、熔点、光谱数据、产品外观、质量、产率等)。
(7) 结果与讨论(结论):在这一部分需要分析和解释实验结果。这是实验报告中最重要的部分。通过这部分内容,学生将理解这个实验而不是仅仅停留在简单地完成这个实验。例如,可写出产物的质量以及如何纯化和表征该产物。在制备实验中,可给出产率。也可讨论本次实验是否达到了预期实验目标,如果没有达到目标,该如何改进。讨论实验中存在的失误和原因,以及对整个反应结果的影响,等等。

1.6.4 实验报告样例

以下是某制备实验的实验报告样本。

1-溴丁烷的制备

实验目的

(1) 学习如何由 1-丁醇制备 1-溴丁烷。
(2) 练习回流、蒸馏及气体吸收装置的安装和操作。
(3) 进一步练习液体产品的纯化方法,包括洗涤、干燥、蒸馏等操作。

实验原理

主反应:(略)
副反应:(略)

实验试剂及物理常数

试剂	分子量	用量	相对密度(d_4^{20})	水中溶解度/(g/100 mL)	沸点/℃
1-丁醇	74	11.2 mL	0.81	7.9	117.7
1-溴丁烷	137		1.28	不溶	101.6
溴化钠	103	15 g		微溶	
浓硫酸	98	18 mL	1.84	易溶	

其他试剂:10%碳酸钠溶液,无水氯化钙。

实验装置

回流装置图(略)。
蒸馏装置图(略)。

实验步骤

(1) 在 100 mL 圆底烧瓶中加入 15 mL 水,加入搅拌磁子,在搅拌下置于冰水浴中冷却。然后在搅拌下小心地加入 18 mL(33 g)浓硫酸(分次加,每次 2~3 mL,每次均需充分混匀并冷却),摇匀后,冷却至室温。再加入 11.2 mL(9.0 g)1-丁醇,混合后加入 15 g 溴化钠,充分搅拌,在烧瓶上安装水冷凝管回流,在冷凝管的上口用软管连接一个漏斗,倒置悬于盛水的烧杯中,如图××。

(2) 小火加热回流 30 min。如果从冷凝管顶部可以看到白色烟雾,则应降低温度。稍冷却后,加入 10 mL 水,改作蒸馏装置,加热蒸馏直至馏出液无油状物。

(3) 将馏出液小心地转入 125 mL 分液漏斗,加入 10 mL 水分液。塞住漏斗口,轻轻振摇几次并放气。注意已形成两层,检验并确定有机层,将两层分开。然后,小心地将有机层转入另一干燥的分液漏斗中,用 5 mL 浓硫酸洗涤。尽量分去硫酸层,有机层依次用水、饱和碳酸氢钠溶液和水各 10 mL 洗涤。产物移入 50 mL 干燥的锥形瓶中,加入无水氯化钙干燥,间歇摇动,直至液体透明。将干燥后的产物小心地转入蒸馏烧瓶中,加热蒸馏,收集 99~103 ℃

的馏分。

实验结果

产量：4.2 g

理论产量 = $\dfrac{0.9}{7.4} \times 137 = 16.4$ （g）

产率 = $\dfrac{4.2}{16.4} \times 100\% = 25\%$

讨论

（略）

第 2 章 有机化学实验基本操作

实验 1 熔 点 测 定

实验目的

(1) 了解测定熔点的基本原理和熔点测定的意义。
(2) 掌握提勒法测定熔点的操作方法。

实验原理

纯晶体物质的熔点是物质从固态转变为液态的温度。大多数固体有机物的熔点较低(50～300℃)，可以比较方便地用简单的仪器测定。纯晶体物质有固定的熔点，从开始熔化(初熔)到完全熔化(全熔)的温度范围称为熔程，熔程一般不超过 0.5～1℃。

物质的固-液两相的蒸气压随温度的变化曲线参见图 2.1。由图可见，固相蒸气压随温度的变化速率比相应的液相大。两曲线相交，交点所对应的温度即熔点，交点处固、液两相共存，这是纯固体有机物有敏锐熔点的原因。

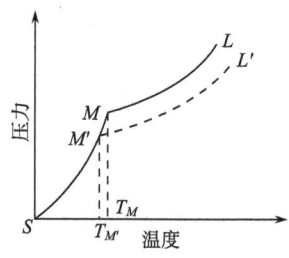

图 2.1 压力-温度相图

根据拉乌尔定律可知，在一定压力和温度下，增加溶质的量会显著稀释溶剂分子，将导致溶剂蒸气压降低。考虑到固体和液体的蒸气压在熔点的位置是相同的，因而蒸气压的降低会导致相应物质熔点的降低。杂质会导致熔点下降，熔程变长。因此，通过测定有机化合物的熔点可以推测物质的纯度。

实验用品

仪器：提勒管、水银温度计(250℃)、酒精灯、毛细熔点管(内径 1.0 mm，长度 60～70 mm，一端封闭)。

试剂：尿素、肉桂酸、尿素与肉桂酸的混合物(质量比 4∶1)、甘油(导热液)。

实验步骤

本实验采用毛细管法，用提勒管测定样品的熔点，具体步骤如下。

1. 熔点管的准备

通常使用内径约为 1.0 mm，长度为 60~70 mm，一端封闭的毛细管[1]。

2. 毛细管的填装

取 0.1~0.2 g 样品，放在干净的表面皿上，用玻璃钉或不锈钢勺研磨成粉末，聚集成堆，将毛细管的开口插入样品堆中，将样品挤入毛细管内（如图 2.2 所示）。然后准备一根长玻璃管，竖立在平整的硬质桌面上，然后将样品管沿着玻璃管扔下，使样品管在玻璃管内做几次跳跃运动，直至样品在毛细管底部填装成 2~3 mm 紧致、均匀的样品柱。该操作需迅速，以防样品吸潮。毛细管外的样品粉末要擦干净，以免污染导热液。

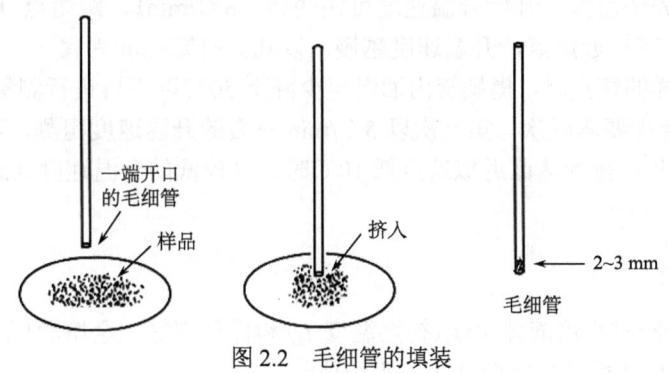

图 2.2　毛细管的填装

3. 安装装置

毛细管法使用提勒管为实验装置。该装置是一个装有导热油的玻璃仪器，使用明火加热，如图 2.3 所示。利用管内液体温度差可以形成液体对流，在不搅拌的情况下维持管内液体温度相对均匀。

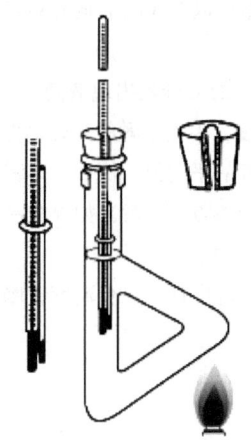

图 2.3　提勒管熔点测定装置

选择合适的导热液作为传热介质，导热液的选择视所需温度而定，同时考虑毒性、可燃性、价格及是否易清洗等因素。导热液必须选择性质稳定的高沸点液体。一般低于 140℃ 可用液体石蜡或甘油;不超过250℃可用浓硫酸(温度超过250℃,浓硫酸产生白烟,影响温度计读数);硅油可加热到250~365℃,但在经济上不可行。

在提勒管内加入导热液,使导热液的液面略低于提勒管上支口。温度计用一个切有沟槽的单孔橡皮塞固定在提勒管的中心轴线上(注意:不要使用密闭的塞子,以防加热时导热液喷出),水银球位于提勒管上、下两支口中间。将填装好的熔点管用橡皮圈套在温度计上[2],样品的中心部位位于温度计水银球的中部。

4. 熔点的测定

使用酒精灯对提勒管中的导热液进行加热。为了得到准确的结果,熔点需与温度计读取的温度尽量保持一致,因此必须小心控制升温速度。

(1)已知样:先快后慢。开始升温速度可快些(5~8℃/min),距熔点10~15℃时,升温速度1~2℃/min,越接近熔点,升温速度越慢,以0.5~1℃/min为宜。

(2)测定未知样的熔点时,提勒管内油温至少降至50℃以下再进行测量(可用吹风机吹冷风加速降温),且至少要测两次。第一次以5℃/min左右的升温速度粗测,可得到一个近似的熔点;第二次测量[3],待到达比近似熔点低10℃时,以较低的升温速度0.5~1℃/min准确测量熔点。

5. 熔点的记录

记录熔点管中刚有小滴液体出现(初熔温度 t_1)和样品刚好完全熔融(全熔温度 t_2)这两个温度点的读数[4]。每个样品测定两次,取平均值[5-7]。

注释

[1] 测定前注意检查是否漏管。方法:将装好样品的毛细管浸入导热液,发现样品溶解或管底渗入液体,说明为漏管,应弃去,另换一根。

[2] 测定前校正温度计。

[3] 重复测定时,导热液温度应降至熔点以下30℃,同时更换并使用新的熔点管,不能使用前面测定时已经使用过的熔点管。

[4] 若物质120℃时开始收缩(坍塌),121℃开始出现液滴,122℃全部液化,熔程应该是121~122℃,且不可把120℃收缩时的温度作为初熔温度。因为熔融的开始仍然以第一滴液体出现为准。

[5] 记录时不能取初熔温度到全熔温度的平均值。例如,熔程为123~125℃,不可记录为熔点124℃。

[6] 有些化合物对热敏感,在熔点前或熔点处分解。如果样品严重变色或产生气体,说明发生了分解。分解温度经常作为一个物理性质代替实际的熔点。

[7] 实验完毕后,不能立即用冷水冲洗温度计,以免温度计爆裂,应待其自然冷却至室温。导热液待充分冷却后才可倒回瓶中。

注意事项

(1)实验过程中需佩戴护目镜。
(2)不要用手触摸热的提勒管,以防烫伤。
(3)毛细管易折断,需小心使用。
(4)实验完毕记得洗手。

思考题

(1) 熔点测定中，若样品研磨不细，对实验结果有什么影响？
(2) 为什么说熔点测定误差太大多数是由加热太快造成的？
(3) 测定熔点时的有机化合物只能使用一次，为什么？

实验 2　简单蒸馏及沸点测定

实验目的

(1) 了解蒸馏和沸点测定的基本原理和意义。
(2) 掌握加热回流和搭建普通蒸馏装置的基本操作。
(3) 通过蒸馏提纯液体并测定其沸点。

实验原理

当液体的蒸气压与外界大气压相等时，液体呈沸腾状态并产生蒸气，此时的温度就是该液体的沸点。通常的沸点是特指在 1 atm 压力下液体的沸腾温度，又称为常压沸点。在一定压力下，纯净的液体具有固定的沸点。通过测定沸点可以鉴别有机化合物。

蒸馏是将液态物质加热到沸腾变为蒸气，又将蒸气冷却为液体并加以收集的过程。由于同一物质的凝结点和沸点相同，因此如果在蒸馏装置中加入温度计，则蒸气在温度计上冷凝时的温度将是该物质的沸点。因此，可用蒸馏操作测定液体的沸点。

如果溶液中有两种不同的液体，沸点低的组分可在另一高沸点组分不被明显蒸出的情况下被蒸出。将低沸点液体蒸发、冷凝并转至另一容器，可以得到纯净的液体，而高沸点的液体或杂质则被留在原来的容器内。蒸馏可以用于从非挥发性产物中除去挥发性溶剂，或将挥发性产物与非挥发性杂质分离，或分离沸点相差较大的液体混合物。但需要注意，具有恒定沸点的液体并不一定都是纯净物。某些有机物往往能与其他组分形成二元或多元恒沸混合物（称为共沸物），它们也和纯液体一样有固定的沸点。

简单蒸馏是在常压下分离沸点相差较大（一般在 30℃ 以上）的液体混合物的一种方法。用简单蒸馏法测定沸点称为常量法，此法用量较大，要求预测物质在 10 mL 以上。若量太小，则可以用微量法测定。

蒸馏时，从接液管滴下第一滴液体时温度计显示的温度称为初馏温度；最后一滴液体从蒸馏烧瓶最低点蒸发的瞬间温度计所显示的温度为末馏温度，两个温度之差为沸程。一定压力下纯净液态有机物的沸程很小，仅 0.5～1.5℃，若有杂质时则沸点降低，沸程增大。因此，用蒸馏的方法测出其沸程，则可定性判断液态有机物的纯度。

实验用品

仪器：圆底烧瓶、蒸馏头、温度计套管、直形冷凝管、接液管、锥形瓶、温度计(150℃)、铁夹、沸石、电热板、铁架台。

试剂：工业乙醇。

实验步骤

1. 搭建蒸馏装置

按照图 2.4 所示搭建蒸馏装置[1, 2]。搭装置时一般采用自下而上、从左到右的原则。其基本步骤如下:

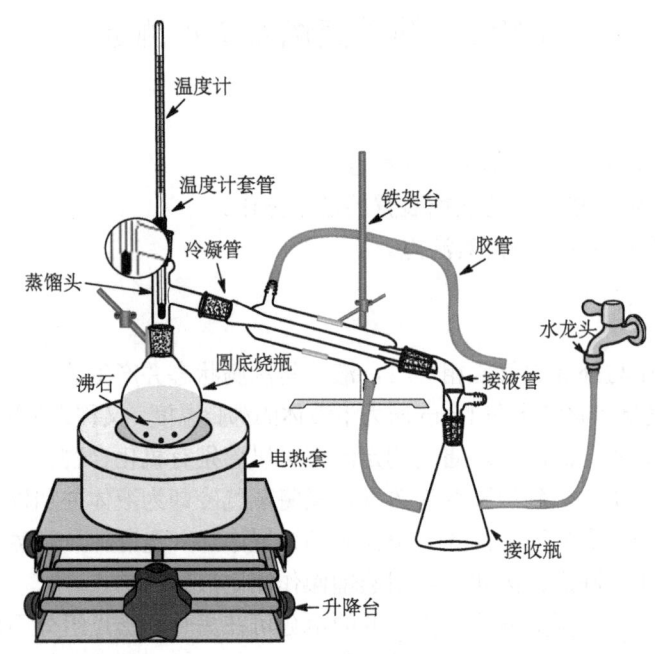

图 2.4 简单蒸馏装置

(1) 选择合适的蒸馏烧瓶,蒸馏烧瓶的大小视待蒸馏液体的体积而定,通常待蒸馏液体的体积占蒸馏烧瓶容量的 1/3~2/3,不要超过 2/3。然后在蒸馏烧瓶中加入一定量的待蒸馏液体。

(2) 在通风橱中,以热源如电热套(或电热板)为基准[3],将圆底烧瓶(蒸馏烧瓶)放在电热套上,烧瓶颈部用烧瓶夹固定在铁架台上,使烧瓶的底部刚刚接触电热套(注意夹子不要夹在烧瓶接头以外的部位,否则容易损坏玻璃仪器)。

(3) 往烧瓶内加入 2~3 颗绿豆大小的沸石[4],然后插入蒸馏头,再插入温度计套管和温度计[5]。为保证蒸馏平稳地进行,在加热之前需要在蒸馏烧瓶中加入沸石或毛细管。如果不加沸石,蒸气难以在液体内生成气泡,导致液体过热,突然放出大量蒸气而造成暴沸。切勿在加热过程中补加沸石,否则会引起液体暴沸冲出瓶外,引发危险。沸石一般由碎瓷片制得,毛细管为两端开口。温度计的位置正确与否直接影响沸点测量结果的准确性。温度计的正确位置是:温度计水银球的上缘应位于蒸馏头支管底缘最高点所在的水平线上(如图 2.4 放大部分所示)。

(4) 另外使用一个铁架台,将冷凝管用夹子固定在铁架台上,在冷凝管的下端连接接液管,可用橡皮筋或夹子固定[6]。

(5) 在接液管下面连接接收瓶(圆底烧瓶或锥形瓶)，用夹子夹住烧瓶接口处固定在铁架台上。确保所有接头处连接紧密，否则装置会漏气，从而影响蒸馏效果。

(6) 用胶管将冷凝管的下支口与水龙头相连，另一胶管的一端与冷凝管的上支口相连，胶管的另一端连入水槽。将电热套与变压器相连，调节变压器到合适的数值。开始加热前检查整个装置是否安装到位。

(7) 先向冷凝管中缓慢通入冷凝水，再开始加热，冷凝水不要开得太大，调节水龙头，使出水管内有细小水流流出即可。控制加热程度，不要太快，使蒸馏速度以每秒滴出 1~2 滴馏出液为宜。

(8) 开始加热，蒸馏烧瓶内的液体开始冒泡。在蒸馏过程中，应使温度计水银球常有被冷凝的液滴润湿，此时的温度计读数就是馏出液的沸点。当温度计读数到达沸点时，收集所需温度范围的馏出液，并记录第一滴馏出液进入接收瓶时的温度[7]。

(9) 继续加热直到蒸出绝大部分液体。**注意：绝对不要蒸干**[8]！

(10) 蒸馏完毕，先拔下电源插头，再移走电热套，然后停止通水，最后拆除蒸馏装置(与安装顺序相反)。

2. 乙醇的简单蒸馏

在蒸馏烧瓶中加入 20 mL 工业乙醇，按照上述步骤搭建简单蒸馏装置，蒸出乙醇并测量其沸点。

注释

[1] 实验过程中要佩戴防护眼镜。

[2] 各磨口玻璃仪器的磨口应配套，连接紧密，蒸馏烧瓶和接收瓶要夹牢，但不要夹得太紧以防夹破。

[3] 使用水浴或电热套，不得使用明火加热！有机液体和蒸气易燃。

[4] 蒸馏任何液体时须在加热前加入 2~3 颗沸石，蒸馏中途严禁加入。万一要在中途补加，则须冷至接近室温方可加入。沸石不能重复使用，因为使用过的沸石的微孔在冷却过程中会充满液体。

[5] 使用温度计要非常小心，水银毒性较大！

[6] 接液管或接收瓶须有一处通向大气，**切勿造成密闭体系！**

[7] 实验目的如为纯化液体，则应将低沸点的前馏分用其他容器接收后弃去，至温度计上显示温度为所需液体沸点时再开始接收。蒸至温度计上显示温度突然变化时，停止收集。本实验中需接收的是 95%乙醇，沸点为 78℃。

[8] 蒸馏操作不得蒸干，以防不可预知的危险发生。当蒸馏瓶中剩下 2~3 mL 液体时停止加热。

思考题

(1) 什么是沸点？沸点和大气压有什么关系？

(2) 为什么必须在蒸馏烧瓶中加入沸石？能否立即将沸石加入将近沸腾的液体中？用过的沸石能否继续使用？

(3) 为什么蒸馏时最好控制馏出液的速度为每秒 1~2 滴为宜？

(4) 如果液体具有恒定的沸点，能否认为它是纯物质？

实验3 重结晶

实验目的

(1) 提纯室温下为固体的有机化合物。
(2) 学习重结晶的原理和方法。
(3) 掌握热过滤、抽滤等基本操作。

实验原理

重结晶是提纯固体有机化合物常用的方法之一。该提纯方法的原理是固体在溶剂中的溶解度与温度的关系,一般是温度升高,溶解度增大;同时利用混合物中被提纯物质与杂质在某溶剂中的溶解度不同。该过程需要找到合适的溶剂,该溶剂对被提纯的成分热时溶解度大,冷时溶解度小。具体方法是:选择一种合适的溶剂,将含有杂质的固体物质溶解在少量热的溶剂中,趁热滤去不溶性杂质,冷却滤液,析出待提纯物质的结晶,可溶性杂质仍留在母液中。从母液中过滤出晶体,洗涤、干燥所得的晶体,最终达到提纯的目的。

重复重结晶操作,可获得纯度更高的晶体。

重结晶主要包括如下几步:选择溶剂、溶解样品、活性炭脱色、热过滤、冷却结晶、抽滤晶体、洗涤和干燥晶体。

1. 选择溶剂

在进行重结晶时,选择理想的溶剂是操作的关键。理想的溶剂应具备以下特征:
(1) 与被提纯物质不发生化学反应。
(2) 被提纯物质在该溶剂中温度较高时易溶,而在室温或更低温度时几乎不溶。
(3) 对杂质的溶解度非常大或非常小。
(4) 容易从提纯产物中除去,溶剂必须具有相对较低的沸点并容易蒸发。
(5) 能给出较好的结晶。
(6) 价廉,低毒,安全。

假设有一种未知化合物需要结晶提纯。要找到一种良好的重结晶溶剂,可以遵循以下步骤。首先,测试其在简单可用溶剂中的溶解性。

1) 单一溶剂法

取 0.1 g 待重结晶固体于试管中,滴入 1 mL 冷溶剂,振荡约 3 min,观察溶解情况。如在室温下能全部或绝大部分溶解,则此溶剂不适合做该固体的重结晶溶剂。如该固体不溶,且加热还不溶,并逐步加大溶剂量至 3 mL,加热至沸腾仍不溶,则该溶剂溶解度太小,也不能用。如该固体能溶于 1~3 mL 沸腾的溶剂中,且冷却时能自行析出大量结晶,则此溶剂适合。

2) 混合溶剂法

如果难以找到合适的单一溶剂,则可以考虑使用混合溶剂。混合溶剂一般由两种可以混溶的溶剂组成,如乙醇-水、丙酮-水、乙酸-水、乙醚-甲醇、乙醚-石油醚、苯-石油醚等。

其中一种溶剂对被提纯固体有较大的溶解度,而另一种对被提纯固体溶解度较小。

本实验重结晶使用单一溶剂水。

2. 溶解样品

将待重结晶固体加入锥形瓶中,加入少量热溶剂和一粒沸石,用电热板加热至接近沸腾,搅拌以溶解固体。若还有未溶解的固体,可继续在搅拌下慢慢加入热溶剂,直至固体刚好全部溶解。若加入溶剂,加热后不见未溶物减少,则可能是不溶性杂质,将在热过滤时除去。不要加太多溶剂,否则溶液不饱和,降低提纯产物的产率。

当使用混合溶剂时,例如,被提纯固体易溶于乙醇而难溶于水。一般先将样品溶于尽量少的沸腾的乙醇中,再趁热逐滴加入温水,滴加水时溶液中会产生浑浊,浑浊在搅拌下最初会消失。当滴加水至溶液中的浑浊不再消失时,补加几滴乙醇,再加热溶液使其变澄清,放置冷却,使结晶析出。不溶性杂质最好在加水前从热乙醇溶液中过滤除去。当样品溶解后,从电热板上取下烧瓶并添加少量溶剂(约为溶液所需量的20%)。

3. 活性炭脱色

如溶液有颜色,可使用活性炭脱色。将热溶液稍冷却后加入少许活性炭(样品质量的1%~5%,加入过多则产品会被活性炭过多吸附而损失)[1],搅拌下加热微沸 5~10 min。**注意**:不能向正在沸腾的溶液中加活性炭,以防溶液暴沸而喷出。

4. 热过滤

这一步在所得溶液颜色正常且无明显不溶物的情况下可以不做。

有机化学实验一般使用重力过滤[如图 2.5(a)所示]和减压过滤(也称抽滤)[如图 2.5(b)所示]两种热过滤方法。

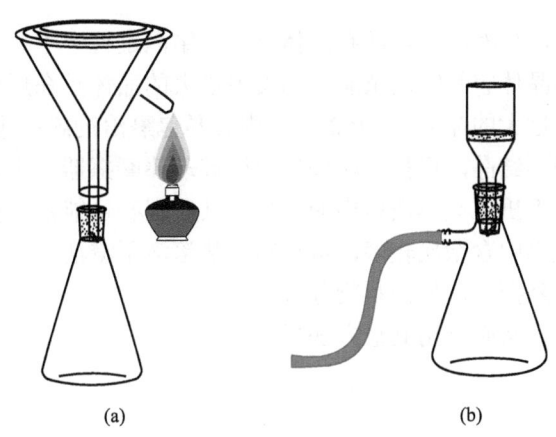

图 2.5 热过滤装置

在过滤的液体量较大的情况下,一般采用重力过滤。重结晶通常使用热过滤装置进行。如图 2.5(a)所示搭好热过滤装置。将折好的菊花滤纸(菊花滤纸有许多折痕,因而表面积大,流速快,其折叠方法如图 2.6 所示)放在一短颈、粗径的玻璃漏斗上,再将玻璃漏斗放在热水漏斗上，热水漏斗夹套中事先冲注热水，并在热滤过程中用酒精灯对热水漏斗侧管加热。在热过滤前，菊花滤纸、漏斗和锥形瓶需要预热以减少在滤纸上析出的晶体。

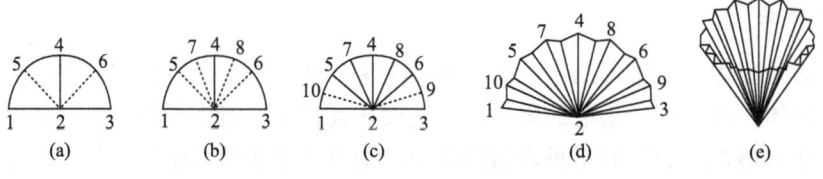

图 2.6　菊花滤纸折叠方法（虚线代表折叠的部位）

趁热将待过滤溶液沿玻璃棒倒入漏斗中进行热过滤。热过滤要准备充分，动作要迅速，每次倾倒液体后，可将锥形瓶放回电热板上保温。倒入漏斗中的液体不要超过滤纸容积的 3/4，液体尽量倒在滤纸中心部位以防止未过滤的液体溢出滤纸从侧面未经过滤直接流下。如滤液中仍含有少量固体杂质，则需要将滤液重新加热并再次热过滤。若有少量晶体析出，可用少量热溶剂洗下，若较多，可用刮刀刮回原瓶，重新热过滤。当过滤的液体量较少时（100 mL 以下），也可用预热后的布氏漏斗进行快速热抽滤。抽滤时使用双层滤纸，以防热抽滤时滤纸被抽破。抽滤后，将热滤液转移至干净的烧杯中。

菊花滤纸的折叠方法：先把圆形滤纸沿 1→3 线对折为半圆形，然后沿 2→4 线对折成扇形，再对折成 1/8 扇形（圆形滤纸折成相等的 8 等分），然后打开，如图 2.6(a) 所示。在 8 等分的每一小格中间再以相反的方向对折成 16 等分，如图 2.6(b)、(c) 所示。打开，继续在 16 等分的每一小格中间再以相反的方向对折，就得到折扇一样的排列，如图 2.6(d) 所示。展开后即得到折好的菊花滤纸，如图 2.6(e) 所示。注意：折叠时不要用手指或指甲直接在折痕上用力，尽量采用压折，否则过滤时滤纸的中央容易破裂。

5. 结晶

滤液在室温下放置，自然冷却，即有晶体析出。结晶过程中不要急速冷却或剧烈搅动，以免晶体过细，过细的晶体容易吸附杂质。当发现过大的晶体正在形成时，应轻轻摇动使其形成较均匀的小晶体，过大的晶体（大于 2 mm）则容易包裹溶剂或杂质。

如果溶液冷却后仍不结晶，可投"晶种"或用玻璃棒摩擦器壁引发晶体的形成。

如果被提纯的物质不析出晶体而析出油状物，其原因有可能是溶质的温度比被提纯物质的熔点高或接近。油状物中含杂质较多，可重新加热溶液至澄清后，让其自然冷却至开始有油状物出现时，立刻剧烈搅拌，使油状物分散。

如果结晶不成功，通常必须用其他方法提纯。

6. 抽滤

如图 2.5(b) 所示装好抽滤装置，剪好滤纸[2]，放在布氏漏斗上，用少量水润湿，开动真空泵使滤纸贴紧布氏漏斗。摇动烧瓶，将晶体和液体倒入布氏漏斗中抽滤。附着在瓶壁的晶体可以用少量冰冷的溶剂冲洗出来。晶体用少量冷溶液洗涤，洗涤前最好关闭真空泵，让洗液停留几秒钟再抽滤。停止抽滤时，需要先打开安全瓶的活塞再关闭真空水泵，否则水泵中的水有可能倒吸至抽滤瓶内。

7. 晶体的干燥

将滤纸上的晶体用刮刀刮下,放在已称量的洁净表面皿上自然晾干或在烘箱中快速烘干。称量并测定干燥产品的熔点,比较提纯后的样品和粗品的熔点。

实验用品

仪器:电子天平、真空水泵、电热板、锥形瓶、量筒、烧杯、玻璃棒、药匙、热水漏斗、玻璃漏斗(短颈)、抽滤瓶、布氏漏斗、酒精灯、表面皿、称量纸、胶管、沸石、滤纸。

试剂:乙酰苯胺(粗品)、蒸馏水、活性炭。

实验步骤

把 2 g 粗乙酰苯胺加入 100 mL 锥形瓶中,加入 40 mL 水,在电热板上加热至接近沸腾,搅拌以溶解固体。若还有未溶解的固体,可继续在搅拌下慢慢加入热水,直至固体刚好全部溶解。将锥形瓶从电热板上取下,再多加 20%左右的溶剂。

如溶液有颜色,将热溶液稍冷却后加入少许(0.2~0.4 g)活性炭,搅拌下加热微沸 5~10 min。趁热用预热后的布氏漏斗和抽滤瓶进行快速热抽滤[3],抽滤时使用双层滤纸,以防热抽滤时滤纸被抽破。

滤液趁热转移至 250 mL 干净烧杯,在室温下放置,自然冷却,即有乙酰苯胺晶体析出。

再安装抽滤装置,将乙酰苯胺晶体和液体倒入布氏漏斗中抽滤,晶体用少量冷水洗涤[4]。抽干后[5],将滤纸上的晶体刮下,放在表面皿上自然晾干或在 80~100 ℃烘箱中烘干,称量产品并记录产量。

乙酰苯胺的熔点为 114 ℃。

$$回收率/\% = 晶体质量/粗品质量 \times 100$$

注释

[1] 活性炭不能加入正在沸腾的溶液中,否则热溶液将暴沸。

[2] 滤纸不能大于布氏漏斗的底面,但须盖住布氏漏斗的所有孔。

[3] 热过滤时,整个操作过程要迅速,否则结晶在滤纸上和漏斗颈部析出。

[4] 用尽量少的溶剂洗涤晶体,以免晶体大量损失。

[5] 停止抽滤时,先将抽滤瓶与泵之间的连接断开,或者将安全瓶上的活塞打开,再关闭泵,防止水倒流入抽滤瓶内。

思考题

(1) 重结晶一般包括哪几个步骤?选用的溶剂应具备哪些条件?

(2) 不纯的物质经一次重结晶后,是否必然纯净?如何决定是否需要再进行一次重结晶?

(3) 重结晶时所用的溶剂为什么不能太多或太少?溶剂量应如何正确控制?

(4) 在布氏漏斗中洗涤晶体,应注意些什么?

实验 4 萃 取

实验目的

(1) 学习萃取的原理和方法。
(2) 掌握萃取和洗涤的操作方法。
(3) 利用萃取技术分离二组分混合物。

实验原理

萃取是提纯或分离有机化合物的重要方法之一。该方法是利用物质在两种不互溶(或微溶)溶剂中溶解度或分配比的不同来达到分离或提纯目的的一种操作。被萃取的物质可以是固体、液体或气体。以除去混合物中的少量杂质为目的的萃取常称为"洗涤"。

萃取的原理是：设溶液由有机化合物 X 溶解于溶剂 A 而成。如果要从其中萃取 X，可以选择一种对 X 溶解度极好，而与溶剂 A 不相混溶且不发生化学反应的溶剂 B。把溶液放入分液漏斗中，加入溶剂 B，充分振荡。静置片刻，A 和 B 因不相溶而分成两层。此时，只要温度保持不变，有机化合物 X 在 A、B 两相中的浓度之比为一常数 K，称为"分配常数"，这种关系称为分配定律，可用下面的公式表示：

$$\frac{X在溶剂A中的浓度}{X在溶剂B中的浓度} = K(分配系数)$$

(注意：分配定律是假设所选用的溶剂 B 不与 X 发生化学反应时才适用。)

当用一定量的溶剂 B 萃取 X 时，总是希望在溶剂 A 中的 X 剩余量越少越好。那么在萃取中，用一定量的溶剂一次萃取好还是分几次萃取好呢？可通过下面的推导说明这个问题。

第一次萃取：设 V 为被萃取溶液的体积(mL)，近似看做与溶剂 A 的体积相等(因溶质量不多，可忽略)；W_0 为被萃取溶液中溶质(X)的总含量(g)；S 为萃取时所用溶剂 B 的体积(mL)；W_1 为第一次萃取后溶质 X 在溶剂 A 中的剩余量(g)；W_2 为第二次萃取后溶质 X 在溶剂 A 中的剩余量(g)；W_n 为经过 n 次萃取后溶质 X 在溶剂 A 中的剩余量(g)。

$W_0 - W_1 =$ 第一次萃取后溶质 X 在溶剂 B 中的含量(g)

$W_1 - W_2 =$ 第二次萃取后溶质在溶剂 B 中的含量(g)

则

$$\frac{W_1/V}{(W_0 - W_1)/S} = K$$

整理得

$$W_1 = \frac{KV}{KV + S} W_0$$

同理

$$\frac{W_2/V}{(W_1 - W_2)/S} = K$$

整理得

$$W_2 = \frac{KV}{KV+S}W_1 = \left(\frac{KV}{KV+S}\right)^2 W_0$$

经过 n 次萃取后的剩余量 W_n 应为

$$W_n = \left(\frac{KV}{KV+S}\right)^n W_0$$

但必须注意，上式只适用于和溶剂 A 几乎不互溶的溶剂。

例如，在 15℃时，正丁酸在水合苯中的分配系数 $K=1/3$，如果每次用 100 mL 苯萃取 100 mL 含 4 g 正丁酸的水溶液，根据以上公式可知，经过一次、二次、三次、四次、五次萃取后，水溶液中剩余的正丁酸的量分别为

$$W_1 = 4 \times \frac{\frac{1}{3} \times 100}{\frac{1}{3} \times 100 + 100} = 4 \times \frac{1}{4} = 1.0(\text{g})$$

$W_2 = 4 \times (1/4)^2 = 0.250$ (g) $W_3 = 4 \times (1/4)^3 = 0.0625$ (g)
$W_4 = 4 \times (1/4)^4 = 0.016$ (g) $W_5 = 4 \times (1/4)^5 = 0.004$ (g)

如果将 100 mL 苯分成三等份，每次用一份萃取上述正丁酸水溶液，萃取三次后水溶液中剩余正丁酸的量为

$$W_3 = 4 \times \left(\frac{\frac{1}{3} \times 100}{\frac{1}{3} \times 100 + \frac{100}{3}}\right)^3 = 4 \times \left(\frac{1}{2}\right)^3 = 0.5(\text{g})$$

计算结果表明：

(1) 用同样体积的溶剂，分多次萃取比用全部溶剂萃取一次的效果好。

(2) 但是，当萃取溶剂的总量保持不变时，萃取次数增加，每次所用溶剂的体积必然减小。如果每次所用溶剂量太少，不仅操作麻烦，而且被萃取物的量增加很小。因此，一般情况下萃取 3～5 次即可。

1. 萃取溶剂的选择

理想的萃取溶剂应具备以下条件：
(1) 不与原溶剂混溶，也不成乳浊液。
(2) 不与溶质或原溶剂发生化学变化。
(3) 对溶质有尽可能大的分配系数。
(4) 沸点较低，易于回收。
(5) 安全，价廉易得。

2. 分液漏斗的使用

液-液萃取常用分液漏斗，其基本使用方法如下：
(1) 选择合适大小的分液漏斗，使待处理液体积为漏斗容积的 1/2~3/4。

(2) 使用前，确保顶部的塞子和底部的旋塞配套，并且是液体密封的[1]。如果漏斗装有玻璃旋塞，确保塞子充分润滑[2]。如果旋塞阀内的塞子转动顺畅（应该有一个夹子帮助固定塞子），则可以进行检漏测试。摇动装有 50%水的漏斗，塞子或旋塞处不应有任何泄漏。

(3) 洗净、烘干分液漏斗。注意不能把旋塞上附有凡士林的分液漏斗放在烘箱内烘干。当使用带有聚四氟乙烯旋塞的分液漏斗时，旋塞不应拧得过紧，以防扭曲；也不要对其快速加热，因为聚四氟乙烯比玻璃膨胀得快得多。

(4) 将分液漏斗固定在铁架台的铁圈上，关闭旋塞。将待萃取的溶液和萃取溶剂倒入分液漏斗。

(5) 盖上塞子，竖直拿住分液漏斗，振荡两三次并倒转漏斗。朝无人处放气[3]，打开旋塞释放可能积聚在分液漏斗内的压力(挥发性溶剂如乙醚等会产生相当大的压力)。

(6) 将分液漏斗放回铁圈内，**打开上口塞子**[4]，使液体静置分层。开启旋塞，将下层液体从下口放出，然后关闭旋塞，将上层液体从上口倒出[5]。

3. 液体分层

在萃取操作中，有时会遇到水层与有机层难分层的现象(特别是当被萃取液呈碱性时，常出现乳化现象，难分层)。此时，应认真分析原因，采取相应的措施。

(1) 若萃取剂与水层的密度比较接近，可能发生难分层的现象。在这种情况下，只要加入一些溶于水的无机盐，增大水层的密度，即可迅速分层。此外，加无机盐(通常用氯化钠)使水溶液饱和后，能显著降低有机物在水中的溶解度，明显提高萃取效果。这就是"盐析作用"。

(2) 若萃取剂与水部分互溶而产生乳化，只要静置较长时间就可以分层。

(3) 若被萃取液中存在少量轻质固体，在萃取时常聚集在两相交界面处使分层不明显，将混合物过滤就能解决问题。

(4) 若因被萃取液呈碱性或含有表面活性剂而产生乳化，可加入少量稀硫酸降低溶液的pH，并轻轻振摇使乳浊液分层。

此外，还可根据不同情况，采用加入醇类化合物降低表面张力、加热破坏乳化等方法。

本实验用乙醚从乙酸水溶液中萃取乙酸，采用以下两种方法：

(1) 一次性用 30 mL 乙醚萃取乙酸。

(2) 进行多次萃取，乙醚用量为每次 10 mL，共 3 次。

实验用品

仪器：分液漏斗、锥形瓶、碱式滴定管。

试剂：冰醋酸和水的混合溶液(冰醋酸：水= 1：19)、乙醚、0.2 mol/L NaOH、酚酞指示剂。

实验步骤

1. 一次萃取

(1) 用移液管准确量取 10 mL 乙酸与水的混合液放入分液漏斗中，用 30 mL 乙醚萃取。

(2) 双手轻轻握紧漏斗，一只手始终紧握塞子，另一只手的手掌抵住旋塞，盖上漏斗塞子。用力振荡溶液，偶尔拧松旋塞，释放由溶剂蒸气或放出的气体产生的压力[图 2.7(a)]。关闭旋塞，水平握住漏斗，摇动漏斗两三次，如前所述倒转漏斗并释放压力。重复此过程，直到打开旋塞不再有压力释放。关闭旋塞，摇动漏斗 15～20 次。

(3) 将分液漏斗置于铁圈中，**打开塞子**，静置液体待其分层。放出下层水溶液于 50 mL 锥形瓶内[参见图 2.7(b)]。控制放液速度，不要太快。当分层界面接近旋塞时，小心控制放液速度。当上层液体刚刚放入旋塞孔内时，关闭旋塞，停止放液。将上层液体从分液漏斗上口倒入第二个烧瓶中。在不能完全确定哪一层是需要保留的液体时，**不要丢弃**任何一层液体[6]。

(4) 加入 3～4 滴酚酞作指示剂，用 0.2 mol/L NaOH 溶液滴定萃取的水溶液，记录用去 NaOH 溶液的体积。

(5) 计算：①留在水中乙酸量及质量分数；②留在乙醚中乙酸量及质量分数。

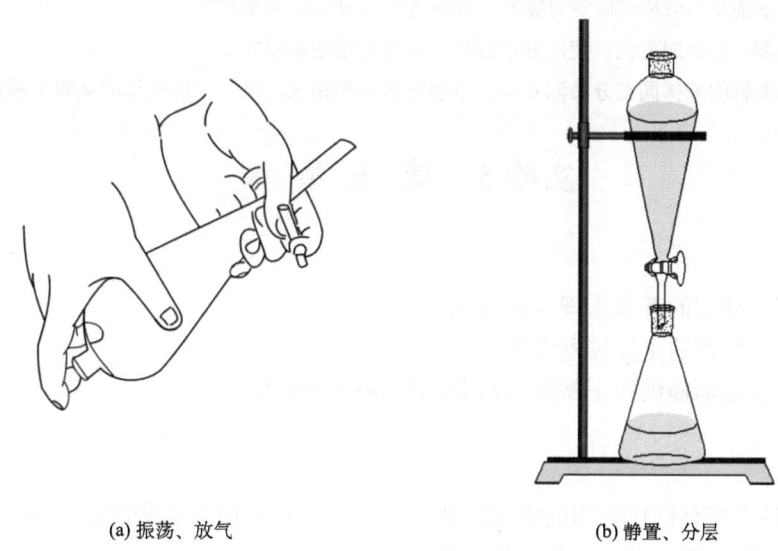

(a) 振荡、放气　　　　　　　(b) 静置、分层

图 2.7　分液漏斗的操作方法

2. 多次萃取

(1) 用移液管准确量取 10 mL 乙酸与水的混合液放入分液漏斗中，用 10 mL 乙醚如上法萃取，分去上层有机层。

(2) 将下层水溶液再用 10 mL 乙醚萃取，分出乙醚溶液。

(3) 将第二次萃取的下层水溶液再用 10 mL 乙醚萃取，如此共 3 次，分出乙醚溶液。

(4) 用 0.2 mol/L NaOH 溶液滴定萃取的水溶液。

(5) 计算：①留在水中乙酸量及质量分数；②留在乙醚中乙酸量及质量分数。

比较两种方法的萃取效果。

注释

[1] 萃取前，一定要对分液漏斗检漏，并检查活塞的灵活性。

[2] 旋塞可以是玻璃的或聚四氟乙烯的。如果旋塞是玻璃的，可在塞子表面涂抹很少量的凡士林以使其

转动顺畅。涂凡士林**不能**沾到容器内部，以免在萃取过程中，因有机溶剂溶解凡士林而产生污染。

[3] 分液漏斗放气时，**不得**将分液漏斗的支口对准任何人！

[4] 从分液漏斗下口放出液体时需要**打开**分液漏斗上口的塞子，否则在放液过程中液体上部的压力下降，内外压力差会降低液体流下的速度直至停止流出。一段时间后，空气会不吸入分液漏斗中，造成已分层的液体再次混合。

[5] 分离液体时，下层液体从下口放出，上层液体从上口倒出。

[6] 在萃取或洗涤时，上、下两层液体都应**保留**到实验完毕。如果中间的操作发生错误，从少量溶液中分离比从废液桶中要容易得多。

思考题

(1) 影响萃取法的萃取效率的因素有哪些？如何才能选择合适的溶剂？

(2) 使用分液漏斗的目的何在？使用分液漏斗时要注意哪些事项？

(3) 两种不相溶解的液体同在分液漏斗中，下层液体从哪里放出来？上层液体应从哪里转移出来？

实验 5 减 压 蒸 馏

实验目的

(1) 学习减压蒸馏的基本原理及其应用。

(2) 熟悉减压蒸馏的主要仪器设备。

(3) 通过减压蒸馏提纯在正常沸点易分解的高沸点液体。

实验原理

减压蒸馏是提纯有机化合物的常用方法之一。它特别适用于某些沸点较高的在常压蒸馏时未达沸点即分解、氧化或聚合的有机化合物。

在低于一个大气压的负压状态下进行的蒸馏称为减压蒸馏。液体的沸点是指它的蒸气压等于外界压力时的温度，因此液体的沸点是随外界压力的变化而变化的。如果借助真空泵降低系统内压力，就可以降低液体的沸点，这就是减压蒸馏操作的原理。

从图 2.8 可以看出，液体的蒸气压随着温度的升高而升高。依据这个关系图，我们发现液体的饱和蒸气压随温度升高逐渐增加，当饱和蒸气压升至 1 atm(101.3 kPa，760 mmHg) 或 101 325 Pa，液体开始沸腾并不断蒸出直至蒸完。

这清楚地说明了降低外部压力可降低液体的沸点这一重要原理。很多高沸点有机化合物在常压下蒸馏时因温度过高而发生分解，因而采用减压蒸馏方法显著降低其沸点可以避免发生这种情况。

液体在常压、减压下的沸点近似关系图参见图 2.9。压力与沸点的关系可以用该图的压力-温度关系图来估算。使用该图时，只要在中间的线段 B 上放一直线，以 B 线段上某温度值为支点，转动该直线与 A、C 线段相交，通过交点的读数即可得到某压力下的沸点近似值。

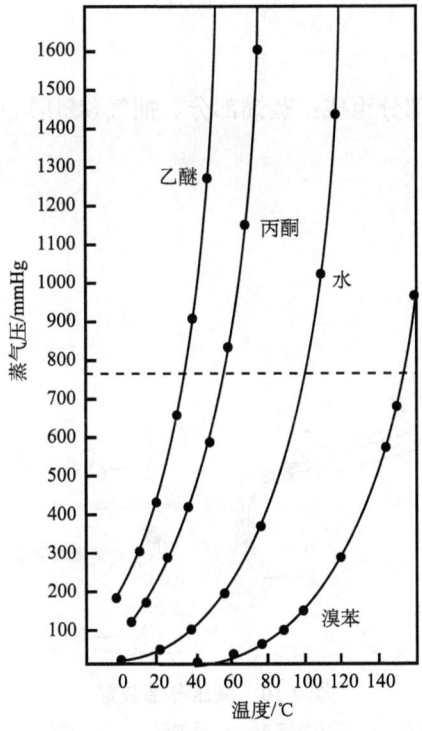

图 2.8 液体的温度与蒸气压关系

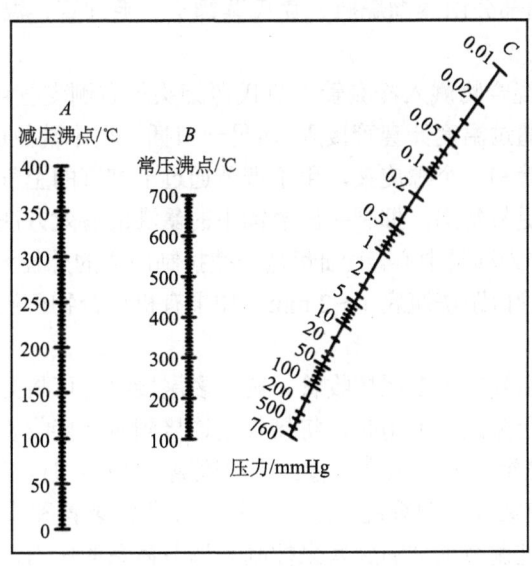

图 2.9 液体在常压、减压下的沸点近似关系图(1 mmHg≈133 Pa)

例如,水杨酸乙酯常压下的沸点为 234℃,减压至 15 mmHg 时的沸点可用图 2.9 求出。在图 2.9 中 B 线上找到 234℃的点,再在 C 线上找到 15 mmHg 的点,通过两点连一直线,该直线与 A 线的交点为 113℃,即近似为水杨酸乙酯在 15 mmHg 的沸点。

实验装置

减压蒸馏装置主要由四部分组成：蒸馏部分、抽气(减压)、安全保护和测压，如图 2.10 所示。

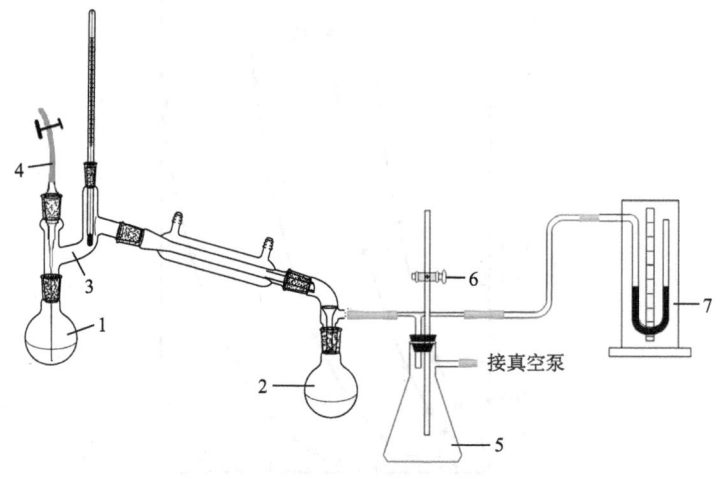

图 2.10 减压蒸馏装置
1. 蒸馏烧瓶；2. 接收瓶；3. 克氏蒸馏头；4. 毛细管；5. 安全瓶；6. 两通旋塞；7. 压力计

(1) 蒸馏部分：蒸馏部分由蒸馏烧瓶、克氏蒸馏头、毛细管、温度计及冷凝管、接收瓶等组成。

克氏蒸馏头可防止混合物溅入冷凝管。克氏蒸馏头的右侧支管可以安装温度计套管(参见图 2.10)，温度计可以通过温度计套管接入。而另一口插一根末端拉成毛细管的厚壁玻璃管，玻璃管上端接一耐压胶管和一个螺旋夹，用于调节通过毛细管的空气量。

液体在减压蒸馏时更易暴沸。抑制减压蒸馏中的暴沸的有效方法之一是通过毛细管导入小气泡至待蒸馏液体，作为沸腾中心。毛细管是一截拉制而成的外径为 3～5 mm 厚壁毛细管，调节毛细管长度，使其管口距烧瓶底 1～2 mm。如果有机化合物可被空气氧化，则可以将毛细管与氮气连接。

进行减压蒸馏时，最好使用多尾接收管。旋转多尾接收管可将几个馏分分别收集到对应的接收瓶中，而不必停止蒸馏。使用时，将接收瓶连接到多尾接收管，将多尾接收管连接到冷凝管。在 140℃以下蒸馏可使用水冷凝管；当温度高于 140℃时，则使用空气冷凝管。

(2) 安全保护：为使减压蒸馏系统安全、可控，在蒸馏装置和真空泵之间安装一些瓶、阱。其中，安全瓶可通过调节顶部的旋塞来控制和调节体系的压力，并防止蒸气倒吸；冷阱可防止挥发物进入油泵；吸收塔安装在冷阱和油泵之间，以防止酸性气体或水蒸气吸入油泵中。冷阱中冷却剂的选择根据需要而定，可用冰-水、冰-盐、干冰-丙酮等。

三个吸收塔(又称干燥塔)安装在接收器和泵之间，用来吸收不同的气体：第一个装无水 $CaCl_2$ 或硅胶，吸收水汽；第二个装粒状 NaOH，吸收酸性气体；第三个装石蜡片，吸收烃类气体。

冷阱和吸收塔装置参见图 2.11。

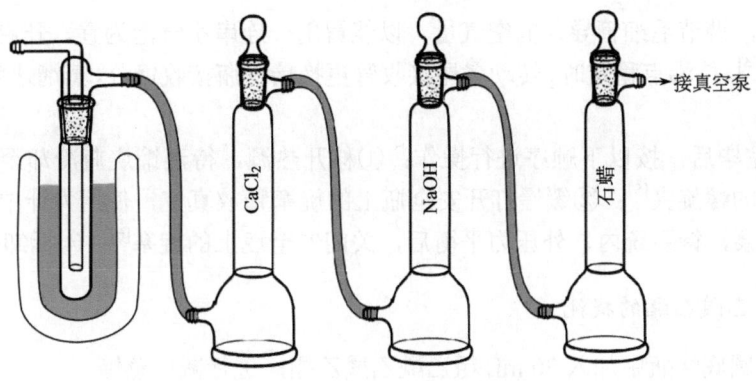

图 2.11　冷阱和吸收塔装置

(3) 抽气(减压)：实验室通常用水泵或油泵进行减压。一般"粗真空"1.333～100 kPa (10～760 mmHg)可用水泵获得；"中真空"0.133～133.3 Pa (0.001～1 mmHg)可用油泵获得；低于 0.133 Pa(0.001 mmHg)的"高真空"需使用扩散泵。

当用油泵进行减压蒸馏时，为除去挥发性杂质(如水汽、溶剂、酸性或碱性化合物)或防止泵油倒吸，以避免其造成真空泵的效率降低或损坏，应在真空系统中安装冷阱。此外，油泵中的油应定期更换。

使用真空泵时，其一般顺序是：先尽量用水泵获得的"粗真空"除去烧瓶中大部分的挥发性馏分；然后改用油泵，将压力降低到较高真空；最后可使用扩散泵使压力进一步降低至"高"真空。

(4) 测压：实验室常用封闭式 U 形管压力计测量减压系统的真空度，压力计两臂汞柱高度之差即为系统的真空度。例如，在封闭式压力计中，如果封闭端的水银柱高度比另一端高 690 mm，则气体压力等于 690 mmHg。使用时应当注意，当减压操作结束时，要小心旋开安全瓶上的两通旋塞，让气体慢慢进入系统，使压力计中的水银柱缓缓复原，以避免因系统内的压力突增使水银冲破玻璃管。

实验步骤

1. 操作方法

(1) 按图 2.10 安装仪器。仪器安装好后，先检查系统是否漏气[1]。必须使用厚壁玻璃仪器[2]，连接真空系统的橡皮管也须是厚壁、耐压的。为了提高真空质量，磨口仪器的所有接口部分都必须涂抹真空油脂。

(2) 检漏：先打开安全瓶上的旋塞，拧紧毛细管上的毛细管夹，然后用泵抽气，再慢慢拧紧安全瓶上的旋塞，夹住连接系统的橡皮管，观察压力计，无变化说明不漏气，有变化即表示漏气。如果漏气，则夹住连接接收瓶和真空系统间的橡皮管，以及接收瓶和压力计间的橡皮管，观察压力计以确定可能漏气的部位。检测到漏气部位后，慢慢打开安全瓶上的旋塞以释放真空，修复漏气部位，然后重新开始第(2)步的操作。如仪器不漏气，则缓慢释放压力，继续进行第(3)步。

(3) 在蒸馏烧瓶中加入待蒸的液体[3](量不要超过蒸馏烧瓶的一半)，关闭安全瓶上的旋塞，开动油泵，调节毛细管导入的空气量，以能冒出一连串小气泡为宜。开启冷凝水，缓慢加热蒸馏烧瓶[4]。待沸点稳定时，转动多尾接收管更换接受瓶接收馏分，蒸馏速度以每秒0.5～1滴为宜。

(4) 蒸馏完毕后，按以下顺序进行操作：①移开热源，待蒸馏烧瓶冷却至室温后，慢慢打开毛细管上的螺旋夹[5]；②缓慢打开安全瓶上的旋塞释放真空，使压力计中的汞柱缓慢复原；③关闭油泵，待系统内、外压力平衡后，关闭安全瓶上的旋塞[6]；④拆卸仪器。

2. 粗乙酰乙酸乙酯的提纯

在 50 mL 圆底烧瓶中加入 20 mL 粗乙酰乙酸乙酯，进行减压蒸馏。

乙酰乙酸乙酯的沸点参见表 2.1。

表 2.1　乙酰乙酸乙酯的沸点

压力/mmHg	760	80	60	40	30	20	18	14	12
b.p. /℃	181	100	97	92	88	82	78	74	71

注释

[1] 在减压蒸馏过程中，特别是读取温度时，必须戴上防护眼镜。

[2] 所用全部玻璃仪器必须是厚壁的，不可使用平底玻璃仪器，如锥形瓶、平底烧瓶等。

[3] 待蒸馏液体应先进行常压蒸馏或水泵减压蒸馏除去易挥发低沸点馏分后，才能用油泵进行减压蒸馏。

[4] 必须在真空度达到要求并稳定后，才能开始加热。否则温度有可能超过所要求的真空度下的沸点，易引起暴沸。

[5] 停止蒸馏时应先移去热源，待蒸馏烧瓶冷却至室温后，才可慢慢打开安全瓶旋塞。否则，高温下大量空气进入，剩余残液易发生氧化引起爆炸。

[6] 关闭泵前一定要先将系统与大气相通(打开安全瓶旋塞)，否则泵油或水易倒吸。

思考题

(1) 什么化合物需要用减压蒸馏进行提纯？

(2) 使用水泵减压蒸馏时，应采取什么防范措施？

(3) 进行减压蒸馏，为什么必须先抽真空后加热？

(4) 当收集完所要的化合物后，应如何停止减压蒸馏？为什么？

实验 6　水蒸气蒸馏

实验目的

(1) 学习水蒸气蒸馏的原理及应用范围。

(2) 掌握水蒸气蒸馏的装置及其操作方法。

(3) 利用水蒸气蒸馏分离对热敏感有机物。

实验原理

水蒸气蒸馏是一种特殊的蒸馏方法。将水蒸气引入蒸馏装置,水蒸气将少量蒸发的化合物带入冷凝烧瓶中,冷凝液相在这里分离,便于收集。这一过程可以有效地在较低温度下蒸馏以提纯温度敏感的有机化合物,减少所需产品的变质。

道尔顿分压定律指出,气体混合物的总压力等于其所有成分的分压之和。因此,将有机化合物 A 溶解在水中而形成的溶液的总压力可由 $p_{总} = p_{H_2O} + p_A$ 得出。当总蒸气压($p_{总}$)等于环境大气压时,溶液沸腾。这表示溶液会在低于纯水沸点(100℃)和低于有机化合物 A 的沸点的温度下沸腾。因此,水蒸气蒸馏可以在较低的温度下蒸馏有机化合物。

水蒸气蒸馏中馏出液组分的计算:假设两组分是理想气体,则根据理想气体定律

$$pV = nRT = WRT/M$$

混合物的组成用下式计算:

$$W_A / W_{H_2O} = M_A p_A / M_{H_2O} p_{H_2O}$$

例如,苯甲醛(b.p. 178℃)和水的混合物在 97.9℃沸腾。这时

$$p_{H_2O} = 703.5 \text{ mmHg}$$

$$p_{C_6H_5CHO} = 760 - 703.5 = 56.5 \text{ (mmHg)}$$

$$M_{C_6H_5CHO} = 106 \text{ g/mol} \qquad M_{H_2O} = 18 \text{ g/mol}$$

代入上式得馏出液的组成为

$$\frac{W_{C_6H_5CHO}}{W_{H_2O}} = \frac{106 \times 56.5}{18 \times 703.5} = 0.473 \text{ (g)}$$

即水蒸气蒸馏蒸出 0.473 g C_6H_5CHO,需蒸出水的量为 1 g。若蒸馏 10 mL C_6H_5CHO(ρ= 1.041 g/mL),需出水量(理论):

$$10 \times 1.041 / 0.473 = 10.41 / 0.473 = 22 \text{ (mL)}$$

注意这个数值为理论值,因为实验时有相当一部分水蒸气来不及与被蒸馏物充分接触便离开蒸馏烧瓶,同时苯甲醛微溶于水,所以实验蒸馏出的水量往往超过计算值,故计算值仅为近似值。

使用水蒸气蒸馏提纯的物质必须具备以下条件:
(1)不溶或难溶于水。
(2)与水一起沸腾时不发生化学变化。
(3)在 100℃左右该物质的蒸气压在 10 mmHg(1.33 kPa)以上。

水蒸气蒸馏常用于以下情况:
(1)从固体多的混合物中分离被吸附的液体物质。
(2)从含有焦油状物质的混合物中分离液体物质。
(3)提纯在常压下蒸馏会发生分解的高沸点有机物质。

实验用品

仪器:水蒸气发生器、圆底烧瓶、克氏蒸馏头、直形冷凝管、锥形瓶、量筒、分液漏斗、

烧杯、接收器。

试剂：苯甲醛、无水 MgSO$_4$。

苯甲醛的物理性质参见表 2.2。

表 2.2 苯甲醛的物理性质

试剂	M_r	m.p./℃	b.p./℃	d_4^{20}	水溶性
苯甲醛	106	−26	179	1.040～1.046	微溶

实验装置

实验室常用的水蒸气蒸馏装置参见图 2.12，包括两部分：水蒸气发生器和蒸馏部分。

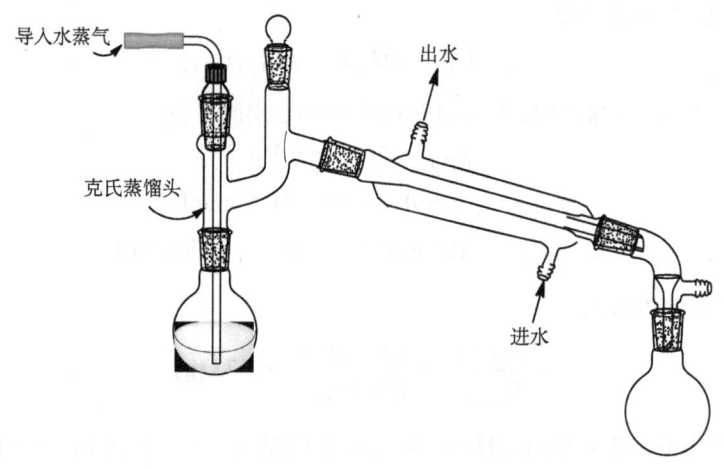

图 2.12 水蒸气蒸馏装置

1. 水蒸气发生器

水蒸气发生器可为金属制的，也可用短颈圆底烧瓶代替。按图 2.13 安装实验装置。水蒸气发生器中插入一根长 1 m、直径约为 5 mm 的玻璃管作为安全管。安全管的底部应位于水蒸气发生器内的水面以下，但不得接触烧瓶底部。水蒸气导出管与一个 T 形管相连，T 形管的支管套上一短橡皮管，橡皮管上用螺旋夹夹住。T 形管的另一端与蒸馏烧瓶的导入管相连，将水蒸气导入蒸馏烧瓶。导入管应尽可能短些，以减少水蒸气的冷凝。T 形管用来排出冷凝下来的水，并保护系统免受过高的压力。如果水蒸气压力继续升高，可关闭安装在 T 形管上的安全夹，以确保水蒸气发生器与大气相通。

2. 蒸馏部分

蒸馏烧瓶常用长颈圆底烧瓶，被蒸馏的液体体积不能超过其容积的 1/3。长颈水蒸气蒸馏烧瓶倾斜，以防止烧瓶中的溶液溅入出口管的入口，从而吹入冷凝管。水蒸气导管的一端弯曲，正对蒸馏烧瓶底部中央，距离瓶底 8～10 mm。或者可以使用带有克氏蒸馏头的短颈圆底烧瓶,以阻止烧瓶中的溶液溅到蒸馏接收器中(如图 2.12 所示)。

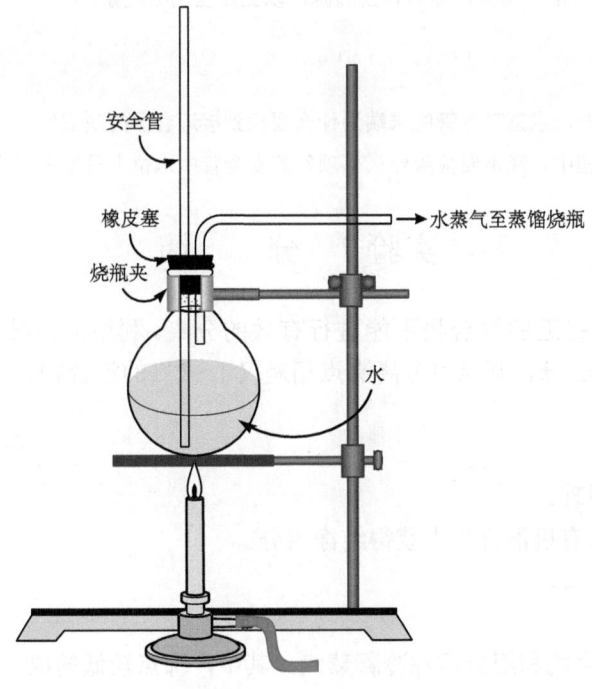

图 2.13 水蒸气发生器

在水蒸气发生器中加入约占容积 3/4 的热水,待检查整个装置不漏气后,旋开 T 形管的螺旋夹,加热至沸腾。当有大量水蒸气从 T 形管的支管冲出时,旋紧螺旋夹,让水蒸气进入蒸馏烧瓶,开始蒸馏。

实验过程中观察水蒸气发生器安全管中水面的高低,判断整个水蒸气蒸馏系统是否通畅。若水面上升很高,说明某部分堵塞,这时应立即旋开 T 形管的螺旋夹,移去热源,拆下装置进行检查和处理,以免造成危险。

蒸馏速度保持在每秒 2~3 滴。

当馏出液无明显油珠时,旋开 T 形管的螺旋夹,移去热源,停止蒸馏。

实验步骤

(1) 按图 2.12 连接装置[1]。
(2) 在圆底烧瓶中加入 15 mL 苯甲醛。
(3) 加热蒸馏:加热前先打开 T 形管的螺旋夹,直到有水蒸气时再关闭螺旋夹,使水蒸气通入圆底烧瓶。必要时可加热圆底烧瓶促使其快速蒸馏,以免水分大量增加[2]。
(4) 当馏出液变澄清时,先打开 T 形管的螺旋夹,再移去热源,停止蒸馏(防止倒吸)[3]。
(5) 馏出液用分液漏斗分出水后,用干燥剂干燥,过滤,收集产品,称量,计算产率。

注释

[1] 安装要正确,连接处要严密。

[2] 调节火焰，控制蒸馏速度每秒 2～3 滴，并时刻注意安全管。

[3] 必须先打开 T 形管的螺旋夹，然后移去热源，以免发生倒吸现象。

思考题

(1) 进行水蒸气蒸馏时，水蒸气导管的末端为什么要插到接近容器的底部？

(2) 在水蒸气蒸馏过程中，经常要检查什么事项？若安全管中水位上升很高，应如何解决问题？

实验 7 分 馏

简单蒸馏对沸点接近的混合物不能进行有效的分离。利用分馏技术可以将沸点相差小于 25℃的液体混合物分开，甚至可分离沸点相差仅 1～2℃的混合物。

实验目的

(1) 了解分馏的原理。

(2) 使用分馏法从有机混合物中获得纯净组分。

实验原理

分馏包括加热混合物和用分馏柱冷凝蒸气。其中，沸点较低的成分通过分馏柱，比沸点较高的成分更早收集。分馏可看做一次运行中的多个简单蒸馏。在典型的分馏过程中，液体混合物在蒸馏烧瓶中被加热，产生的蒸气上升到分馏柱，并在柱内冷凝。柱内冷凝的液体不断地被上升的蒸气重新沸腾。由于每次再沸都会使滴下的冷凝液中挥发性较低的组分富集，当液体滴回烧瓶时，其挥发性较低的组分即可富集。同时，每次再沸都会使上升的蒸气中的挥发性组分富集。当蒸气冷凝并收集在接收瓶中时，冷凝液富含挥发性更强的组分。这些连续的沸腾—冷凝—再沸操作使混合物分离。蒸气经过这样冷凝和再蒸发的循环次数称为"理论塔板数"。

需要注意的是，由于共沸混合物具有恒定的沸点，与蒸馏一样，分馏操作也不可用来分离共沸混合物。这是因为当共沸物沸腾时，蒸气中各组分的含量与未沸腾液体中的含量相同。例如，乙醇-水的混合物经分馏产生的液体中含约 95.5%体积的乙醇。一旦达到该比例，液体和蒸气的组分含量将保持不变，因此也不再发生乙醇和水的分离。

分馏装置包括五个部分：热源、圆底烧瓶、分馏柱、冷凝管和接收器，如图 2.14 所示。分馏装置与蒸馏装置不同的只是蒸馏头和蒸馏烧瓶间插一支分馏柱。分馏的效率取决于分馏柱的使用。

分馏常用的分馏柱有球形分馏柱、刺形分馏柱[又称韦氏(Vigreux)分馏柱]和填充式分馏柱，它们均提高了蒸气与冷凝液接触的面积。球形分馏柱分离效率较差。韦氏分馏柱比同样长度的填充式分馏柱分离效率低，适合分离少量且沸点差距较大的液体。填充式分馏柱效率较高，适合分离一些沸点差距较小的化合物。

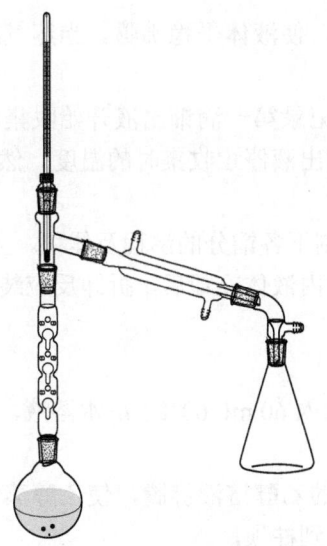

图 2.14 分馏装置

分馏柱的效率与柱的长度和填料有关。用表面积大、长度足够的填料填充柱体,可获得良好的效率。增加蒸气必须经过的表面积会增加馏出液的纯度,因为有效地增加了蒸馏的次数。蒸气接触的表面积越大,分离组分的效率越高。

柱中填料不同,效率不同。例如,玻璃管填料(6~20 mm 的小段)性质稳定,但分馏效率较低;金属丝绕成固定形状效率较高,但会与某些有机物如卤代烃等发生反应。在填装填料时,填料要比较松散,为液体和蒸气的流动留出一定的空间,但不要太松散,否则会影响其效率。

分馏柱通常是绝热的,以保持柱温基本恒定。如果柱温波动很大,就很难保持缓慢、恒定的蒸馏速度。因此,一个好的选择是使用真空夹层玻璃柱与冲压金属件填充,商品名为"Pro-Pak"。分馏柱不需要真空夹层,但在这种情况下,必须至少用玻璃棉绝热。

实验装置

分馏的实验室装置通常包括下列装置(参见图 2.14):热源(如油浴锅或电热套),蒸馏烧瓶(通常是圆底烧瓶),接收瓶(圆底烧瓶或锥形瓶),分馏柱(韦氏柱),蒸馏头,温度计和温度计套管,冷凝管,接液管,沸石等。

注意:整个温度计水银球的上缘须与分馏柱支管接口下缘相平。确保整个装置没有完全密闭,装置须与大气相通,否则加热时装置会被喷开。控制加热的速度,使温度计显示低沸点组分的沸点。

实验步骤

1. 分馏的一般方法

(1)将待分馏物质装入圆底烧瓶[1],放入 2~3 粒沸石防止暴沸[2],然后依序安装分馏柱、温度计、冷凝管、接液管及接收瓶[3],参见图 2.14。

(2) 接通冷凝水，开始加热，使液体平稳沸腾。当蒸气缓缓上升时，注意控制温度，使馏出速度维持在 2~3 s 一滴[4]。

(3) 用接收瓶收集馏出液。记录第一滴馏出液开始收集时的温度。继续蒸馏直至温度计读数显著上升或下降时，记录馏出液停止收集时的温度。然后根据需要升高加热温度，更换接收瓶开始第二馏分的收集。

(4) 依此类推，收集并记录剩下各馏分的沸程及体积。当蒸馏瓶内剩下 1~2 mL 液体时，移开热源，停止蒸馏。待蒸馏瓶内液体冷却后，拆卸反应装置。

2. 乙醇-水混合物的分馏

(1) 在 100 mL 圆底烧瓶中加入 60 mL 60%乙醇水溶液，加入两粒沸石，安装分馏装置，打开冷凝水。

(2) 加热圆底烧瓶，至瓶内的乙醇溶液沸腾，使乙醇蒸气慢慢升入分馏柱，此时需严格控制加热温度，使蒸气缓慢上升到柱顶。

(3) 当温度计读数达到 78℃时，开始收集馏出液，并保持馏出液的速度为 2~3 s 一滴。

(4) 外界条件不变的情况下继续加热液体，当温度计温度持续下降时，停止加热。

(5) 记录馏出液的馏出温度范围和体积。

(6) 用酒精比重计测量乙醇馏出液的质量分数。

注释

[1] 待蒸馏液体的体积不要超过蒸馏烧瓶体积的 2/3。

[2] 待蒸馏液中须加入沸石，以防止暴沸。**切勿**在加热的液体中直接加入沸石，否则会有暴沸的危险。另外，沸石不能重复使用。

[3] 分馏柱的外围可用玻璃棉包住，减少柱内热量的散发，使柱温基本保持恒定。

[4] 分馏一定要缓慢进行，控制恒定的蒸馏速度(2~3 s 一滴)，这样可以得到较好的分馏效果。

思考题

(1) 分馏和简单蒸馏有哪些异同？如果是两种沸点很接近的液体组成的混合物，能否用分馏提纯？

(2) 若加热太快，用分馏分离液体混合物的效率会显著下降，为什么？

实验 8　柱　色　谱

色谱法是有机分析中最重要的技术之一，它既可以用来分离复杂混合物，又可以用来提纯和鉴定物质，尤其适用于少量物质的分离、纯化和鉴定，其分离效果远比萃取、蒸馏、分馏、重结晶好。

色谱法是一种物理分离方法，待分离的组分分布在两相之间，一相是固定相，另一相是流动相。流动相通过毛细或重力作用经过固定相时，由于与固定相的亲和力强度不同，在固定相中滞留时间不同，与流动相亲和力大的组分流动速度快，而与固定相亲和力大的则流动慢，从而达到分离的目的。

色谱法的种类较多。其中按固定相种类，可分为分配色谱(液体作固定相涂渍在惰性载体上)、吸附色谱(固定相为多孔性吸附剂)、离子交换色谱(固定相为离子交换树脂)和凝胶渗透色谱(固定相为多孔惰性凝胶)等。按流动相状态，可分为气相色谱、液相色谱和超临界流体色谱。按固定相的外形可分为柱色谱和平板色谱，后者又可分为薄层色谱和纸色谱。

实验目的

(1) 学习柱色谱的原理和应用。
(2) 掌握柱色谱分离有机化合物的操作方法。

实验原理

柱色谱是分离和提纯固体和液体最有用的方法之一。它属于固-液色谱，其中固定相是固体，流动相是液体。色谱柱内装有固体吸附剂(固定相)，通常为硅胶(SiO_2)或氧化铝(Al_2O_3)。被提纯样品从柱顶加入，然后从柱顶加入流动相(溶剂)作为洗脱溶剂，并通过重力或外部压力(快速色谱法)向下流过柱。化合物的分离是通过固定相和流动相的不同吸附和相互作用实现的。不同的组分被固定相保留，并以不同的速度通过色谱柱。最后各组分先后随溶剂流出，分别收集，从而实现分离。

主要试剂

甲基橙和亚甲基蓝混合液(1 mg 甲基橙和 5 mg 亚甲基蓝溶解在 2.2 mL 95%乙醇中)，中性氧化铝。

实验装置

柱色谱装置参见图 2.15。

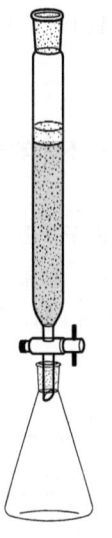

图 2.15 柱色谱装置

实验步骤

1. 装柱

选择一支合适的色谱柱,洗净干燥后垂直固定在铁架台上。用镊子取少许脱脂棉(或玻璃棉)放入色谱柱中,并用长玻璃棒将脱脂棉推到底部,轻轻压紧,再在脱脂棉上铺一层 0.5 cm 厚的海石砂,用套有橡皮塞的玻璃棒轻敲柱子,使砂子上层水平,关闭活塞。向色谱柱内加入 95%乙醇至柱高的 3/4 处,打开活塞,控制流出速度为 1 滴/s,然后通过一干燥的长颈漏斗慢慢地加入中性氧化铝,边加边轻敲柱身,使其装填均匀紧密,装入量约为柱长的 3/4。再在氧化铝顶部加盖一层水平的 0.5 cm 厚的海石砂[1]。将顶部多余的溶剂放出,当柱内液面降至比海石砂面还高 1 cm 处时,关闭活塞。

2. 上样

用长滴管从柱顶加入 2 mL 样品溶液,用另一干净滴管吸取尽量少的洗脱剂洗涤色谱柱内壁上的样品。打开活塞,使液面降至氧化铝上表面。关闭活塞,塞住色谱柱,静置 2 min。

3. 洗脱

在色谱柱上安装滴液漏斗,用 95%乙醇进行洗脱。打开活塞,按 1 滴/s 的速度流出,用锥形瓶接收洗脱液。洗脱过程中要连续不断地加入洗脱剂,不使氧化铝表面的溶液流干[2]。当保护砂层上方仍有超过 2 cm 的溶剂时,将溶剂加满。

极性大的化合物在吸附剂上的吸附力更强,因而停留时间更长,而极性小的物质流动更快。在本实验中,亚甲基蓝极性小,向下移动快;甲基橙极性大,留在柱的上端。当蓝色带流近出口时,收集蓝色带洗脱液。当蓝色带全部从柱中流出,洗脱液呈无色时,更换接收瓶,加入 10 mL 水为洗脱剂。当甲基橙流近出口时,用接收瓶收集,直至洗脱液无色为止[3]。

注释

[1] 色谱柱应装填得尽可能紧密、均匀,柱中不可有气泡、裂缝或断层,否则会严重影响分离效果。吸附剂顶端要平,否则影响分离效果。

[2] 在整个操作过程中,都要控制溶剂**不能流干**,否则易使色谱柱产生气泡或裂缝,影响分离效果。

[3] 在洗脱过程中,应控制洗脱液的流出速度。如果流速太慢,扩散过程会导致谱带变宽,有时会发生样品分解。如果流速太快,则没有足够的时间达到平衡,化合物快速流出,造成拖尾。

思考题

(1) 实验中先用极性小的洗脱剂再用极性大的洗脱剂洗脱,能否反过来?

(2) 为什么在加样前需要把液面调到与吸附剂表面相平,而吸附剂又不能超出液面?

实验 9　薄 层 色 谱

实验目的

(1) 学习薄层色谱的原理和应用。
(2) 掌握薄层色谱分离有机化合物的操作方法。

实验原理

薄层色谱(TLC)属于分配色谱,是快速分离和定性分析少量物质、监测反应进程的一种技术。最典型的 TLC 是在玻璃板上均匀铺上一薄层吸附剂(固定相),制成薄层板,用毛细管将样品溶液点在起点处,然后在盛有展开剂(流动相)的密闭容器中展开。由于样品中各组分在固定相和流动相中的分配系数不同,样品在两相中不断分配,分布在薄层板的不同位置上。物质在薄层板上的移动距离可用比移值(R_f)表示。

R_f 值是表示被分离物质在薄层板上上升的高度与溶剂上升的高度的比值,按下式计算(参见图 2.16):

$$R_f = \frac{原点中心至斑点（层析点）的距离}{原点中心至溶剂前沿的距离}$$

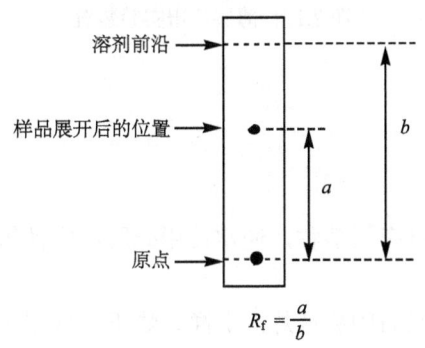

图 2.16　比移值(R_f)的计算

R_f 值与色谱条件有关:溶剂体系、吸附剂的厚度、物料斑点的量和温度等。在这些实验条件固定的情况下,对每一种化合物来说,R_f 值是一个特定数值,因此可通过测定 R_f 值对未知物进行定性鉴定。要鉴定某一具体化合物,应在相同实验条件下,与已知标准物质的 R_f 值进行对比。

薄层色谱一般应用于以下四个方面:
(1) 微量样品的分离和鉴定。
(2) 常量样品的精制。
(3) 监测反应进程。
(4) 为柱色谱摸索最佳分离条件。

主要试剂

硅胶 GF254、0.5%羧甲基纤维素钠溶液(CMC)、环己烷、乙酸乙酯、石油醚、乙醚、乙醇、正丁醇、固体碘、偶氮苯的乙醇饱和溶液、1%间苯二酚的乙醚溶液、1%邻氨基苯酚的乙醚溶液、苏丹Ⅲ的乙醇饱和溶液、1%圆珠笔油的乙醇溶液。

实验装置

薄层色谱实验装置如图 2.17 所示。

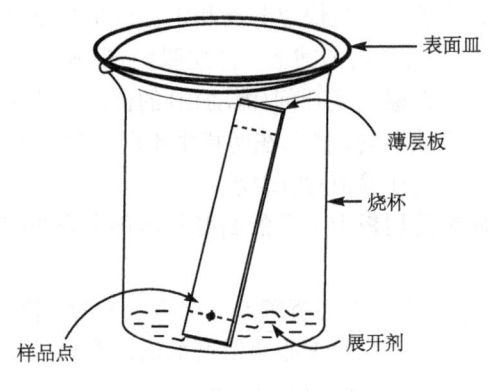

图 2.17　薄层色谱实验装置

实验步骤

1. 薄层板的制备

用于制备薄层板的吸附剂有很多种,通常使用硅胶和中性氧化铝,其中硅胶的应用最为广泛。

(1)取 4 片 7 cm×3 cm 左右的载玻片,洗净、烘干。取用时拿住载玻片的两边,不能触摸载玻片表面。

(2)称取 3 g 硅胶 GF254 于小烧杯中,边搅拌边慢慢加入 8 mL 0.5%羧甲基纤维素钠溶液,调成均匀的糊状。

(3)铺制薄层板的方法有多种,包括倾注法、浸涂法、平铺法和喷涂法。本实验采用倾注法。将调好的糊状硅胶均分到 4 块载玻片上,用玻璃棒抹平,使硅胶在载玻片上铺开。然后将载玻片在桌边上轻轻敲震,使硅胶薄层表面均匀光滑[1]。最后将载玻片放在水平架或台面上,自然晾干 45 min。将晾干的薄层板放入烘箱中进行活化,在 105～110℃下加热 30 min。

2. 点样

如图 2.18 所示,在距薄层板底端 1 cm 处用铅笔轻轻画一条横线作为起始线,并用铅笔轻轻画出所要点样的位置,注意不要破坏薄层板表面。若在同一薄层板上点几个样品,则样品在起始线上的间距为 1～1.5 cm。用毛细管蘸取样品溶液,垂直轻轻地点在薄层板的起始线上。点样后斑点直径不超过 2 mm。如果溶液太稀,一次点样不够,待前一次溶剂挥发后再重新在同一位置点样[2]。

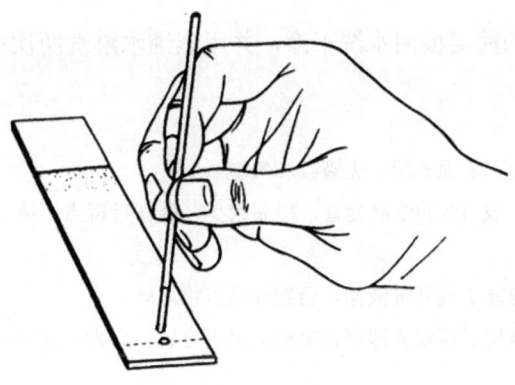

图 2.18 点样

3. 展开

薄层色谱的展开需在密闭容器中进行，可以是带盖的玻璃缸，或在烧杯上盖表面皿代替。先将选择的展开剂倒入展开缸中(液面高度约为 0.5 cm)，使展开缸内溶剂蒸气饱和 5~10 min。再将点好样品的薄层板放入展开缸中，盖好盖子进行展开。注意展开剂的液面不得高于样品点的高度[3]。当展开剂前沿上升至距薄层板上端 1 cm 处时，取出薄层板，并用铅笔尽快标记出展开剂前沿的位置。

一般情况下，选择的展开剂使样品的 R_f 值分布在 0.2~0.8 为宜。

4. 显色

如果薄层板有色斑，用铅笔轻轻地圈起来。大多数样品没有颜色，需要用紫外灯或碘熏法显色。当使用紫外灯时，将灯放在薄层板上，并圈出看到的任何点。注意紫外线会灼伤眼睛和皮肤，因此实验过程中一定要佩戴防护眼镜和手套，且不得直接注视紫外灯。

另一种无色有机化合物在薄层板上显色的方法是将其置于碘蒸气(I_2)中，测试其对碘蒸气的吸收[4]。将带有无色斑点的薄层板放入装有少量碘的密封容器中，约 10 min 后，无色斑点逐渐变为深棕色。

5. 用 TLC 分析混合物的组成

(1) 求出下列物质的 R_f 值。

(i) 偶氮苯的乙醇饱和溶液(展开剂：环己烷 : 乙酸乙酯=9 : 1，体积比)。

(ii) 苏丹Ⅲ的乙醇饱和溶液(展开剂：环己烷 : 乙酸乙酯=9 : 1，体积比)。

(iii) 1%间苯二酚的乙醚溶液(展开剂：石油醚 : 乙醚=1 : 2，体积比；用碘熏法显色)。

(iv) 1%邻氨基苯酚的乙醚溶液(展开剂：石油醚 : 乙醚=1 : 1，体积比；用 254 nm 紫外灯显色)。

(2) 1%圆珠笔油的乙醇溶液(展开剂：正丁醇 : 乙醇 : 水=8 : 5 : 2，体积比)。

6. 结束整理

实验完毕,将用过的薄层板用水洗干净,并用蒸馏水淋洗两次后烘干,备用。

注释

[1] 薄层板要求厚度均匀,表面光滑,无裂纹和气泡。

[2] 样品浓度要适中,一般 1%的浓度为宜。样品太少,展开后斑点不清;样品太多或直径太大,展开时会造成拖尾、扩散现象。

[3] 薄层板点样的位置应高于展开剂液面,否则样品点将溶解。

[4] 碘熏法显色,取出薄层板后要立即用铅笔标记斑点位置,因为碘在空气中挥发后,斑点消失。

思考题

为什么薄层板要活化后才能使用?

实验10 纸 色 谱

纸色谱是一种以滤纸为载体的色谱方法,主要用于分离和检测多官能团或高极性的亲水化合物,如醇类、羟基酸、氨基酸、糖类和黄酮类等。

实验目的

(1)学习纸色谱的原理和应用。
(2)掌握纸色谱分离有机化合物的操作方法。

实验原理

纸色谱的原理比较复杂,一般认为属于分配色谱。它是以纸为载体,固定相为结合于滤纸纤维中的的水,形成液-液分配色谱固定液;流动相为展开缸中被水饱和的有机溶剂。

样品点在滤纸的一端,将该端边缘浸入展开剂。由于纸纤维的毛细作用,展开剂沿纸向另一端移动,并带动样品前进。样品组分在前进过程中在水相和有机相之间分配,利用各组分在两相中的分配系数不同(亲脂性的组分大多分配在流动相中,前进的速度比亲水性的组分快),从而达到分离的目的。

实验用品

仪器:夹子、表面皿、烧杯、玻璃棒、滤纸(6 cm ×15 cm)、电吹风或烘箱、毛细管、喷雾器、尺子、铅笔。

试剂:0.5%脯氨酸水溶液、0.5%亮氨酸水溶液、0.5%茚三酮乙醇溶液、脯氨酸和亮氨酸的混合溶液、展开剂(正丁醇:乙酸:乙醇:水=4:1:1:2,体积比)。

实验步骤

1. 层析滤纸的选择

滤纸应不含杂质，具有适当的厚度、吸附性和保水性。根据不同分离需求可以选择不同的滤纸。快速滤纸适用于大多数分离，当需要较高分离效果时，可以选择慢速滤纸。实验室常用新华 1 号滤纸。本实验将滤纸切成 6 cm×15 cm 的纸条。纸色谱装置参见图 2.19。

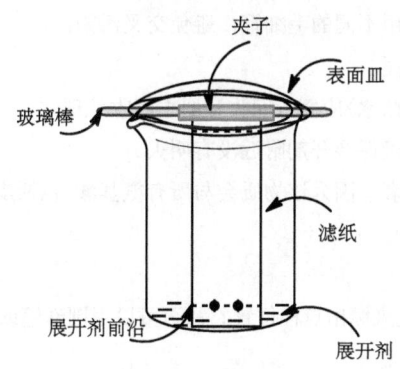

图 2.19　纸色谱装置

2. 点样

用铅笔在滤纸一端距底边 2～3 cm 处轻轻画一条线[1]。在线上标出三个点，各点间距离约为 1.5 cm，并用铅笔标明各点对应样品的代号。用毛细管吸取少量 0.5%脯氨酸水溶液、0.5%亮氨酸水溶液、两种氨基酸的混合溶液，在对应点进行点样[2]。样点必须尽可能小，没有任何扩散，以便更好地分离。最好将样品逐滴点样，用热风(或电吹风)干燥，以蒸发溶剂。干燥后，可再滴一滴并干燥。当另一个样品点样时，一定要更换毛细管。

3. 饱和与展开

将点好样品的滤纸悬吊于装有展开剂的层析缸中，用盖盖好，如图 2.19 所示。点样端向下，将滤纸的样点以下部分垂直浸入展开剂中，展开剂液面在点样点以下约 1 cm 处[3]，用盖盖好，展开约 1 h。

样品展开时间主要取决于被分离物质的性质、展开剂的性质、滤纸的质量和环境温度。为了获得良好的分离效果，R_f 值控制在 0.4～0.8 比较合适。

当溶剂前沿升到接近滤纸顶端时，取出滤纸，立即用铅笔画出溶剂前沿所在位置，吹干[4,5]。

4. 显色

用喷雾器向滤纸均匀喷洒显色剂 0.5%茚三酮乙醇溶液[6]，以滤纸基本打湿为宜。然后用电吹风热风缓缓吹干并加热滤纸，直到显示出紫色斑点为止(或将滤纸放入 90℃烘箱烘干显色)。

5. 测量 R_f 值与鉴定样品

用铅笔将所有斑点的轮廓描出来。分别量出点样点到溶剂前沿和各斑点位置的距离,计算各斑点的 R_f 值。比较各斑点的 R_f 值大小,确定两个斑点各是什么物质。

注释

[1] **不得**用手直接接触滤纸表面,因为皮肤上的油渍会在纸色谱上显色。
[2] 对每种氨基酸和未知物使用不同的毛细管,避免交叉污染。
[3] 样品点必须高于展开剂液面。
[4] 使用过的展开剂(丁醇/乙酸/水)应倒入回收容器中回收处理。
[5] 丁醇是易燃液体,使用时确保展开剂附近没有明火。
[6] 使用茚三酮溶剂时应戴手套,因为该物质会与所有氨基酸(包括皮肤中的氨基酸)形成有色化合物。

思考题

为什么要用铅笔在滤纸上画线来标出点样位置?是否可以用圆珠笔或钢笔标记?

第3章 有机化合物制备实验

实验11 环己烯的制备

实验目的

(1) 掌握环己醇分子内脱水制环己烯的合成方法。
(2) 学习分馏反应、盐析、干燥和分液等基本操作。
(3) 熟悉水浴加热、蒸馏、洗涤等操作。

实验原理

本实验主反应和副反应分别如下。
主反应：

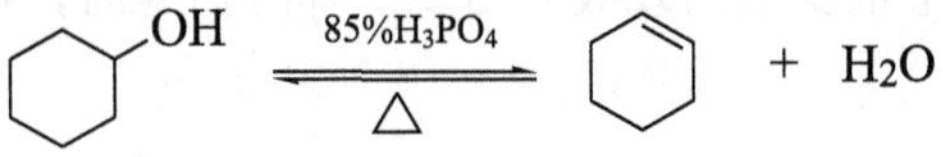

$$\text{C}_6\text{H}_{11}\text{OH} \underset{\triangle}{\overset{85\%\text{H}_3\text{PO}_4}{\rightleftharpoons}} \text{C}_6\text{H}_{10} + \text{H}_2\text{O}$$

副反应：

$$2\,\text{C}_6\text{H}_{11}\text{OH} \underset{\triangle}{\overset{85\%\text{H}_3\text{PO}_4}{\rightleftharpoons}} (\text{C}_6\text{H}_{11})_2\text{O} + \text{H}_2\text{O}$$

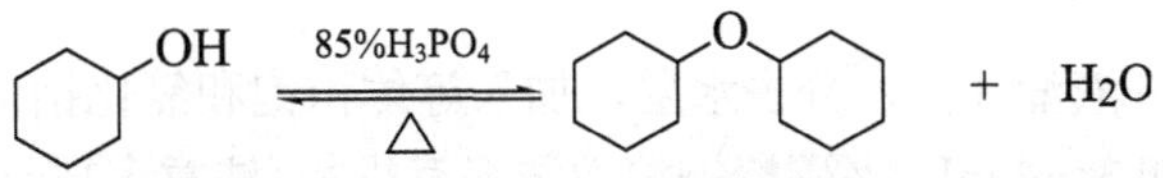

由于该反应可逆，为了促进正向反应转化率，采用在反应的同时蒸出产物环己烯与水的二元共沸物(共沸点70.8℃，含水10%)的方法。但是需要注意，原料环己醇也可与水形成二元共沸物(共沸点97.8℃，含水80%)。为了让产物环己烯和水以共沸物形式蒸出而又不夹带走原料环己醇，本实验采用分馏反应法。通过应用这种方法，连续地从反应体系中除去产物，可以大大提高转化率。控制柱顶端馏出温度不超过90℃，以减少原料损失。

实验用品

仪器：圆底烧瓶、分馏柱、锥形瓶、直形冷凝管、温度计、温度计套管、蒸馏头、接液管、电热套、分液漏斗。

试剂：85%磷酸溶液、5%碳酸钠溶液、饱和氯化钠溶液、无水氯化钙、环己醇。

实验步骤

1. 环己烯的制备

取50 mL干燥的圆底烧瓶，加入10 mL(9.6 g，0.096 mol)环己醇[1]和4 mL 85%磷酸(或1 mL浓硫酸[2])，充分振摇使其混合均匀[3]。再加几粒沸石。搭建分馏反应装置(如图3.1所示)：圆底烧瓶上加装一小段分馏柱，分别接上蒸馏头、温度计、冷凝管、接液管和锥形瓶，并将锥形瓶浸在冷水中冷却。

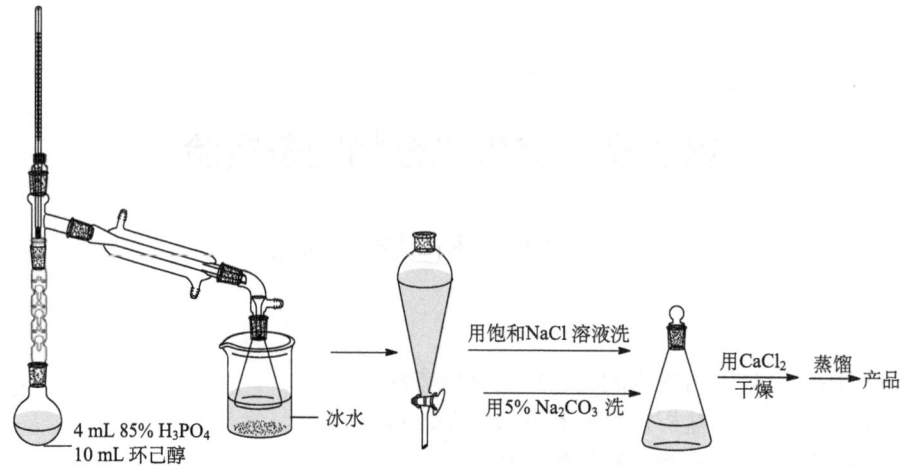

图 3.1　由环己醇制备环己烯的实验流程

开启电热套电源,用小火缓缓加热至沸,控制加热功率使分馏柱顶端的馏出温度不超过 90℃[4],馏出液为浑浊液体(环己烯及少量水的混合物)。当无液体蒸出时,可适当加大火力促使反应完全。当烧瓶内只留下很少量的残液并出现一阵阵白雾时,停止蒸馏[5]。反应时间约需 40 min。

2. 后处理

将馏出液转移至分液漏斗内,分去水层,加入等体积饱和氯化钠溶液充分振摇,静置分层,分去水层。然后加 3～4 mL 5%碳酸钠溶液洗涤有机相(注意多放气)。静置分层,将水相放出(须尽量将水分离完全),有机相转移至干燥锥形瓶内,用适量无水 $CaCl_2$ 干燥[6]。静置溶液,干燥约 15 min。将干燥后液体滤入干燥的圆底烧瓶中,加入几粒沸石后利用水浴加热蒸馏,收集 80～85℃的馏分于干燥的锥形瓶(事先已称量)内。产量为 4～5 g。

环己烯的红外光谱和核磁共振氢谱分别如图 3.2 和图 3.3 所示。

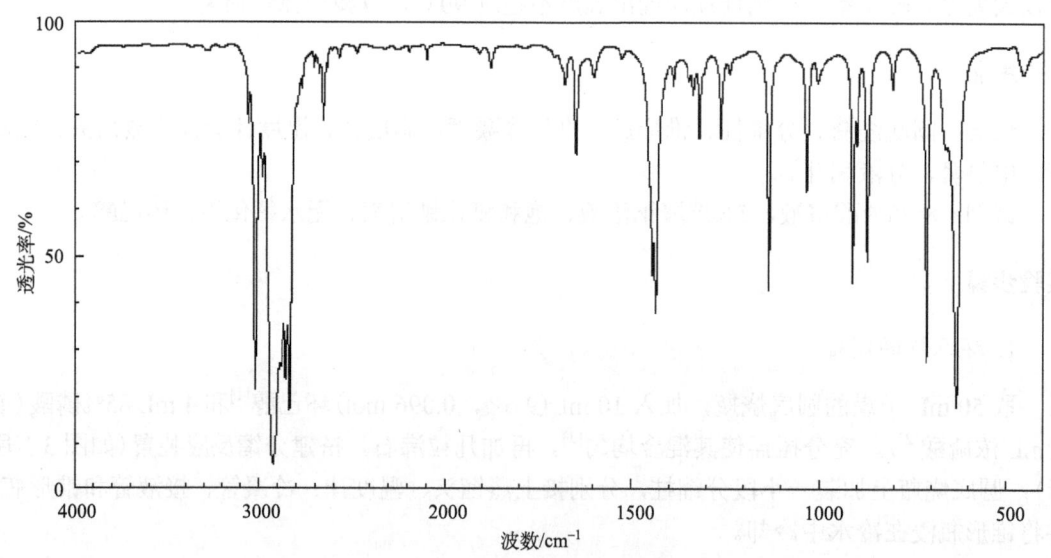

图 3.2　环己烯的红外光谱

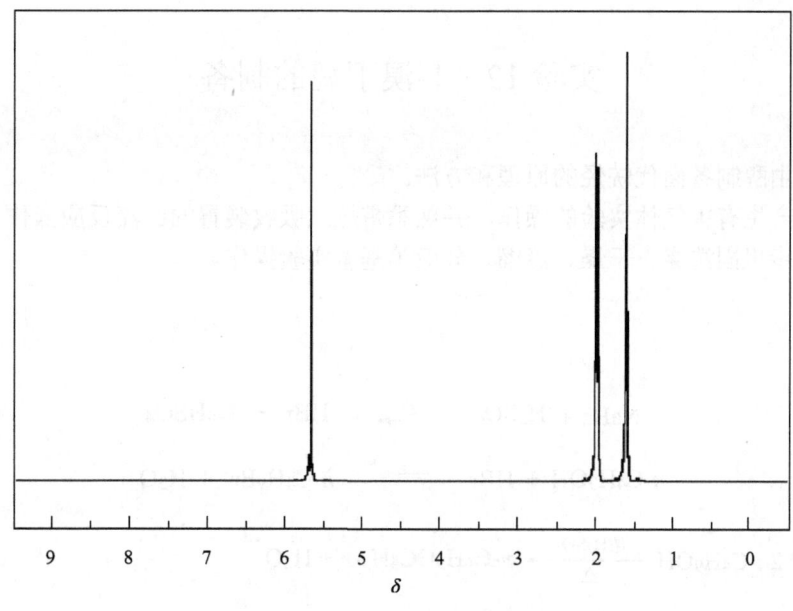

图 3.3　环己烯的核磁共振氢谱

注释

[1] 环己醇在常温下为黏稠液体，因而量取时应特别注意转移过程中的损失。

[2] 本实验也可用 1 mL 95%浓硫酸作催化剂，其余步骤相同。但由于磷酸的氧化能力较硫酸弱得多，可减少氧化等副反应的发生。

[3] 环己醇与磷酸要混合充分，否则加热过程中可能会局部炭化，体系变黑。

[4] 最好采用空气浴或油浴加热使蒸馏烧瓶受热均匀。由于反应过程中环己烯与水形成共沸物，而环己醇也能与水形成共沸物，因此加热温度切不可过高，以免环己醇也被蒸出。有文献报道分馏柱顶端温度控制在 73℃左右可减少原料的损失。但是在该温度下反应速度太慢，时间过长。本实验为加快蒸出速度，可控制温度在 90℃以下。

[5] 反应终点的判断可参考以下几个参数：

(i) 分馏反应进行 40 min 左右。

(ii) 分馏出的环己烯和水的共沸物已达到理论计算的量。

(iii) 烧瓶中出现一阵阵白雾。

(iv) 分馏柱顶端温度下降后又升到 85℃以上。

[6] 干燥前，水层应尽量分离完全，否则无水 $CaCl_2$ 的用量会增加，导致产物吸附损失增加。本实验用无水 $CaCl_2$ 干燥较适合，因它还可除去少量环己醇。无水 $CaCl_2$ 的用量视粗产品中的含水量而定。

思考题

(1) 分馏反应过程中，为什么要控制分馏柱顶端温度不超过 90℃？

(2) 停止蒸馏前，烧瓶内出现的阵阵白雾是什么物质？

(3) 本实验后处理时，为什么要加入氯化钠使水层饱和？

(4) 后处理时，干燥前为什么要将水层尽量分离完全？

实验 12 1-溴丁烷的制备

实验目的

(1) 学习由醇制备卤代烷烃的原理和方法。
(2) 掌握产生有害气体实验的操作,并熟悉带尾气吸收装置的回流反应操作。
(3) 进一步巩固洗涤、干燥、蒸馏、分液等基本实验操作。

实验原理

主反应:

$$NaBr + H_2SO_4 \xrightarrow{\triangle} HBr + NaHSO_4$$

$$n\text{-}C_4H_9OH + HBr \rightleftharpoons n\text{-}C_4H_9Br + H_2O$$

副反应:

$$2n\text{-}C_4H_9OH \xrightarrow[\triangle]{\text{浓}H_2SO_4} n\text{-}C_4H_9OC_4H_9\text{-}n + H_2O$$

$$n\text{-}C_4H_9OH \xrightarrow[\triangle]{\text{浓}H_2SO_4} CH_2=CHCH_2CH_3 + CH_3CH=CHCH_3 + H_2O$$

$$2HBr + H_2SO_4 \longrightarrow Br_2 + SO_2 + 2H_2O$$

实验用品

仪器:带有尾气吸收的回流装置、分液漏斗、75°弯管、锥形瓶、量筒。
试剂:1-丁醇、溴化钠(无水)、浓硫酸、10%碳酸氢钠溶液、无水氯化钙、5%氢氧化钠溶液。

实验步骤

1. 1-溴丁烷的制备

如图 3.4 所示,在 100 mL 圆底烧瓶中先加入 10 mL 水,在不断振荡下慢慢加入 12 mL (0.22 mol)浓硫酸(务必小心!)。冷却至室温后,再依次加入 7.5 mL(0.08 mol)1-丁醇和 10 g (0.10 mol)研细的溴化钠(注意:瓶口内不要沾上固体颗粒)[1],充分振荡后加入几粒沸石。安装带尾气吸收的回流装置,用 5%氢氧化钠溶液作吸收剂(注意:漏斗部分没入水面,部分翘起,以防止倒吸)[2]。在电热套上加热至微沸,反应至固体全溶。然后调节温度保持混合物平稳地回流,反应需 30~40 min。待反应液冷却后,补加 10 mL 水,改为蒸馏装置,利用水蒸气蒸馏蒸出粗产物(注意蒸馏终点判断)。

2. 粗产物的提纯

将馏出液转移至分液漏斗中,加入 10 mL 水洗涤,静置分层后(产物在下层),将产物转入另一干燥的分液漏斗中,再用 5 mL 浓硫酸洗涤,除去粗产物中少量未反应的 1-丁醇及副产物正丁醚、1-丁烯、2-丁烯等。尽量分去硫酸层(下层)后,有机相再依次用 10 mL 水(除硫酸)、10%碳酸氢钠溶液(中和未除尽的硫酸)和水(除去残留的碱)洗涤[3],转移至干燥的锥形瓶中,加入 1~2 g 无水 $CaCl_2$ 干燥[4],间歇摇动锥形瓶,直到液体透明为止。

将干燥彻底的粗产物转移至 50 mL 圆底烧瓶内蒸馏,收集 99~103℃的馏分。

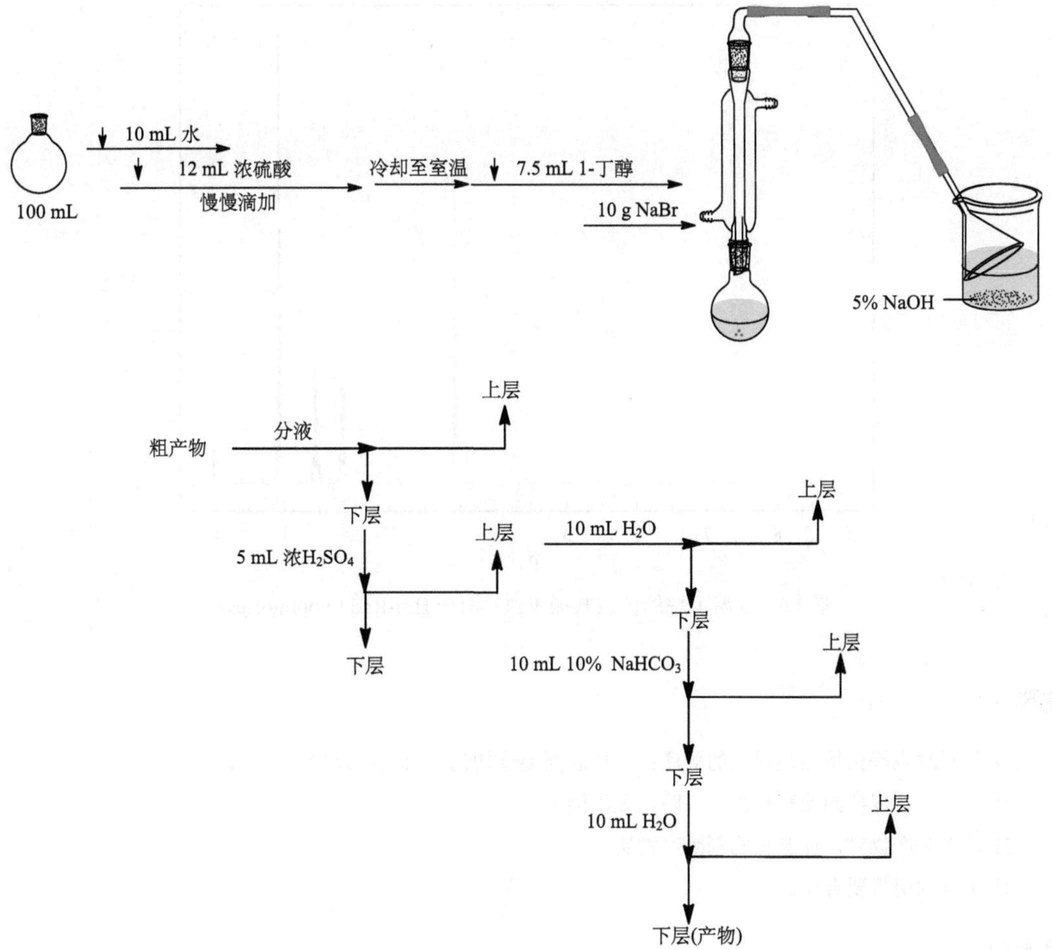

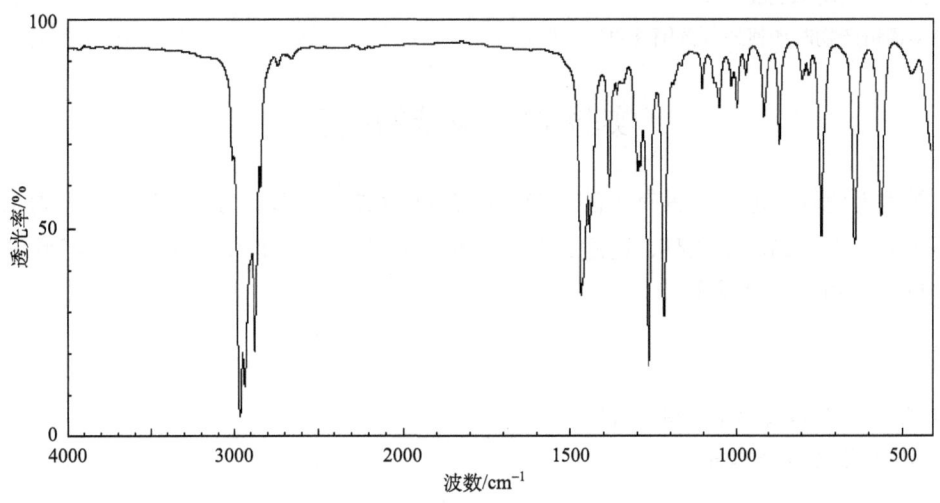

图 3.4　由 1-丁醇制备 1-溴丁烷的实验流程

纯 1-溴丁烷为无色透明液体，沸点为 101.6℃，n_D^{20} 为 1.4401。1-溴丁烷的红外光谱和核磁共振氢谱分别如图 3.5 和图 3.6 所示。

图 3.5　1-溴丁烷的红外光谱

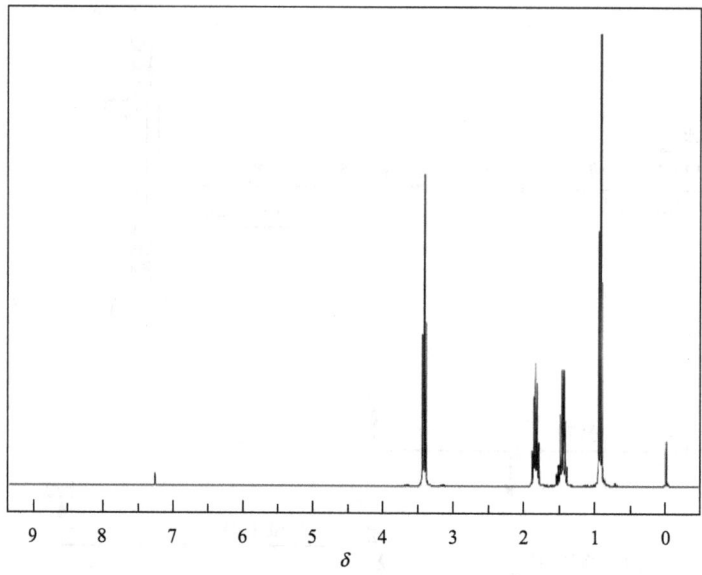

图 3.6　1-溴丁烷的核磁共振氢谱（来源: Bio-Rad Laboratories）

注释

[1] 投料时应严格按照教材上的顺序；投料后要混合均匀，再进行反应。
[2] 反应时，保持回流平稳进行，防止发生倒吸。
[3] 洗涤粗产物时，注意正确判断产物层。
[4] 干燥剂用量要合理。

思考题

(1) 1-溴丁烷的制备实验为什么要采用回流反应装置？
(2) 在 1-溴丁烷制备实验开始时，为什么要用一定浓度的硫酸而不直接用浓硫酸？
(3) 为什么要用浓硫酸洗涤粗产物？
(4) 蒸馏粗产物时如何控制蒸馏终点？

实验 13　溴苯的制备

溴苯为无色油状液体，具有苯的气味，熔点为 –30.7℃，沸点为 156.2℃，n_D^{20} 为 1.5590。该物质不溶于水，易溶于乙醚、丙酮、苯、甲醇、四氯化碳等多数有机溶剂，可用作有机溶剂、分析试剂和有机合成等。

实验目的

(1) 学习制备溴苯的原理和方法。
(2) 巩固蒸馏等基本操作。

实验原理

在铁屑的存在下,苯与溴发生亲电取代反应制得溴苯:

$$\text{C}_6\text{H}_6 + \text{Br}_2 \xrightarrow{\text{Fe}} \text{C}_6\text{H}_5\text{Br} + \text{HBr}$$

在本反应中,真正起催化作用的是三溴化铁。反应对水汽敏感,因为三溴化铁易水解失效。另外,在该体系中,生成的溴苯有可能继续溴代生成二溴苯(主要为对二溴苯)。

$$2\,\text{C}_6\text{H}_5\text{Br} + 2\text{Br}_2 \xrightarrow{\text{Fe}} p\text{-C}_6\text{H}_4\text{Br}_2 + o\text{-C}_6\text{H}_4\text{Br}_2 + 2\text{HBr}$$

苯、溴苯和对二溴苯尽管沸点差别较大,但经一次蒸馏难以分开,因此采用二次蒸馏的方法进一步提纯粗产物。

实验用品

仪器:带有尾气吸收的回流装置、分液漏斗。

试剂:苯、液溴、铁屑、10%氢氧化钠溶液。

实验步骤

1. 溴苯的制备

如图 3.7 所示,在 100 mL 三颈烧瓶上分别安装回流冷凝管和恒压滴液漏斗,冷凝管的顶端加装一套溴化氢气体吸收装置[1]。烧瓶内加入 6 mL(5 g,0.06 mol)无水苯和 0.2 g 铁屑,滴液漏斗中事先盛 2.5 mL(8 g,0.05 mol)液溴(注意:液溴有强挥发性和腐蚀性,转移时务必小心!)[2]。先往瓶内滴入 1~2 滴液溴,反应随即开始(必要时可加热),可观察到有溴化氢气体放出。然后,在搅拌下慢慢滴加其余的液溴,使反应液保持微沸状态。约需 20 min 滴加完毕。调节温度使反应液保持温和回流,直至无溴化氢气体逸出为止,反应过程约需 10 min。

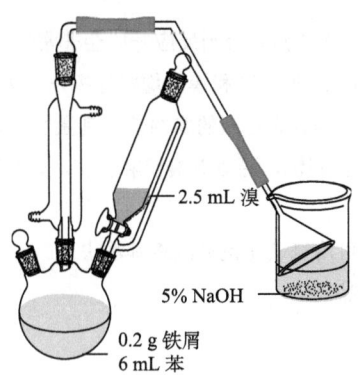

图 3.7 由苯制备溴苯的反应装置

2. 粗产物的提纯

将 15 mL 水加入上述烧瓶内，充分振摇后，将瓶内混合物倾出抽滤。将滤液转移至 50 mL 分液漏斗中，依次用 25 mL 水、6 mL 10% NaOH 溶液[3]、25 mL 水洗涤。将有机相转移至干燥的锥形瓶内，加适量的无水 $CaCl_2$ 干燥后得粗产物。之后，采用蒸馏对粗产物进行提纯。先蒸除混合液中的苯，当温度升至 135℃时，换空气冷凝管，收集其中 140～170℃的馏分，最后将此馏分再重新蒸馏提纯一次，收集 150～160℃的馏分[4]。所得纯品产量为 4～5 g，产率为 55%～60%。本实验需 5～6 h。

纯溴苯为无色透明液体，沸点为 156.2℃，n_D^{20} 为 1.5590，其红外光谱见图 3.8。

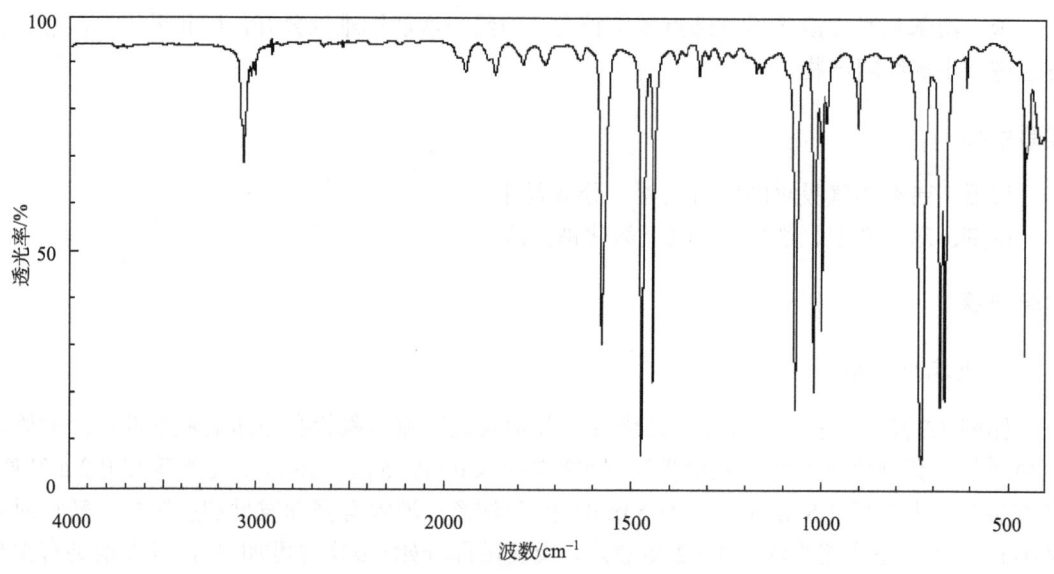

图 3.8 溴苯的红外光谱

注释

[1] 本实验中，所有反应仪器须完全干燥，否则反应开始会非常慢，甚至不能发生。因此，实验开始时，需仔细检查仪器是否干燥。另外，由于液溴、苯和溴苯均极易挥发，需仔细检查仪器的密闭性。

[2] 处理液溴要非常小心，因为液溴具有强烈的腐蚀性和刺激性，量取、转移液溴时必须在通风良好的通风橱内进行，并戴上防护手套和防护眼镜，避免液溴接触皮肤。如不慎触及皮肤，应立即用大量水洗，再用甘油按摩后涂上油膏。

[3] 由于溴在水中溶解度不大，可用氢氧化钠溶液将其洗去。

[4] 也可采用水蒸气蒸馏提纯产品。

思考题

(1) 使用液溴等具有强腐蚀性和刺激性的药品时，应注意什么事项？

(2) 本实验应如何尽量减少副产物对二溴苯的生成？

(3) 本实验为什么要用二次蒸馏进行提纯？

实验 14 2-甲基-2-丁醇的制备

实验目的

(1) 学习并熟悉格氏试剂的制备方法、应用范围及反应条件。
(2) 学习机械搅拌回流实验的操作。
(3) 进一步巩固回流、萃取、蒸馏等基本操作。

实验原理

先利用溴乙烷与镁屑反应制备乙基溴化镁，然后乙基溴化镁与丙酮发生加成反应得 2-甲基-2-丁氧基镁盐，最后在稀盐酸中水解镁盐可制得 2-甲基-2-丁醇。相关反应式如下：

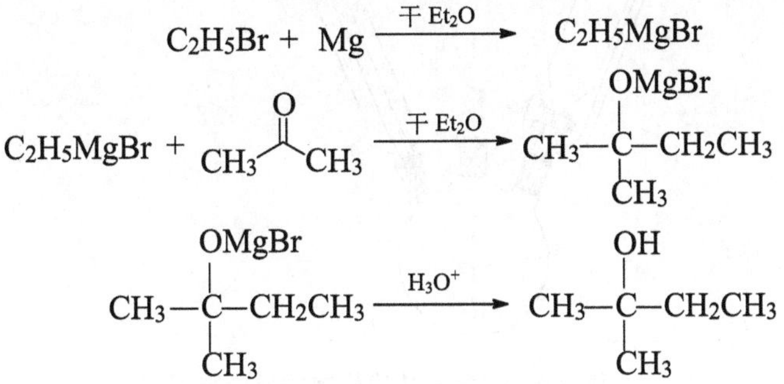

$$C_2H_5Br + Mg \xrightarrow{\text{干} Et_2O} C_2H_5MgBr$$

$$C_2H_5MgBr + CH_3COCH_3 \xrightarrow{\text{干} Et_2O} \underset{\underset{CH_3}{|}}{\overset{\overset{OMgBr}{|}}{CH_3-C-CH_2CH_3}}$$

$$\underset{\underset{CH_3}{|}}{\overset{\overset{OMgBr}{|}}{CH_3-C-CH_2CH_3}} \xrightarrow{H_3O^+} \underset{\underset{CH_3}{|}}{\overset{\overset{OH}{|}}{CH_3-C-CH_2CH_3}}$$

实验用品

仪器：机械搅拌器、恒压滴液漏斗、三颈烧瓶、蒸馏头、球形冷凝管、直形冷凝管、干燥管、接液管、圆底烧瓶。

试剂：镁屑、溴乙烷、丙酮、无水乙醚、5%碳酸钠溶液、20%硫酸、乙醚、无水碳酸钾、无水氯化钙。

实验步骤

1. 乙基溴化镁的制备

如图 3.9 所示，在 100 mL 充分干燥的三颈烧瓶上分别安装机械搅拌器、球形冷凝管和恒压滴液漏斗，并在冷凝管的上口装 $CaCl_2$ 干燥管[1]。在三颈烧瓶内先加入 1.7 g (0.07 mol) 镁屑及一小粒碘。事先在恒压滴液漏斗中盛入 6.5 mL (14.6 g, 0.13 mol) 溴乙烷和 10 mL 无水乙醚，并摇匀。反应开始时，先从滴液漏斗中滴入约 1/3 混合液于三颈烧瓶内，溶液逐渐浑浊并呈微沸状态[2]，随即碘的颜色消失(若不消失，可用温水浴温热)。之后，开动搅拌器，继续滴加剩余混合液，滴加速度以维持反应液呈微沸状态为宜。若发现反应物呈黏稠状态，可补加适量的无水乙醚。滴加完毕后，于温水浴中继续搅拌回流约 30 min，直至镁屑基本消失为止。

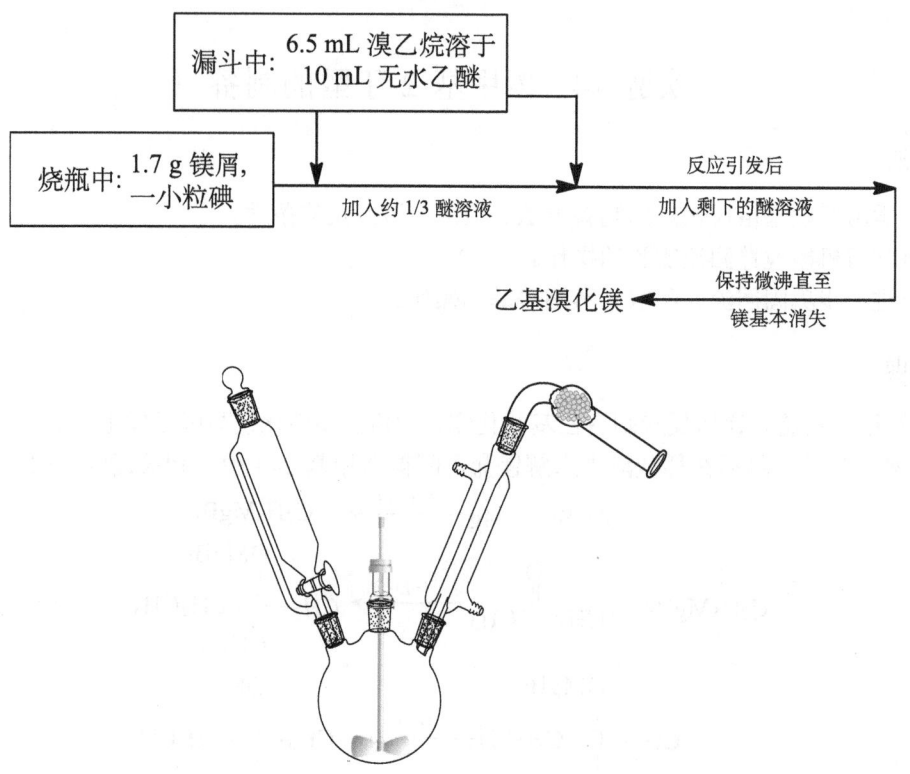

图 3.9 制备乙基溴化镁的反应装置

2. 与丙酮的加成反应

将烧瓶置于冰水浴中,在机械搅拌下自滴液漏斗慢慢滴加 5 mL(3.95 g,0.07 mol)丙酮与 5 mL 乙醚混合液,滴加完毕,于室温下继续搅拌约 15 min,此过程烧瓶内有灰白色黏稠状固体析出。

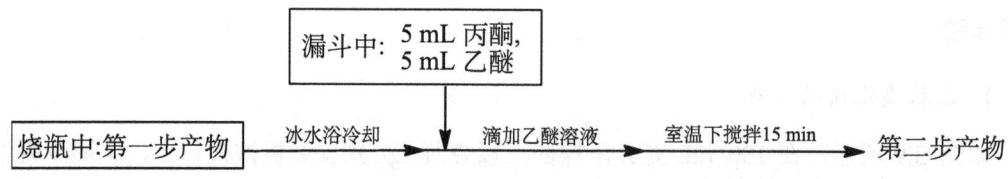

3. 后处理

将烧瓶置于冰水中冷却,搅拌下从滴液漏斗小心滴入 30 mL 20%硫酸溶液(事先配好,于冰水中冷却透彻)水解粗产物。然后分液,收集醚层,水层用乙醚萃取两次(每次 10 mL)。合并醚层,用 8 mL 5%碳酸钠溶液洗涤一次,再用无水 K_2CO_3 干燥[3]。用水浴蒸除乙醚后[4],改为蒸馏装置进行蒸馏,收集 95~105℃馏分。称量,计算产率。

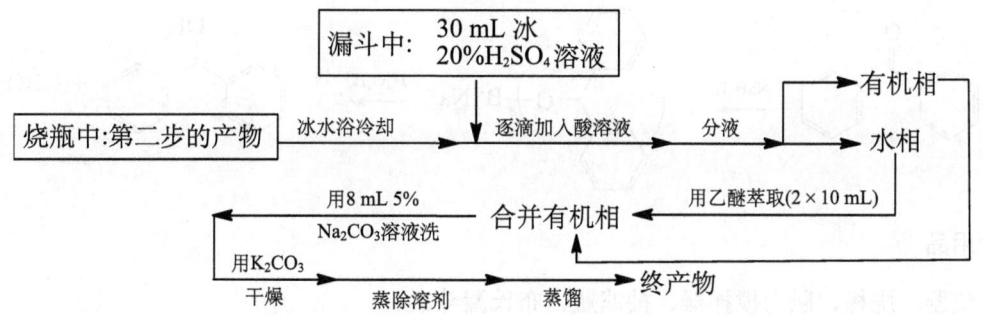

纯 2-甲基-2-丁醇为无色液体，沸点为 102.5℃，n_D^{20} 为 1.4025。

注释

[1] 格氏反应对水高度敏感，因此所用的仪器和药品须经过彻底干燥处理。冷凝管上需接干燥管以隔绝空气中的水分。

[2] 滴加溴乙烷时须控制恰当速度，保持体系微沸状态为宜。

[3] 2-甲基-2-丁醇和水可形成共沸物，采用无水 K_2CO_3 干燥粗产物时一定要充分。

[4] 乙醚易挥发、易燃易爆，仪器接口处需紧密且须远离火源。

思考题

(1) 本实验成败的关键是什么？
(2) 采取哪些措施可以提高本实验的产率？
(3) 为什么加入少量碘可促进格氏反应的引发？
(4) 粗产物为什么用 K_2CO_3 而不用 $CaCl_2$ 干燥？

实验 15 二苯甲醇的制备

实验目的

学习以硼氢化钠还原法由酮制备仲醇的原理和方法。

实验原理

二苯甲醇主要用于有机合成，是一些农药、医药和其他产品的重要中间体。例如，二苯甲醇可作为苯甲托品和苯海拉明的中间体。

还原反应是有机化学中最重要的反应类型之一。醇经常可由相应的醛或酮还原生成。二苯甲酮可以通过多种还原剂得到二苯甲醇，如锌-碱、硼氢化钠、氢化锂铝等。硼氢化钠是负氢试剂，能选择性地还原羰基，且操作方便。反应可以在含水醇中进行。微型实验中，硼氢化钠是十分理想的试剂。1 mol 硼氢化钠理论上能还原 4 mol 醛、酮，但在实际反应中常用过量的硼氢化钠。反应式为

$$\underset{\text{(二苯甲酮)}}{\text{Ph}_2\text{C}=\text{O}} \xrightarrow{\text{NaBH}_4} (\text{Ph}_2\text{CH}-\text{O})_4\text{B}^-\text{Na}^+ \xrightarrow{\text{H}_2\text{O, H}^+} \text{Ph}_2\text{CHOH} + \text{H}_3\text{BO}_3$$

实验用品

仪器：烧杯、磁力搅拌器、抽滤瓶、布氏漏斗。

试剂：二苯甲酮、硼氢化钠、95%乙醇、石油醚(30~60℃)、10%盐酸。

实验步骤

在 200 mL 烧杯中加入 1 g (5.5 mmol) 二苯甲酮、3 mL 水和 20 mL 95%乙醇，搅拌使固体物全部溶解，在搅拌下小心地分批加入 0.4 g (10.5 mmol) 硼氢化钠[1]。此时可观察到有气泡产生，溶液变热。硼氢化钠加入速度以反应温度不超过 50℃ 为宜。待硼氢化钠加毕，继续搅拌 20 min 左右，直到有沉淀物出现[2]，冷至室温后，加入 50 mL 冷水分解过量的硼氢化钠，然后逐滴加入 10%盐酸，直至反应停止。

当反应液冷却后，抽滤，用水洗涤所得固体，干燥后得粗产物。粗产物用石油醚(30~60℃)重结晶，得二苯甲醇针状晶体[3]。纯二苯甲醇熔点为 69℃。

注释

[1] 硼氢化钠具有很强的腐蚀性，不能与皮肤、眼睛和鼻腔接触。

[2] 若无沉淀出现，可在水浴上蒸去大部分乙醇，冷却后将残液倒入 6 g 碎冰和 1 mL 10%盐酸的混合液中。

[3] 也可以用己烷代替石油醚进行重结晶。

思考题

(1) 硼氢化钠和氢化锂铝都是负氢还原剂，试说明它们在还原性及操作上有何不同。

(2) 试提出合成二苯甲醇的其他方法。

(3) 反应完成后，为什么要加入 10%盐酸？

实验 16 1,1'-联二萘酚的合成（文献实验）

实验目的

(1) 学习英文文献的查阅方法，按照文献合成目标产物。

(2) 学习并掌握固相研磨合成方法。

(3) 学习合成简单的手性化合物，并了解手性配体的性质。

实验原理

有的反应可在无溶剂存在的环境下进行，而且相对于溶液中的反应能耗更低、效果更好、选择性更高。同时，因为不需要溶剂，反应对环境友好。

本实验通过固相研磨合成方法促使 β-萘酚在三氯化铁存在下发生氧化偶联反应,合成 1,1′-联二萘酚(BINOL)的外消旋体。

$$2\ \text{C}_{10}\text{H}_7\text{OH} + 2\text{FeCl}_3 \cdot 6\text{H}_2\text{O} \xrightarrow{\text{固相研磨}} \text{BINOL (R+S)}$$

实验用品

仪器:烘箱、研钵、研杵、水浴锅、抽滤装置。

试剂:β-萘酚、六水合三氯化铁、5%盐酸、乙醇。

实验步骤

1. 固相研磨反应

将 1.0 g β-萘酚与 3.8 g 六水合三氯化铁置于研钵内混合均匀,并充分研磨足够长的时间(约 0.5 h)[1,2]。然后将研钵连同混合物置于烘箱中加热至 50℃,并保温使继续反应约 2 h。

2. 后处理

取出研钵,冷却至接近室温。在搅拌下加入 5%盐酸溶液水解粗产物。冷却,过滤,滤饼用少量蒸馏水洗涤三次,以除去残留的铁离子或亚铁离子。充分抽干粗产物。该粗产物可用乙醇重结晶,得白色晶体。纯 1,1′-联二萘酚的熔点为 214~217℃。

注释

[1] 固相研磨应尽量充分,使固体颗粒尽量小,增大原料颗粒的比表面,以有利反应。

[2] 三氯化铁极易吸水,研磨时会出现潮解,体系出现糊状。这不影响实验结果,实验可继续进行。

思考题

(1) 固相有机合成相对于传统有机溶剂中的合成有何优点?

(2) 本实验中,三氯化铁起什么作用?用三氯化铁有何优势?

(3) 固体反应混合物为什么要研磨充分?

(4) 反应后为什么要用稀 HCl 处理混合物?

参考文献

Toda F, Tanaka K, Iwata S. 1989. Oxidative coupling reaction of phenols with FeCl$_3$ in solid state. J Org Chem, 54: 3007-3009.

实验 17 正丁醚的制备

实验目的

(1) 掌握醇分子间脱水制备醚的原理和方法。

(2) 学习带分水器的回流操作。

实验原理

主反应:

$$2CH_3CH_2CH_2CH_2OH \xrightarrow[134\sim135℃]{浓H_2SO_4} CH_3CH_2CH_2CH_2OCH_2CH_2CH_2CH_3 + H_2O$$

副反应:

$$CH_3CH_2CH_2CH_2OH \xrightarrow[>135℃]{浓H_2SO_4} CH_3CH_2CH=CH_2 + CH_3CH=CHCH_3 + H_2O$$

为了促使平衡右移、提高转化率,用分水器把回流反应生成的水不断从体系中分离出来。

实验用品

仪器:二颈烧瓶、球形冷凝管、分水器、温度计、分液漏斗、圆底烧瓶。

试剂:1-丁醇、沸石、浓硫酸、饱和氯化钠溶液、5%氢氧化钠溶液、饱和氯化钙溶液、无水氯化钙。

实验步骤

1. 正丁醚的制备

在 50 mL 二颈烧瓶中加入 15.5 mL(12.5 g,0.17 mol)1-丁醇、2.5 mL 浓硫酸和几粒沸石。摇匀上述混合物后,搭建分水回流反应装置(参见图 3.10)。在分水器内装满水,直到水位刚好到达支管,然后从分水器底部放出 2.0 mL 水[1]。烧瓶一口中插入球形冷凝管,另一口中插入温度计(其水银球浸入反应液面)。然后,将混合物加热至沸腾,并在温和回流下继续加热。反应中产生的水经冷凝后收集在分水器的下层,蒸气中含有反应溶剂,水从反应瓶中向上进入冷凝器冷凝。不混溶的液体分成两层(下面是水,上面是溶剂)。当上层有机相积至分水器支管时,即可流回烧瓶继续反应,而水层留在分水器中。继续加热直到反应液温度升至 134~136℃[2]。当分水器全部被水充满时停止加热。若继续加热,则反应液变黑并有较多副产物生成。反应约需 1 h。

2. 后处理

将反应液冷却到室温,转移至事先盛有 25 mL 水的分液漏斗中,充分振摇,静置分层,弃去下层水相。上层的粗产物依次用 10 mL 水、8 mL 5%氢氧化钠溶液[3]、8 mL 水和 8 mL 饱和氯化钙溶液洗涤,然后用 1~2 g 无水 $CaCl_2$ 干燥。干燥后的产物倾入 50 mL 干燥圆底烧瓶中进行蒸馏提纯,收集 140~144℃馏分。纯正丁醚为无色液体,沸点为 142.4℃,n_D^{20} 为 1.3992。

注释

[1] 本实验反应体系产生的水理论体积约为 1.5 mL,但实际分出水的体积应略大于理论值,否则产率偏低。

[2] 制备正丁醚的较适宜温度是 130~140℃。但开始回流时,这个温度很难达到,因为正丁醚可与水形成共沸物(沸点 94.1℃,含水 33.4%)。另外,正丁醚与水及 1-丁醇形成三元共沸物(沸点 90.6℃,含水 29.9%,

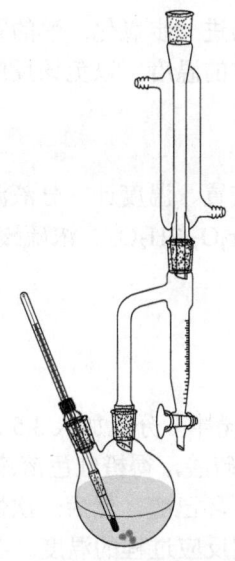

图 3.10 制备正丁醚的反应装置

1-丁醇 34.6%)，正丁醇也可与水形成共沸物(沸点 93℃，含水 44.5%)，故在 100～115℃反应约 0.5 h 后温度才能达到 130℃以上。

[3] 用氢氧化钠溶液洗的过程中，不可过于剧烈地振荡分液漏斗，否则容易乳化，造成分离困难。

思考题

(1) 采取哪些措施能提高本实验的产率？
(2) 为什么分水器中要事先加一定量的水而不是先全部空着或者加满？
(3) 能否在酸催化下由乙醇和异丙醇制备乙基异丙基醚？如果不行，应采用什么方法？

实验 18　环己酮的制备

实验目的

(1) 学习采用铬酸氧化醇制备酮的原理和方法。
(2) 了解氧化反应的操作技术。
(3) 学习反应温度及速率的控制方法。

实验原理

酮是一类重要的有机合成原料。仲醇的氧化或脱氢是合成酮的重要方法。工业上多采用催化氧化或催化脱氢法，在银、铜、铜-铬合金等过渡金属催化下，相应的仲醇在较高的温度(250～350℃)下合成酮。实验室一般采用化学氧化法合成酮。其中酸性重铬酸钠(钾)是常用的氧化剂。本实验用重铬酸氧化环己醇合成环己酮。反应式如下：

$$\text{环己醇} \xrightarrow[50\sim55°C]{Na_2Cr_2O_7, 浓H_2SO_4} \text{环己酮}$$

酮对氧化剂较稳定,一般不容易进一步氧化。醇的氧化是放热反应,但又需一定温度以使反应完全,因此须严格控制反应时的温度,以免该反应过于剧烈。

实验用品

仪器:烧杯、圆底烧瓶、蒸馏装置、温度计、分液漏斗。

试剂:环己醇、重铬酸钠($Na_2Cr_2O_7 \cdot 2H_2O$)、浓硫酸、乙醚、精盐、无水硫酸镁。

实验步骤

1. 环己酮的制备

取 50 mL 烧杯加 20 mL 水,在搅拌下分批加入 3.5 g 重铬酸钠,搅拌至完全溶解。然后边搅拌边用注射器慢慢加入 3 mL 浓硫酸,得橙红色溶液,冷却至 30℃以下备用。

在 50 mL 圆底烧瓶中加 3.5 mL 环己醇,然后一次性加入上述制好的铬酸溶液,振摇使反应液充分混合均匀。用温度计监测反应过程的温度。当温度上升到 55℃时,立即用冷水浴冷却。保持反应温度为 50~55℃[1]。约 15 min 后,移去冷水浴,于室温下放置 15 min。期间间歇振摇反应瓶,促进反应完全。体系最后显墨绿色。

2. 后处理

加 20 mL 水和几粒沸石于上述烧瓶中,改为简单蒸馏装置(实际为简易水蒸气蒸馏)。将环己酮随水一起蒸出,环己酮与水形成共沸混合物(沸点为 95℃),直至无油滴蒸出为止,收集约 15 mL 馏出液。

馏出液用约 3 g 精盐饱和[2],然后转移至分液漏斗内,静置分层,分离出有机层。水层用 5 mL 乙醚萃取一次,合并有机层,用无水 $MgSO_4$ 干燥。清液过滤至圆底烧瓶内,加沸石,于水浴上蒸除乙醚后,蒸馏收集 151~155℃的馏分。纯环己酮的沸点为 155.7℃。

注释

[1] 本实验是放热反应,必须严格控制反应温度。

[2] 加入精盐是起盐析作用,降低环己酮在水中的溶解度,从而有利于分层。水馏出量不宜过多,否则盐析后仍会有少量环己酮因溶于水而损失。31℃时环己酮在水中的溶解度为 2.4 g。

思考题

(1) 环己醇用铬酸氧化得到环己酮,用高锰酸钾氧化则得到己二酸,试解释原因。

(2) 本实验中温度过高或过低有什么不好?可能有哪些副产物?写出有关反应方程式。

(3) 为什么后处理时,馏出液中先要加盐处理?

实验 19 苄叉丙酮和二苄叉丙酮的制备

实验目的

(1) 学习采用羟醛缩合增长碳链的原理和方法。

(2) 学习利用原料的投料比控制产物的方法。

实验原理

羟醛缩合是指两分子具有活泼 α-氢的醛/酮在稀酸或稀碱的催化下发生分子间缩合生成 β-羟基醛/酮；或者进一步脱水生成 α,β-不饱和羰基化合物。羟醛缩合是制备 α,β-不饱和羰基化合物的重要方法，也是有机合成中增长碳链的重要策略。

羟醛缩合分为自身羟醛缩合和交叉羟醛缩合。无 α-氢的芳醛与含 α-活泼氢的醛/酮之间的交叉羟醛缩合称为克莱森-施密特反应。该反应是合成侧链上含两种官能团的芳香族化合物及含几个芳环的脂肪族体系中间体的重要方法。在苯甲醛和丙酮的交叉羟醛缩合中，通过改变反应物的投料比可得到两种不同的产物。

若苯甲醛过量，反应式为

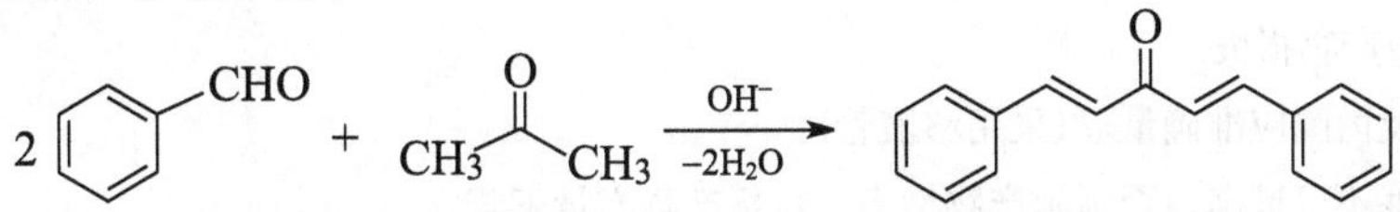

$$2\,C_6H_5CHO + CH_3COCH_3 \xrightarrow[-2H_2O]{OH^-} C_6H_5CH=CHCOCH=CHC_6H_5$$

若丙酮过量，反应式为

$$C_6H_5CHO + CH_3COCH_3 \xrightarrow[-H_2O]{OH^-} C_6H_5CH=CHCOCH_3$$

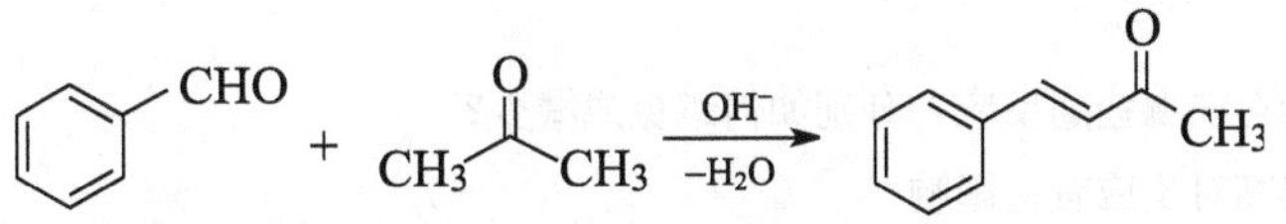

苄叉丙酮(又称 5-苯基-3-丁烯-2-酮)具有类香豆素的气味，可作为合成香料的原料和花香香精的变调剂、染料工业的媒染剂及电镀工业的光亮剂。

实验用品

仪器：圆底烧瓶、布氏漏斗、抽滤瓶、磁力搅拌器、锥形瓶。

试剂：苯甲醛、丙酮、95%乙醇、无水乙醇、冰醋酸、10%氢氧化钠溶液、乙醚、1∶1盐酸、饱和氯化钠溶液、无水硫酸镁。

实验步骤

1. 二苄叉丙酮的制备

在 250 mL 圆底烧瓶中加搅拌磁子。在磁力搅拌下，将 5.3 mL(0.05 mol)新蒸苯甲醛、1.8 mL(0.025 mol)丙酮、40 mL 95%乙醇和 50 mL 10%氢氧化钠溶液依次加入圆底烧瓶中。继续室温搅拌 20 min 后，停止反应。

将上述反应混合物抽滤，用少量水洗涤固体，抽干水分。用 1 mL 冰醋酸和 25 mL 95%乙醇配成的混合液浸泡滤饼，抽滤，再用少量水洗涤一次[1]。将固体转移到 100 mL 锥形瓶中，用无水乙醇进行重结晶[2]。

纯二苄叉丙酮为淡黄色片状晶体，熔点为 110~111℃(113℃时分解)。

2. 苄叉丙酮的制备

在 100 mL 三颈烧瓶中加搅拌磁子，并分别装上滴液漏斗、球形冷凝管和温度计。在磁力搅拌下依次加入 22.5 mL 10%氢氧化钠溶液和 4 mL(0.054 mL)丙酮[3]。然后从滴液漏斗逐滴加入 5.3 mL(0.05 mol)新蒸苯甲醛,控制滴加速度,使反应物温度保持在 25~30℃[4]。滴加完毕继续反应 30 min。

通过滴液漏斗加入适量 1:1 盐酸,直至反应液 pH 为 7.0。用分液漏斗分出黄色油层,水层用乙醚萃取三次(每次 10 mL)。将萃取液与前述油层合并,用 10 mL 氯化钠饱和溶液洗涤,然后用无水硫酸镁干燥,过滤。滤液用水浴蒸馏,回收乙醚,可得产物约 5 g。

纯苄叉丙酮为白色或淡黄色晶体,熔点为 42℃。

注释

[1] 制备二苄叉丙酮时,洗涤、浸泡的步骤均可在布氏漏斗上进行。

[2] 制备二苄叉丙酮时,若溶液颜色呈棕红色,可加入少量活性炭脱色。重结晶过滤时,温度要降低至 0℃左右,以减少产物损失。

[3] 苯甲醛及丙酮应准确量取(采用移液管)。

[4] 反应温度不可过高,否则副产物增多,目标产物产量下降。

思考题

(1) 本实验中可能存在哪些副反应?分别如何避免或减少?

(2) 若碱的浓度偏高对反应有何影响?

(3) 若反应所得产品显示红棕色,是什么原因?遇到这种情况应如何处理?

实验 20　安息香的制备

实验目的

(1) 了解并掌握安息香缩合的原理。

(2) 学习并掌握采用维生素 B_1 作催化剂进行安息香缩合的实验操作。

(3) 进一步巩固冷却、回流、重结晶等操作。

实验原理

安息香缩合反应是两个醛之间发生的偶合反应,生成安息香的母体。传统方法采用氰离子催化,需要使用剧毒的氰化钠。

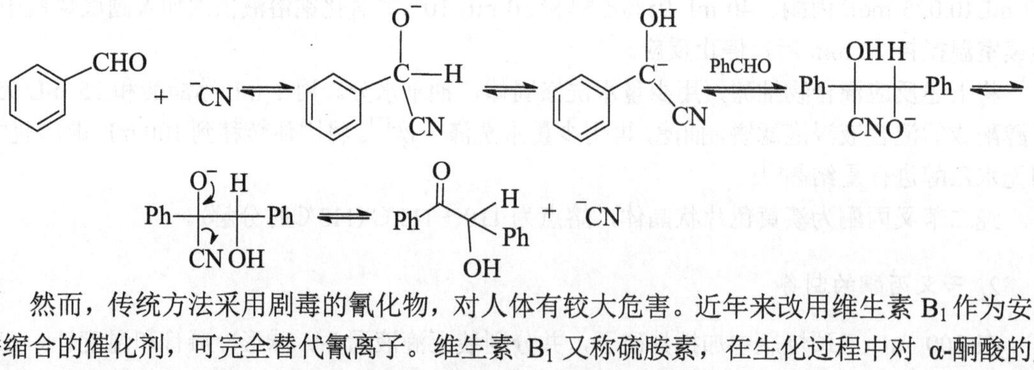

然而,传统方法采用剧毒的氰化物,对人体有较大危害。近年来改用维生素 B_1 作为安息香缩合的催化剂,可完全替代氰离子。维生素 B_1 又称硫胺素,在生化过程中对 α-酮酸的脱羧和生成偶姻(α-羟基酮)等酶促反应发挥辅酶的作用。维生素 B_1 的结构如下:

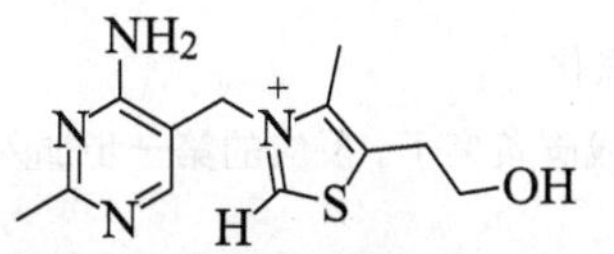

维生素 B_1 分子噻唑环上的 S 和 N 之间的氢原子有较强的酸性,在合适碱的作用下可生成碳负离子,从而催化安息香缩合反应,机理如下:

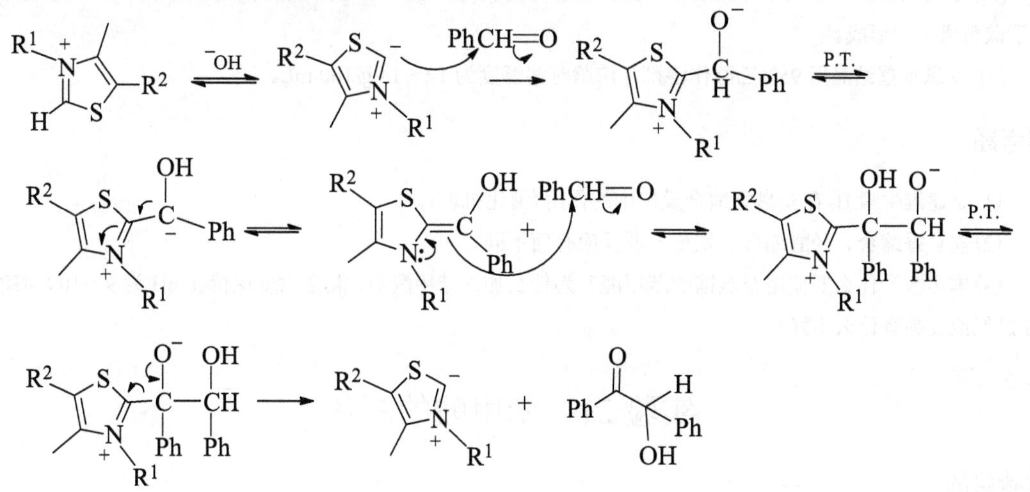

实验用品

仪器:圆底烧瓶、冷凝管、布氏漏斗、抽滤瓶、温度计、熔点仪。

试剂:维生素 B_1、蒸馏水、新蒸苯甲醛、95%乙醇、10%氢氧化钠溶液。

实验步骤

取 50 mL 圆底烧瓶,加入 1.0 g 维生素 B_1[1]、2 mL 蒸馏水和 8 mL 95%乙醇。不时振摇烧瓶使维生素 B_1 溶解。然后塞紧瓶口,将烧瓶放在冰水浴中冷却,同时取 2 mL 10%氢氧化钠溶液于一支试管中并在冰水浴中冷却[2]。在不断摇动下,将冷却的 10%氢氧化钠溶液逐滴滴入烧瓶内。调节 pH 至 9~10(此时溶液呈黄色)。撤去冰水浴,滴入 5 mL(5.2 g,0.05 mol)新蒸苯甲醛[3],加入几粒沸石,装上冷凝管,将混合物置于 60~75℃水浴中温热约 1.5 h(后期温度可升到 80~90℃)。期间注意适当摇动反应瓶并保持 pH 为 9~10(必要时可适当补加 10%氢氧化钠溶液以维持该 pH 范围)[4]。此时,反应混合物呈橘黄或橘红色均相溶液。移去水浴,待烧瓶冷却至室温后放在冰水浴中使其结晶析出。抽滤并用 20 mL 冷水洗涤两次,干燥,可得淡黄色安息香粗产物。

粗产物用 95%乙醇重结晶(必要时可加活性炭脱色)[5],干燥,称量(产量约 2 g),用熔点仪测熔点。

纯安息香为白色针状晶体,熔点为 137℃。

注释

[1] 维生素 B_1 在反应中作为催化剂使用,其质量对反应产生直接的影响。维生素 B_1 通常在酸性条件下稳定,在碱性溶液中噻唑环易开环失效;易吸水,在水溶液中易被氧化而失效;同时光和某些金属离子(如

[2] 氢氧化钠能促进维生素 B_1 形成碳负离子。反应的第一步加入冰冷的氢氧化钠，其目的是防止噻唑环发生开环反应。

[3] 苯甲醛易被空气氧化而含有苯甲酸，所以反应应该用新蒸馏的苯甲醛。

[4] 投完原料后，必须保持反应体系呈弱碱性，调节 pH 为 9~10（采用精密 pH 试纸）。因为碱性条件有利于碳负离子的形成。

[5] 安息香重结晶用 95%乙醇作溶剂，沸腾时溶解度为 12~14 g/100 mL。

思考题

(1) 试述维生素 B_1 在安息香缩合反应中的作用（催化机理）。

(2) 安息香缩合、羟醛缩合、坎尼扎罗反应有何不同？

(3) 本实验为什么要使用新蒸馏的苯甲醛？为什么加入苯甲醛后，混合物要保持在 pH 为 9~10？溶液的 pH 过低或过高有什么不好？

实验 21　苯甲酸的制备

实验目的

(1) 了解和学习甲苯光促卤代反应制备苯甲酸的原理和方法。

(2) 巩固回流、抽滤、干燥等基本操作。

实验原理

本实验采用甲苯在光照条件下溴化，再经水解和酸化，合成苯甲酸。

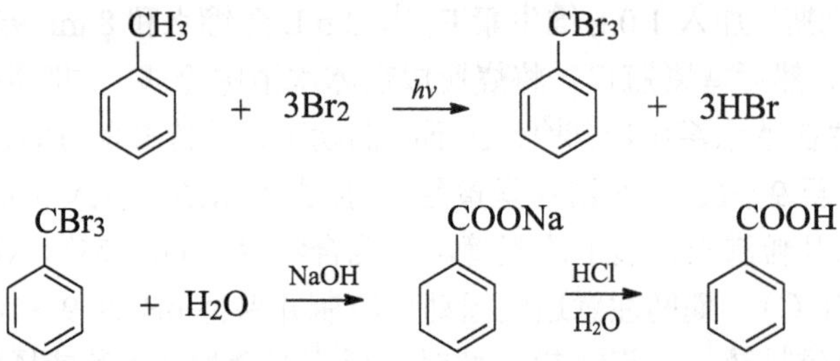

$$C_6H_5CH_3 + 3Br_2 \xrightarrow{h\nu} C_6H_5CBr_3 + 3HBr$$

$$C_6H_5CBr_3 + H_2O \xrightarrow{NaOH} C_6H_5COONa \xrightarrow[H_2O]{HCl} C_6H_5COOH$$

实验用品

仪器：紫外灯、磁力搅拌器、红外灯、恒压滴液漏斗、三颈烧瓶、塑料盆、橡皮管、玻璃塞、接收瓶。

试剂：甲苯、溴、沸石、20% NaOH、10% HCl。

实验步骤

1. 光溴化反应合成卤代中间体

如图 3.11 所示，在一个已用油漆涂黑的盒子或用金属箔覆盖窗子的通风橱中搭建装置。

在 500 mL 三颈烧瓶中放入搅拌磁子。然后在烧瓶内加入 2 mL(0.0188 mol)甲苯。开动磁力搅拌，并开启水泵[1]。

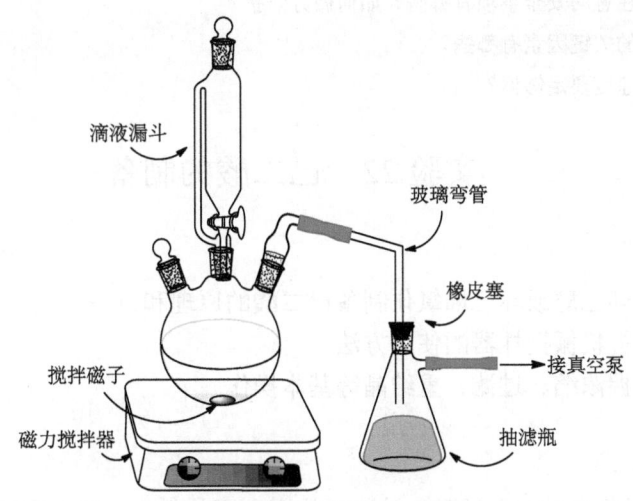

图 3.11　光促反应合成苯甲酸的实验装置

在滴液漏斗中小心加入 3 mL(0.0585 mol)溴[2]，保持滴加速度为每 4～5 秒一滴。将紫外灯尽量靠近烧瓶，并照在烧瓶中的甲苯上[3]。溴加完后，继续光照反应直至溶液呈淡黄色为止。关闭紫外灯。

2. 碱性水解反应

移去烧瓶上的滴液漏斗与水泵的连接。在烧瓶内加入 3 粒沸石及 20 mL 20% NaOH 溶液，混匀。用电热套加热混合物，回流反应约 2 h，并不时振摇反应瓶。然后，将烧瓶冷却至室温，补加 20 mL 水。将反应混合物抽滤，所得滤液备用。

3. 酸化

将上述滤液一边搅拌一边滴加 10% HCl 溶液，直至使蓝色石蕊试纸显酸性。抽滤，压干，收集结晶，放入 80℃烘箱中进行干燥。

采用微焰灯作热源，用升华法进一步提纯苯甲酸。把产物放在预先称量的小瓶内，计算产率。

注释

[1] 为了促进反应进行，需采用水泵移除反应中生成的溴化氢。

[2] 溴有强腐蚀性和刺激性，操作时务必小心。戴上橡胶手套，并在通风橱中操作。不可将液溴洒出、直接接触和吸入。

[3] **不要**直视紫外灯。该反应只能在漆黑的盒子或覆有金属箔的窗子的通风橱中进行。

思考题

(1) 本实验需要注意的安全事项有哪些？如何做好防护？
(2) 做好本实验的关键因素有哪些？
(3) 为什么加溴速度要足够慢？

实验 22　己二酸的制备

实验目的

(1) 掌握利用环己醇或环己烯氧化制备己二酸的原理和方法。
(2) 学习并掌握机械搅拌器的使用方法。
(3) 进一步掌握浓缩、过滤、重结晶等基本操作。

实验原理

己二酸是一种重要的二元羧酸。该物质为白色固体粉末，略带酸味，微溶于水，易溶于乙醇、甲醇、异丙醇和醚等。工业上利用己二酸和己二胺的缩聚反应制备尼龙-66。己二酸也是制备聚氨酯、热熔胶等的中间体。己二酸还可与醇类反应制备己二酸酯，后者可用作制备聚氯乙烯及氯乙烯共聚物、聚苯乙烯、乙基纤维素和硝化纤维素等聚合物的增塑剂。

目前实验教材中通常用 50% HNO_3 或 $KMnO_4$ 氧化环己醇制备己二酸。但是，这些反应均释放出有毒有害的副产物，而且存在反应时间长或后处理复杂等缺点。为了克服这些缺点，降低污染，本实验采取 H_2O_2(30%) 作为氧化剂，并利用 Na_2WO_4/H_3PO_4 为催化剂，对环己烯进行催化氧化制备己二酸。该方法具有条件温和、易于控制等优点，反应时间相对较短，且产物易分离，过程中无毒害性物质产生，是一种较好的绿色合成方法。反应式如下：

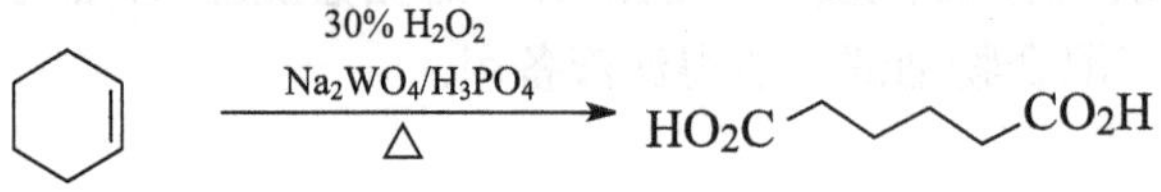

$$\text{环己烯} \xrightarrow[\triangle]{\substack{30\%\ H_2O_2 \\ Na_2WO_4/H_3PO_4}} HO_2C\text{～～～}CO_2H$$

实验用品

仪器：机械搅拌器、三颈烧瓶、烧杯、冷凝管、恒压滴液漏斗、尾气吸收装置、水浴锅、抽滤装置。

试剂：环己醇、50%硝酸、钒酸铵、环己烯、30% H_2O_2、钨酸钠、磷酸。

实验步骤

1. 钒酸铵催化硝酸氧化环己醇

在 100 mL 三颈烧瓶中加入 8 mL (0.085 mol) 50%硝酸和一小粒钒酸铵[1]。三颈烧瓶的三个口分别安装机械搅拌器、恒压滴液漏斗和冷凝管。冷凝管上端接气体吸收装置，用碱液吸收反应中产生的二氧化氮气体。滴液漏斗中加入 2.7 mL (0.025 mol) 环己醇。将三颈烧瓶在水浴中预热到 50℃左右，移去水浴，搅拌下先滴入 5～6 滴环己醇。反应开始后，瓶内反应物温度升高，并有红棕色气体放出[2]。慢慢滴入剩余的环己醇，调节滴加速度[3]，使瓶内反应物保持微沸状态。温度过高或过低时,可借助冷水或热水浴加以调节。滴加完毕(约需 15 min),再用沸水浴加热 10 min,至几乎无红棕色气体逸出为止。将混合物小心倾入外部用冷水冷却的烧杯中,晶体析出完全后,抽滤收集固体,用极少量的冷水洗涤晶体[4]。粗产物干燥后质量为 2~2.5 g,熔点 149~155℃。用水重结晶后,熔点 151~152℃,产量约 2 g。纯己二酸为白色棱状晶体,熔点 153℃。本实验需 3~4 h。

2. 钨酸钠/磷酸催化过氧化氢氧化环己烯

过氧化物的制备:在 100 mL 三颈烧瓶中依次加入搅拌磁子、0.88 g 钨酸钠、0.6 g(约 0.5 mL)磷酸及 22.3 mL 30% H_2O_2。分别在三颈烧瓶上装冷凝管、温度计和恒压滴液漏斗。在 50~60℃下搅拌回流 0.5 h。注意加热温度不可超过 70℃,且搅拌时间不可过久[5]。

己二酸粗产物的制备:打开恒压滴液漏斗缓慢滴加 5 mL(0.049 mol)环己烯,加完后,升温至 102℃搅拌回流 1 h。当溶液显墨绿色时,停止加热[6]。

己二酸的提纯:将上述反应液冷至室温,倾入 100 mL 烧杯中,用 10%盐酸溶液酸化至 pH 为 1~2。加热浓缩至 15 mL。再用冰水浴冷却,抽滤,收集固体,干燥,称量,计算产率。测熔点检测纯度。

注释

[1] 硝酸有强腐蚀性和刺激性,量取、转移硝酸要特别小心。
[2] 要等反应引发成功后,才能加入其余的环己醇。
[3] 环己醇滴加速度以维持体系微沸状态为宜。
[4] 洗涤用的冰水不可太多,否则会有溶解损失。
[5] H_2O_2 在较高温度下分解。
[6] 升温时,因反应初期环己烯-水-H_2O_2 形成共沸物,初期温度会在 70℃停留 20~60 min。当反应温度达到 102℃时,若溶液呈浅灰色不必停止反应,此时 Mo^{6+} 被环己烯还原为 Mo^{5+} 或 Mo^{4+}。但当溶液呈墨绿色时,催化剂活性已有所抑制,此时反应可停止。

思考题

(1) 比较两种己二酸制备方法的优缺点。
(2) 第一个反应中,为什么要用碱液吸收尾气?
(3) 第二个反应中,为什么反应液显墨绿色即可停止反应?

参考文献

任冬梅, 王健, 赵岩, 等. 2016. 己二酸绿色合成实验的教学改进. 科技展望, 01(01): 185.

实验 23 肉桂酸的制备

实验目的

(1) 掌握用珀金反应制备肉桂酸的原理和方法。
(2) 进一步掌握水蒸气蒸馏的原理及操作方法。

实验原理

在碱性催化剂的作用下,无 α-H 的芳香醛(如苯甲醛)与酸酐可以发生缩合反应,称为珀金反应。

珀金反应常用的催化剂有:反应中使用的酸酐相应羧酸的钠盐或钾盐、碳酸钾和叔胺等。催化剂的主要作用是促使酸酐烯醇化,生成酸酐碳负离子。

苯甲醛和乙酸酐在乙酸钾存在下发生珀金反应,生成肉桂酸。

反应式:

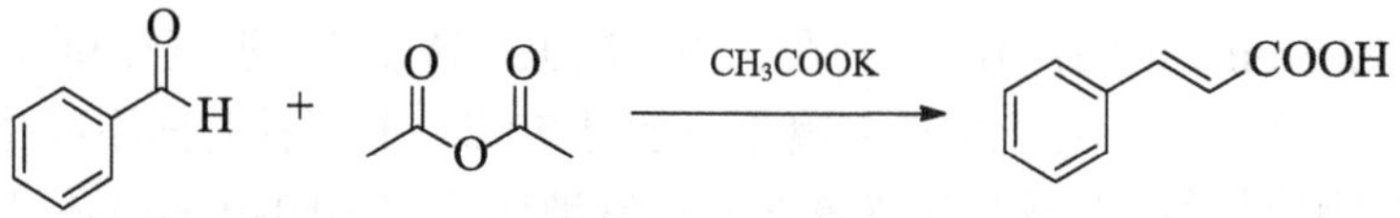

$$PhCHO + (CH_3CO)_2O \xrightarrow{CH_3COOK} PhCH=CHCOOH$$

反应机理:

(反应机理图)

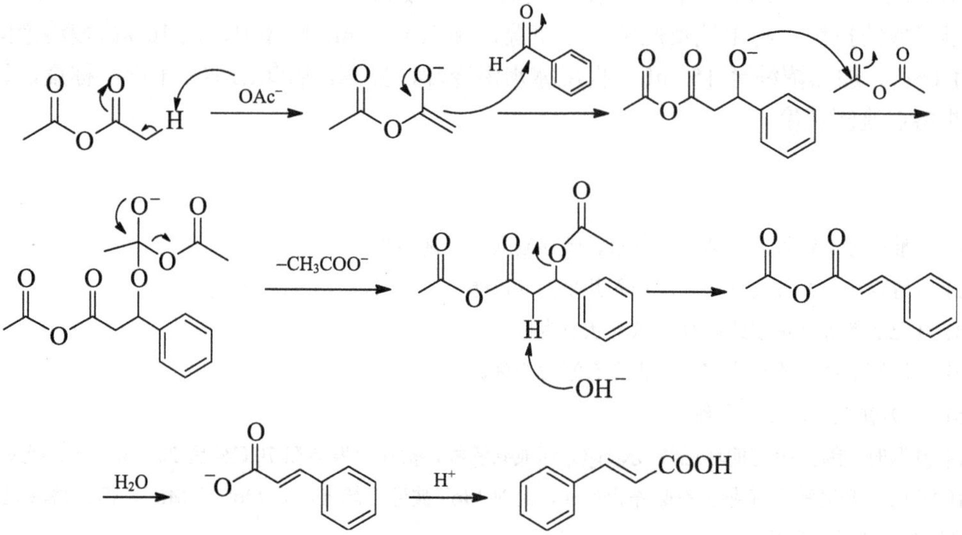

实验用品

仪器:三颈烧瓶、空气冷凝管、烧杯、温度计(250℃)、接液管、锥形瓶、水蒸气蒸馏装置、抽滤瓶、布氏漏斗、电热套。

试剂:新蒸苯甲醛、乙酸酐、无水碳酸钾、10%氢氧化钠溶液、活性炭、浓盐酸、乙醇、刚果红试纸。

反应物和产物的物理性质参见表 3.1。

表 3.1 反应物和产物的物理性质

物质	M/(g/mol)	ρ/(g/mL)	m.p./℃	b.p./℃
PhCHO	106.12	1.04	−26	179
K_2CO_3	138.21	2.428	891	333.6
PhCH=CH—COOH	148.17	1.245	135~136	300

实验步骤

1. 制备

在 100 mL 三颈烧瓶中加入 1.5 mL(0.015 mol)新蒸苯甲醛、4 mL(0.036 mol)乙酸酐和 2.2 g(0.016 mol)研细的无水碳酸钾,振荡混匀。按图 3.12 所示安装实验装置[1]。将混合物加热回流 40 min,温度保持在 150~170℃[2]。待反应液稍冷后(如果有固体析出,需加入 5~10 mL 热水溶解),移去空气冷凝管,改为水蒸气蒸馏装置,蒸馏除去未反应完的苯甲醛,直至馏出液无油珠为止。

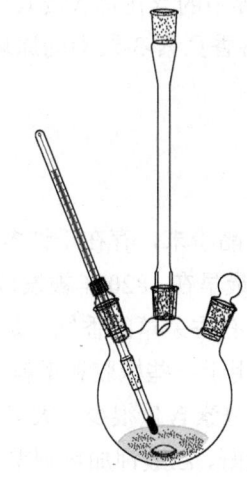

图 3.12 制备肉桂酸的反应装置

2. 提纯

待烧瓶冷却后,在搅拌下加入 10 mL 10%氢氧化钠溶液调至 pH 为 9~10,使肉桂酸成钠盐而溶解。若仍有少量固体不溶,在搅拌下加入适量热水直至固体溶解。如果溶液有颜色,加入少许活性炭,加热煮沸 2~3 min 脱色,趁热过滤。将滤液转入 250 mL 烧杯中,在搅拌下用 1∶1 盐酸酸化至刚果红试纸变蓝[3]。冷却,抽滤,并用少量水洗涤晶体,粗产物可在 30%乙醇中重结晶。在 85℃以下烘干,称量,计算产率。

纯肉桂酸为白色片状晶体,熔点为 135~136℃(反式)。

注释

[1] 由于乙酸酐遇水即水解成乙酸,无水碳酸钾也极易吸潮,因此所用仪器须充分干燥。

[2] 加热回流时保持回流蒸气柱高度为冷凝管的 1/3~1/2,反应时间约为 40 min。

[3] 用盐酸酸化时,要酸化至溶液使试纸明显变色。

思考题

(1) 酸化时能否用浓硫酸？

(2) 试写出丙酸酐与苯甲醛在丙酸钾存在下的反应式。

(3) 本实验用水蒸气蒸馏是为了除去什么？能否用简单蒸馏代替？

实验24 香豆素-3-羧酸的制备

实验目的

(1) 了解香豆素类化合物在自然界中的存在形式及其生物学意义。

(2) 掌握用 Knoevenagel 反应制备香豆素-3-羧酸的原理和方法。

实验原理

1. 简介

香豆素是一种白色斜方晶体或结晶粉末，存在于许多天然植物中，化学名称为 1,2-苯并吡喃酮或邻羟基肉桂酸内酯。香豆素最早在 1820 年被发现并从香豆的种子中获得，它也存在于薰衣草、桂皮的精油中。香豆素具有香茅草的香气，是重要的香料，常用作定香剂，可用于配制香水、花露水、香精等，也应用于一些橡胶和塑料制品中，其衍生物还可以用作农药、杀鼠剂、医药等。由于天然植物中香豆素含量很少，大多数是通过合成获得的。1868年，珀金采用邻羟基苯甲醛(水杨醛)与乙酸酐、乙酸钾加热制得香豆素，该方法也称为珀金合成法。其反应过程如下：

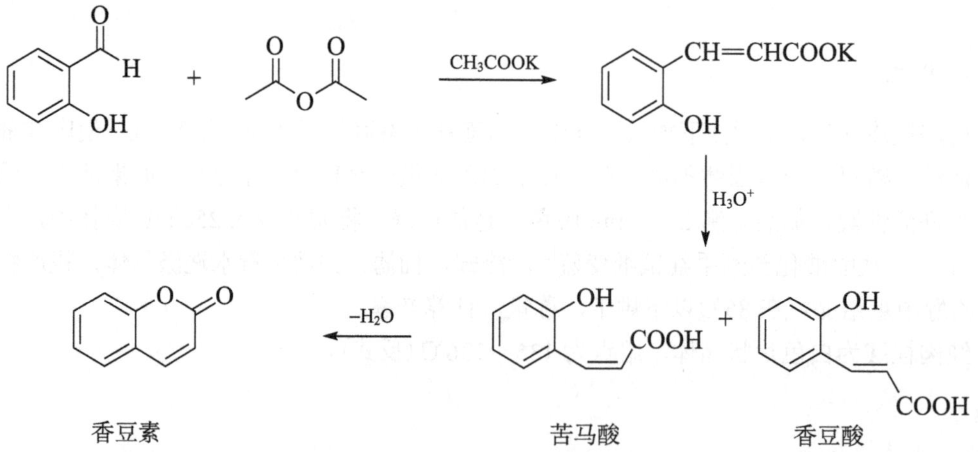

但是，珀金法有许多缺点，如反应时间长、反应温度高、产率不稳定等，因此本实验不用珀金法。

2. 利用 Knoevenagel 反应制备香豆素

低温下，以水杨醛和丙二酸二乙酯为原料，在有机碱催化下合成香豆素-3-羧酸。这种在有机碱作用下发生的改性羟醛反应称为 Knoevenagel 反应。水杨醛和丙二酸二乙酯在哌啶催化下发生 Knoevenagel 反应得到香豆素-3-羧酸乙酯，经过氢氧化钠水解、盐酸酸化和内酯化一系列后续反应，生成香豆素-3-羧酸。

该方法用一个或两个吸电子基团的活泼亚甲基化合物取代珀金法中的酸酐，增加了亚甲基氢的活泼性。同时，使用较弱的有机碱可以避免醛的自身缩合，扩大了缩合反应的原料使用范围。

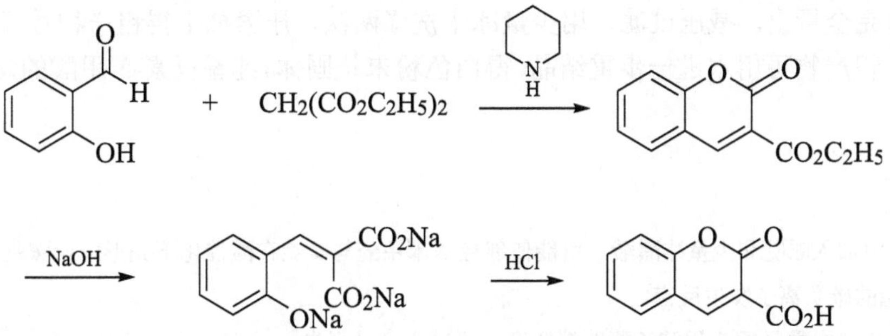

实验用品

仪器：水浴锅、圆底烧瓶、冷凝管、干燥管、锥形瓶、减压过滤装置、熔点测定仪。

试剂：水杨醛、丙二酸二乙酯、无水乙醇、95%乙醇、哌啶、冰醋酸、氢氧化钠、浓盐酸、无水氯化钙。

试剂的物理性质参见表 3.2。

表 3.2 试剂的物理性质

物质	M/(g/mol)	ρ/(g/mL)	m.p./℃	b.p./℃
水杨醛	122.12	1.17	-7	197
丙二酸二乙酯	160.17	1.0551	-50	199.3
哌啶	85.15	3.0	-7	106
香豆素-3-甲酸乙酯	218.2054	1.289	92～93	378
香豆素-3-甲酸	190.15	—	190	—

实验步骤

1. 香豆素-3-甲酸乙酯的制备

于 50 mL 干燥的圆底烧瓶中加入 1.3 mL（1.8 g，0.015 mol）水杨醛、2.6 g（0.016 mol）丙二酸二乙酯、10 mL 无水乙醇、10 滴哌啶和 1 滴冰醋酸[1]。加入几粒沸石，装上回流冷凝管，冷凝管上口需接氯化钙干燥管。将反应体系水浴加热回流 0.5 h。

待反应液稍冷却后，将所得混合液转入锥形瓶内，加 8 mL 水，冰水浴冷却，使产物结晶析出。减压过滤，晶体用冷却的 50%乙醇洗涤两次（每次 3～5 mL）[2]，压紧抽干。粗产物香豆素-3-甲酸乙酯为白色晶体。干燥，称量，计算产率。粗产物可用 25%乙醇重结晶。可通过测定粗产物的熔点检测其纯度。纯香豆素-3-甲酸乙酯的熔点为 92～93℃。

2. 香豆素-3-甲酸的制备

在 50 mL 圆底烧瓶中加入 2.0 g 香豆素-3-甲酸乙酯、1.5 g 氢氧化钠、10 mL 95%乙醇、5 mL 水和 1~2 粒沸石。装上回流冷凝管,加热回流。当酯和氢氧化钠全部溶解后,再继续加热回流 15 min。

将反应液趁热倒入稀盐酸(由 7.5 mL 浓盐酸和 25 mL 水混合制得)中进行酸化,冰水浴冷却使晶体完全析出,减压过滤,用少量冰水洗涤两次,压紧抽干得粗产物香豆素-3-甲酸。干燥,称量。粗产物可用水进一步重结晶,得白色粉末状固体。纯香豆素-3-甲酸的熔点为 190 ℃(分解)。

注释

[1] 反应中加入哌啶和少量冰醋酸,可能使邻羟基苯甲醛与哌啶在酸催化下先形成亚胺基化合物,再与丙二酸二乙酯的碳负离子发生反应。

[2] 降低乙醇中香豆素-3-甲酸乙酯的溶解度,可减少产品损失。

思考题

(1) 反应中加入冰醋酸的目的是什么?

(2) 如何利用香豆素-3-甲酸制备香豆素?

实验 25 呋喃甲醇和呋喃甲酸的制备

实验目的

(1) 了解通过坎尼扎罗反应用呋喃甲醛制备呋喃甲醇和呋喃甲酸的原理和方法。

(2) 巩固洗涤、萃取、简单蒸馏、减压过滤和重结晶等基本操作。

实验原理

坎尼扎罗反应是歧化反应。它是指在浓碱的存在下,两分子无 α-H 的醛发生氧化还原反应,生成等物质的量的伯醇和羧酸。呋喃甲酸和呋喃甲醇可以通过坎尼扎罗反应,利用呋喃甲醛制备。

反应式:

$$2 \text{ furan-CHO} \xrightarrow{\text{NaOH}} \text{furan-CH}_2\text{OH} + \text{furan-COONa}$$

$$\text{furan-COONa} \xrightarrow{\text{HCl}} \text{furan-COOH} + \text{NaCl}$$

反应机理:

实验用品

仪器：烧杯、磁力搅拌器、分液漏斗、圆底烧瓶、直形冷凝管、温度计(250℃)、电热套。
试剂：呋喃甲醛、40%氢氧化钠溶液、乙醚、盐酸、无水硫酸镁。
试剂的物理性质参见表 3.3。

表 3.3　试剂的物理性质

物质	M/(g/mol)	ρ/(g/mL)	b.p./℃	溶解性
呋喃甲酸	112.0835	—	—	
呋喃甲醇	98.10	1.1296	171	溶于乙醚
呋喃甲醛	96.09	2.428	161.7	溶于乙醚

实验步骤

1. 制备

称取 4 g 氢氧化钠，溶于盛有 6 mL 水的小烧杯中，并于冰水浴下冷却至 5℃ 左右。采用磁力搅拌器，边搅拌边滴加 6.6 mL 呋喃甲醛于氢氧化钠溶液中(约需 10 min)[1]，维持反应温度在 8～12℃。加完后，将反应体系置于室温下，继续搅拌反应 20 min[2]，得到黄色浆状物。然后边搅拌边加入适量的水(约 5 mL)使其溶解[3]，此时溶液呈暗红色。

2. 分离

用乙醚萃取反应液三次(每次 10 mL)，分为水相与有机相。

3. 提纯呋喃甲醇

向有机相中加入 2 g 无水硫酸镁干燥。将乙醚溶液过滤至蒸馏烧瓶中，水浴加热除去乙醚。用电热套加热残液，收集 169～172℃ 馏分，得到呋喃甲醇。

4. 提纯呋喃甲酸

向水相中加入浓盐酸酸化至 pH=3[4]。冷却，析出晶体，抽滤得到呋喃甲酸。用少许水洗涤晶体，干燥，称量。

纯呋喃甲醇为无色透明液体，沸点为 171℃，n_D^{20} 为 1.4868。

纯呋喃甲酸为白色针状晶体，熔点为 133～144℃。

注释

[1] 需采用新蒸馏的呋喃甲醛。
[2] 体系是在两相间反应的，因此必须充分搅拌。
[3] 加水溶解反应所得沉淀物时，需控制加水的量，否则呋喃甲酸溶于水，造成产品的损失。
[4] 酸化水相时要加入足量的酸，使呋喃甲酸完全析出。

思考题

(1) 为什么呋喃甲醛要重新蒸馏？长期放置的呋喃甲醛可能含有哪些杂质？如果杂质不事先除去，对实验有什么影响？
(2) 本实验是将呋喃甲醛滴加到氢氧化钠溶液中，若滴加顺序相反，反应有何变化？对产率是否有影响？
(3) 影响产率的关键因素是什么？

实验26 乙酸乙酯的制备

实验目的

(1) 了解制备乙酸乙酯的原理和方法。
(2) 进一步巩固回流、蒸馏和萃取等基本操作。

实验原理

主反应：

$$CH_3COOH + C_2H_5OH \xrightleftharpoons[120℃]{浓硫酸} CH_3COOC_2H_5 + H_2O$$

副反应：

$$2C_2H_5OH \xrightarrow[140℃]{浓硫酸} CH_3CH_2OCH_2CH_3 + H_2O$$

$$C_2H_5OH \xrightarrow[170℃]{浓硫酸} H_2C=CH_2 + H_2O$$

实验用品

仪器：圆底烧瓶、温度计(100℃)、温度计套管、分液漏斗、蒸馏头、锥形瓶、直形冷凝管、电热套。

试剂：无水乙醇、冰醋酸、浓硫酸、饱和碳酸钠溶液、饱和氯化钠溶液、饱和氯化钙溶液、无水硫酸镁(或无水碳酸钾)。

试剂的物理性质参见表 3.4。

表 3.4 试剂的物理性质

物质	M/(g/mol)	ρ/(g/mL)	m.p./℃	b.p./℃
乙醇	46.07	0.789	−114.1	78.3
乙酸乙酯	88.11	0.902	−84	77
乙烯	28.06	—	−169	−103.7

实验步骤

1. 乙酸乙酯的制备

在 50 mL 干燥的圆底烧瓶中加入 12 mL (0.2 mol)无水乙醇和 6 mL (0.1 mol)冰醋酸[1]，一边振荡一边向其中缓慢加入 0.5 mL 浓硫酸[2]，摇匀[3]，加入 1～2 粒沸石，反应体系加热回流 0.5 h。待反应物冷却后，将回流装置改成蒸馏装置。加热蒸出产物乙酸乙酯，直到馏出液体积约为反应物体积的 1/2 为止。

2. 提纯

在馏出液中加 10 mL 水，然后转入分液漏斗中分液。弃去水相，保留有机相。依次用 10 mL 10%碳酸钠溶液[4]、5 mL 饱和氯化钠溶液、5 mL 饱和氯化钙溶液和 5 mL 水洗涤有机相[5]。最后将有机相转入干燥的具塞锥形瓶中，用无水硫酸镁干燥[6]。

将干燥后的乙酸乙酯过滤除去硫酸镁，滤液倒入 25 mL 蒸馏烧瓶中蒸馏[7]，收集 73～78℃馏分，称量，计算产率。

纯乙酸乙酯的沸点为 77℃，具有果香味，n_D^{20} 为 1.3723。

乙酸乙酯的红外光谱和核磁共振氢谱分别见图 3.13 和图 3.14。

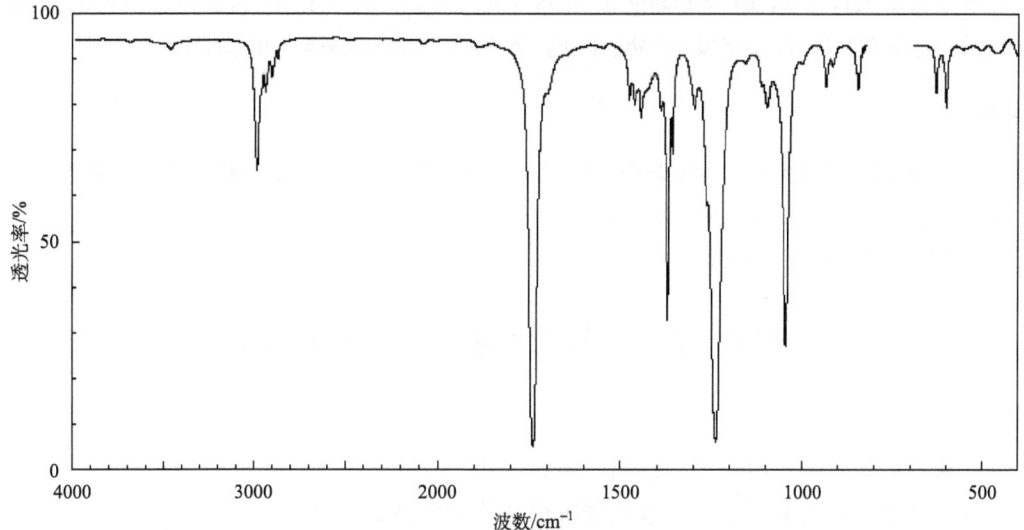

图 3.13 乙酸乙酯的红外光谱

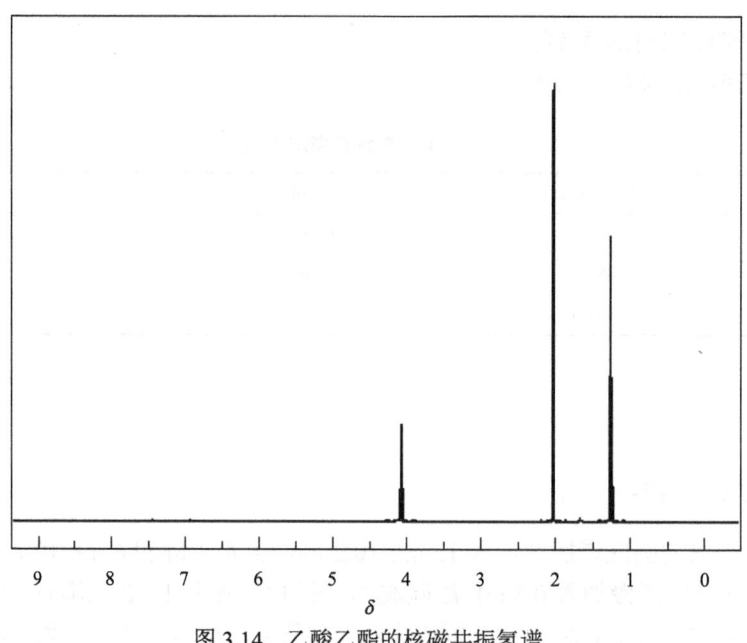

图 3.14 乙酸乙酯的核磁共振氢谱

注释

[1] 酯化反应所用的仪器必须干燥,防止水影响反应平衡。

[2] 浓硫酸的使用非常危险,如有溢出,应立即用大量水清洗,并用小苏打中和溢出物。

[3] 加热前一定将反应混合物混合均匀,否则容易炭化。

[4] 用10%碳酸钠溶液洗有机相时会产生二氧化碳,应注意及时给分液漏斗放气。

[5] 馏出液一定要先用饱和氯化钠溶液洗,然后用饱和氯化钙溶液洗,否则会产生碳酸钙沉淀,给分液带来困难。

[6] 有机相蒸馏前要充分干燥,否则会产生较多的前馏分。乙酸乙酯与水可形成沸点为70.4℃的二元共沸混合物(含水8.1%);乙酸乙酯、乙醇和水可以形成沸点为70.2℃的三元共沸混合物(含乙醇8.4%,水9%)。

[7] 加热前记得加沸石。为了减少产物的损失,蒸馏时用到的接收瓶可以用冰水冷却。

思考题

(1) 酯化反应的特征是什么?本实验采用什么措施可以使酯化反应尽量向产物生成方向移动?

(2) 本实验有哪些可能的副反应?如何避免?

(3) 乙酸是否可以过量?为什么?

实验 27　邻苯二甲酸二丁酯的制备

实验目的

(1) 学习制备邻苯二甲酸二丁酯的原理和方法。

(2) 学习分水器的使用方法,掌握减压蒸馏的操作技能。

实验原理

邻苯二甲酸二丁酯是一种有机化合物，常用作增塑剂。它通常由邻苯二甲酸酐和正丁醇在强酸（如浓硫酸）催化下反应而得。反应式如下：

邻苯二甲酸酐 + n-C_4H_9OH $\xrightarrow{H_2SO_4}$ 邻苯二甲酸单丁酯（$-OC_4H_9$-n, $-COOH$）

邻苯二甲酸单丁酯 + n-C_4H_9OH $\xrightarrow{H_2SO_4}$ 邻苯二甲酸二丁酯

实验用品

仪器：二颈烧瓶、圆底烧瓶、冷凝管、分水器、锥形瓶、温度计、布氏漏斗、抽滤瓶、分液漏斗。

试剂：邻苯二甲酸酐、正丁醇、浓硫酸、饱和 NaCl 溶液、无水 Na_2SO_4、5% Na_2CO_3 溶液。

实验步骤

将 1.5 g（0.010 mol）邻苯二甲酸酐、3.3 mL（0.036 mol）正丁醇、2 滴浓硫酸依次加入 100 mL 二颈烧瓶，充分搅拌。按图 3.10 所示搭好装置，分水器中放入计算好的水。

将反应物加热回流，直到温度达到 140℃，约需 2 h[1]。

将反应液冷却至 70℃以下，移入分液漏斗中，用 20～30 mL 5% Na_2CO_3 溶液中和。用 20 mL 温热饱和 NaCl 溶液洗涤两次，直至呈中性[2]。最后有机层用无水 Na_2SO_4 干燥。

将粗产物常压蒸馏除去正丁醇，然后减压蒸馏，收集纯产物，称量，计算产率。

纯邻苯二甲酸二丁酯为无色油状液体，沸点为 340℃（175～176℃/666.5 Pa），n_D^{20} 为 1.4911。

注释

[1] 当反应温度升到 140℃便可停止反应。

[2] 为防止皂化反应，中和过程使用的碱浓度不宜过高，中和温度应低于 70℃。

思考题

(1) 浓硫酸为催化剂，正丁醇在加热下有哪些反应？

(2) 在反应过程中加热过快有何不良影响？

实验 28　乙酰乙酸乙酯的制备

实验目的

(1) 掌握克莱森酯缩合反应制备乙酰乙酸乙酯的原理和方法。
(2) 巩固回流和减压蒸馏等基本操作。

实验原理

乙酰乙酸乙酯(EAA)主要用作生产各种化合物的化学中间体。它是通过乙酸乙酯的克莱森酯缩合反应制备的。克莱森酯缩合是两个酯或一个酯与另一个羰基化合物在强碱存在下发生的碳-碳键形成反应，生成 β-酮酯或 β-二酮。本实验在乙醇钠(首先由钠和乙酸乙酯中的微量乙醇反应生成)存在下，通过两分子乙酸乙酯的缩合反应制备乙酰乙酸乙酯。

反应式：

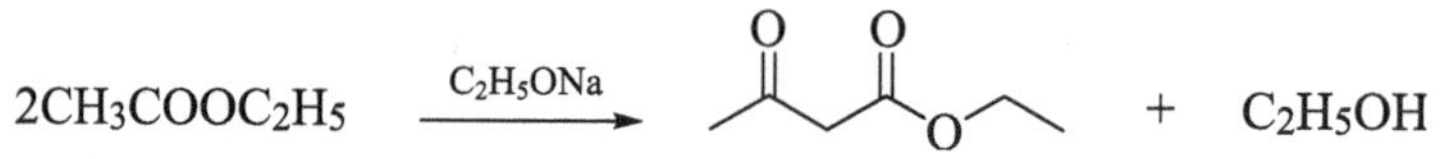

$$2CH_3COOC_2H_5 \xrightarrow{C_2H_5ONa} CH_3COCH_2COOC_2H_5 + C_2H_5OH$$

实验用品

仪器：圆底烧瓶、冷凝管、干燥管、分液漏斗、锥形瓶、温度计、电热套、布氏漏斗、抽滤瓶、压钠机。

试剂：乙酸乙酯、金属钠、50%乙酸、氯化钙、饱和氯化钠溶液、无水硫酸钠。

实验步骤

在干燥的 50 mL 圆底烧瓶中加入 10 mL 干燥的乙酸乙酯[1]，快速加入 0.9 g 新鲜压制的金属钠丝[2]，迅速装上带有氯化钙干燥管的冷凝管，仪器接头处事先涂抹少量真空油脂。

小火加热，保持微沸状态回流，直至金属钠全部消失。稍冷后，慢慢加入 50%乙酸并不断振荡，直至反应液呈弱酸性(pH=5~6)。

将反应液转移至分液漏斗中，加入等体积的饱和氯化钠溶液，振荡，静置，分出有机层，用无水硫酸钠干燥。将干燥过的液体滤到圆底烧瓶中，先常压蒸馏除去乙酸乙酯，然后改用减压蒸馏，在相应压力下蒸出乙酰乙酸乙酯[3]。称量得到的纯品。

纯乙酰乙酸乙酯为无色液体，沸点为 180.4℃(分解)，n_D^{20} 为 1.4194。

注释

[1] 减压蒸馏前，所用仪器和乙酰乙酸乙酯都必须彻底干燥。
[2] 钠很容易与氧气和水反应，并释放大量的热。当空气中的温度足够高时，钠发生自燃，使用时应十分小心。
[3] 乙酰乙酸乙酯在常压蒸馏下易分解，因此应采用减压蒸馏提纯。

思考题

(1) 如果减压蒸馏时仪器和乙酰乙酸乙酯不干燥会有什么影响？

(2) 加入 50%乙酸和饱和氯化钠溶液的目的是什么?
(3) 乙酰乙酸乙酯沸点并不是非常高,为什么要用减压蒸馏提纯?

实验 29　乙酰水杨酸(阿司匹林)的制备

实验目的

(1) 以水杨酸和乙酸酐为原料合成阿司匹林。
(2) 熟悉重结晶和减压过滤等基本操作。

实验原理

乙酰水杨酸又称阿司匹林,是一种非常普遍的治疗感冒的药物,具有镇痛、退热及抗风湿等功效。水杨酸是一个具有双官能团的化合物,酚羟基和羧基都可以发生酯化反应。水杨酸经乙酸酐酰化反应,酚羟基上的氢原子被乙酰基取代生成乙酰水杨酸。通常用少量硫酸作催化剂。

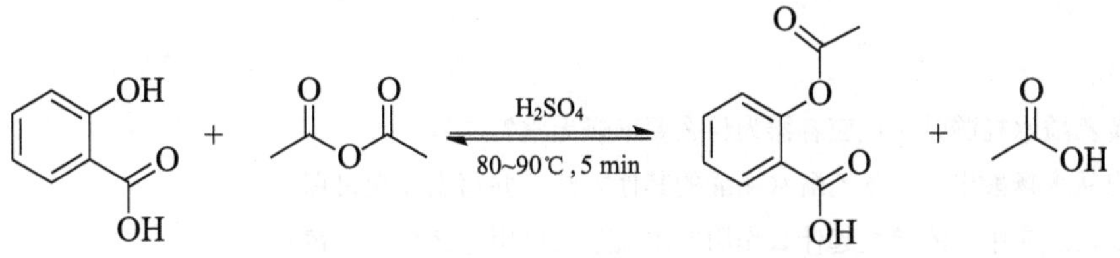

水杨酸　　　乙酸酐　　　　　　　　　乙酰水杨酸　　乙酸
　　　　　　　　　　　　　　　　　　 (阿司匹林)

实验用品

仪器：锥形瓶、温度计、布氏漏斗、抽滤瓶、烧杯、表面皿。
试剂：水杨酸、乙酸酐、乙酸乙酯、浓硫酸、浓盐酸、饱和碳酸氢钠溶液、1%三氯化铁溶液。

实验步骤

在 150 mL 锥形瓶中加入 2.0 g(0.014 mol)水杨酸、5 mL(0.05 mol)新蒸乙酸酐和 5 滴浓硫酸[1,2],充分摇动。在 80~90℃水浴中加热 5 min,并不断搅拌[3]。固体全部溶解。

取出锥形瓶,边摇边滴加 1 mL 冷水。然后快速倒入 50 mL 冷水,冷却至室温。当乙酰水杨酸晶体析出后,将锥形瓶放入冰水浴中进一步冷却。若无晶体出现,可用玻璃棒摩擦内壁促进结晶。抽滤,用 5 mL 冷水洗涤晶体,抽干,得到乙酰水杨酸粗产品。

乙酰水杨酸粗产品用重结晶提纯。将粗产品转移到 100 mL 烧杯中,边搅拌边慢慢加入 25 mL 饱和碳酸氢钠溶液,继续搅拌直到无二氧化碳气体逸出。抽滤除去不溶性杂质,将滤液倒入盛有 10 mL 水和 5 mL 浓盐酸溶液的 150 mL 烧杯中,搅拌均匀,用冰水冷却,确保溶液呈酸性,乙酰水杨酸沉淀完全。减压过滤,用冷水洗涤两次(每次 5 mL),抽干水分。

将晶体转移至表面皿上,自然干燥。然后用热乙酸乙酯(2~3 mL)重结晶。在蒸汽浴上持续加热,确保产品完全溶解。将溶液冷却至室温,然后放入冰浴中。抽滤,收集产品,并用几毫升冷石油醚冲洗烧瓶。干燥[4]，称量，测定熔点(文献值 135~136℃)，并计算产率。

氯化铁试验：取几粒晶体加入盛有 5 mL 水的试管中，加入 1~2 滴 1%三氯化铁溶液，观察颜色变化，检验产品纯度[5]。含有产品的试管在测试后只应变为非常浅的紫色，表明纯度很高。

纯乙酰水杨酸为白色针状晶体，熔点为 135~136℃。

注释

[1] 制备实验中,所用仪器应经过干燥处理,试剂也要事先经过干燥处理。乙酸酐需重新蒸馏,收集139~140℃的馏分。

[2] 注意加样顺序，如果先加水杨酸和浓硫酸，水杨酸就会被氧化。

[3] 因为乙酸酐具有强刺激性，实验在通风橱中进行。

[4] 乙酰水杨酸受热易分解，分解温度为 126~135℃，因此重结晶时不宜长时间加热。粗产品可用乙醇-水或苯-石油醚(60~90℃)重结晶提纯。

[5] 如果粗产品中有未反应的水杨酸，三氯化铁溶液与水杨酸反应显紫色。

思考题

(1)制备乙酰水杨酸时，反应容器为什么要干燥无水？
(2)给出从水杨酸出发制备乙酰水杨酸的其他方法，并写出反应机理。
(3)在酰化反应中，浓硫酸起什么作用？浓硫酸可以用什么物质代替？
(4)本实验的副反应有哪些？
(5)为什么水杨酸能溶于碳酸氢钠溶液，而其聚合物却不溶？
(6)在浓硫酸存在下，水杨酸与过量的乙醇作用将得到什么产物？
(7)用玻璃棒在器皿内壁上摩擦可促进结晶，为什么？除此之外，还有什么方法可以加快结晶？

实验 30　乙酰苯胺的制备

实验目的

(1)用苯胺和乙酸酐反应制备乙酰苯胺。
(2)学习用重结晶提纯乙酰苯胺。
(3)掌握分馏的操作技术。

实验原理

乙酰苯胺是较早使用的解热镇痛药，具有镇痛、退热作用，有"退热冰"之称。乙酰苯胺本身是重要的药物，而且是磺胺类药物(含有磺胺官能团的药物)合成中重要的中间体。乙酰苯胺可由苯胺与乙酰化试剂如乙酰氯、乙酸酐或乙酸等发生乙酰化反应制备。反应活性顺序是乙酰氯＞乙酸酐＞乙酸。由于乙酰氯和乙酸酐的价格较贵且有腐蚀性，本实验用乙酸作为乙酰化试剂。

反应式：

$$\underset{}{\text{C}_6\text{H}_5\text{NH}_2} + \text{CH}_3\text{COOH} \xrightarrow{\text{Zn}} \text{C}_6\text{H}_5\text{NHCOCH}_3 + \text{H}_2\text{O}$$

本实验除了在合成中具有重要意义外，还有保护芳环上氨基的作用。芳环上的氨基易氧化，通常先乙酰化将其保护，然后在芳环上引入其他所需基团，再利用酰胺能水解成胺的性质，恢复氨基。

实验用品

仪器：圆底烧瓶、韦氏分馏柱、接液管、电热套、温度计(150℃)、抽滤瓶、布氏漏斗、量筒、滤纸、烧杯、水泵。

试剂：新蒸苯胺、冰醋酸、锌粉、活性炭。

实验步骤

在 50 mL 圆底烧瓶中加入 5 mL (0.055 mol) 新蒸苯胺[1]、7.5 mL (0.13 mol) 冰醋酸和少许锌粉[2]。按照图 3.15 搭好实验装置。缓慢加热至反应物沸腾，回流 45 min，维持温度计读数在 105℃ 左右[3]。接收馏出液约 4 mL。当温度计读数不断下降时，停止加热。

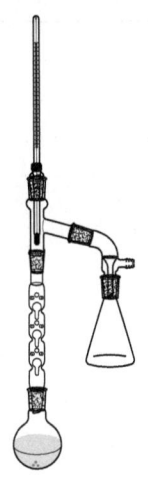

图 3.15 用乙酸制备乙酰苯胺的反应装置

将反应液趁热倒入装有 100 mL 冷水的烧杯中，边倒边搅拌（注意锌粉不要倒进去）[4]，此时有细粒状固体析出。冷却后抽滤，并用冷水洗涤固体两次。抽干水分，得到白色或淡黄色粗产品。

将粗产品转移到 250 mL 烧杯中，加入 100 mL 水，在搅拌下加热至沸腾。继续加热水并搅拌，直到晶体全溶。如果溶液颜色较深或含有油状物，稍冷后加入 0.5 g 活性炭[5]，并煮沸 2 min。在保温漏斗中趁热过滤除去活性炭等不溶性杂质[6]。将滤液自然冷却至室温，再用冰水冷却，待结晶完全后[7]，进行抽滤。用少量冷水洗涤滤饼两次，压紧抽干。将结晶转移至预先称量的表面皿中，自然晾干，称量，计算产率。

纯乙酰苯胺为白色鳞片状晶体，熔点为 114.3℃。

乙酰苯胺的红外光谱和核磁共振氢谱分别见图 3.16 和图 3.17。

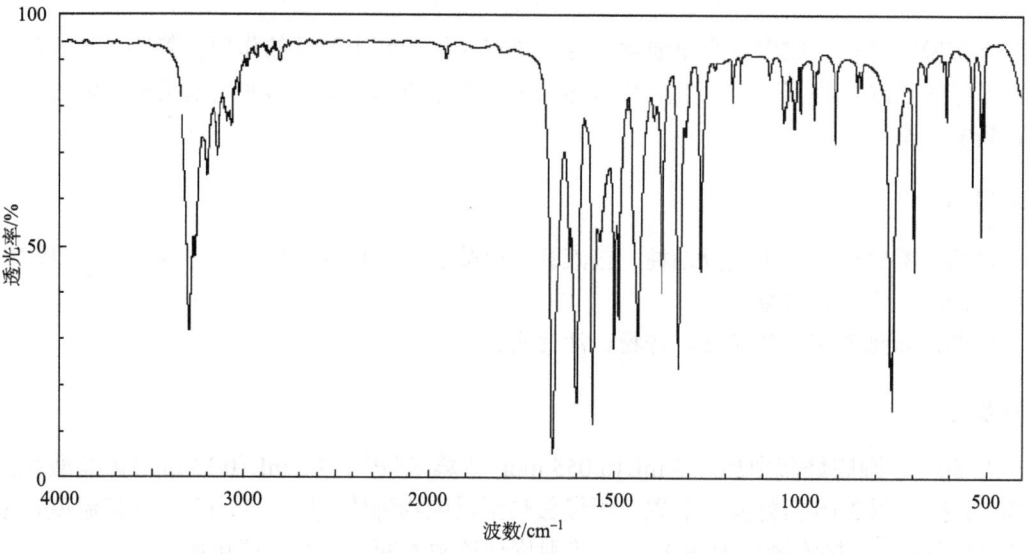

图 3.16　乙酰苯胺的红外光谱

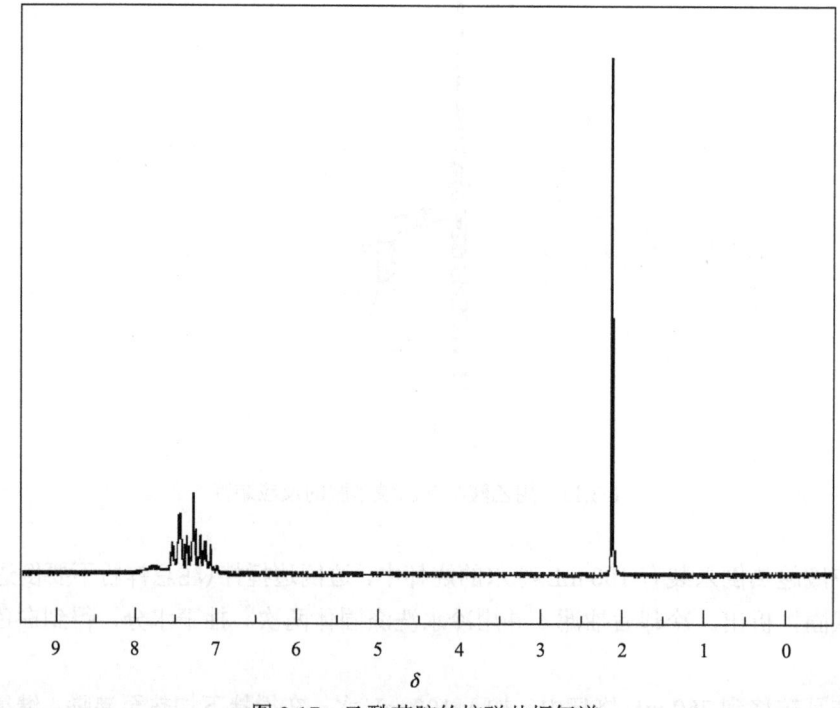

图 3.17　乙酰苯胺的核磁共振氢谱

注释

[1] 久置的苯胺因为氧化而颜色较深，使用前要重新蒸馏。因为苯胺的沸点较高，蒸馏时选用空气冷凝管冷凝或采用减压蒸馏。

[2] 锌粉的作用是防止苯胺氧化,只要少量即可,加得过多,会出现不溶于水的氢氧化锌。
[3] 反应时分馏温度不能太高,以免大量乙酸蒸出而降低产率。
[4] 反应液冷却时会析出乙酰苯胺固体,粘在烧瓶壁上不易处理,所以应趁热将反应液倒出。
[5] 不能向沸腾的溶液中加入活性炭,否则会引起暴沸。
[6] 趁热过滤时,布氏漏斗要预热,否则产品析出,堵塞滤纸的毛孔,粘在漏斗内部。
[7] 重结晶时,如果溶液过饱和,则可用玻璃棒摩擦烧杯内壁或加入晶种使晶体析出。

思考题

(1) 用乙酸作酰化试剂时,如何提高乙酰苯胺的产率?
(2) 反应温度为什么控制在105℃左右?反应温度过高或过低对实验有什么影响?
(3) 根据反应式计算,理论上能产生多少水?为什么实际收集的液体量多于理论量?
(4) 反应结束时,温度计读数为什么下降?

实验 31 对乙酰氨基苯磺酰氯的制备

实验目的

(1) 学习磺酰化反应的原理和反应的一般步骤。
(2) 熟练掌握重结晶操作。

实验原理

磺酰氯是制备一系列磺胺类药物的基本原料。对乙酰氨基苯磺酰氯与氨或氨的衍生物反应是制备磺胺类药物的关键一步。

氯磺酰基可以通过称为氯磺化的芳基亲电取代反应一步引入乙酰苯胺中乙酰胺基团的对位。首先可能是 SO_3 作为亲电试剂进攻苯环,形成磺酸。由于电子效应和空间效应的综合作用,取代以对位为主。然后生成的磺酸与过量的氯磺酸反应,转化为对乙酰氨基苯磺酰氯,同时生成硫酸作为副产物。

反应式:

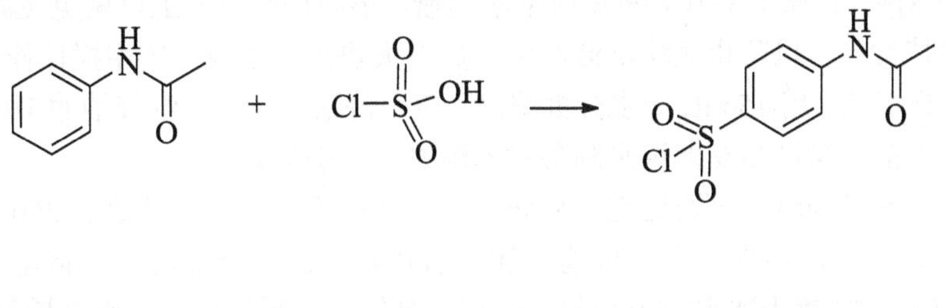

实验用品

仪器:圆底烧瓶、恒压滴液漏斗、弯管接头、克氏蒸馏头、磁力搅拌器、温度计套管、冰水浴、电热套、温度计(150℃)、抽滤瓶、布氏漏斗、弯管、滤纸、烧杯、水泵。

试剂:乙酰苯胺、氯磺酸、5% NaOH 溶液。

实验步骤

在装有搅拌磁子的 100 mL 干燥圆底烧瓶中加入 5 g 干燥的乙酰苯胺。将烧瓶置于电热套中加热使固体熔化。取下烧瓶，缓缓旋转烧瓶，然后将其塞住并置于冰水浴中。此时在烧瓶底部形成一层薄膜[1]。在恒压滴液漏斗中加入 13 mL 氯磺酸，塞上塞子并仔细检查以确保没有泄漏（在通风橱中操作并戴上手套！）[2]。当反应烧瓶冷却后，将烧瓶与克氏蒸馏头连接，然后在克氏蒸馏头上方装滴液漏斗，在侧方连接弯管接头，接头另一端连接弯管。弯管的另一端通过温度计套管插入装有 5% NaOH 溶液的气体吸收烧瓶中。该装置如图 3.18 所示。

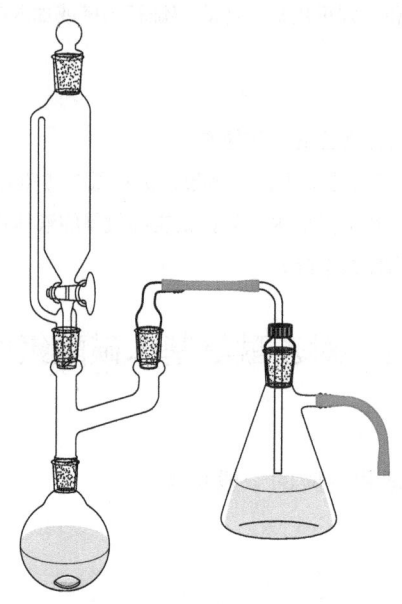

图 3.18　制备对乙酰氨基苯磺酰氯的反应装置

将反应烧瓶置于冷水浴中，温度维持在 10～15℃。然后从滴液漏斗中将氯磺酸快速地一次性加入烧瓶中。快速搅拌反应溶液，并使反应瓶所在的冷水浴中的水温低于 20℃[3]。

当大部分乙酰苯胺溶解，并且初始放热反应消退后，移除冷却浴使溶液在搅拌下升至室温。然后将反应瓶在 70～80℃ 的热水浴中加热 20 min 以完成反应。

将反应混合物在冰浴中冷却至室温以下。然后将反应混合物小心倒入装有 100 g 碎冰的 500 mL 烧杯中（注意：戴上手套并将反应溶液缓慢、小心地倒在冰上以避免飞溅）。倒入过程中需使用搅拌棒搅拌，以防止大块沉淀产生。如有大块沉淀生成，需用搅拌棒小心地搅散。

抽滤，收集沉淀，用 15 mL 冷水分批洗涤，直至滤液呈中性。继续抽滤 10 min 进一步干燥滤渣，得粗产品。粗产品立即用来制备对乙酰氨基苯磺酰氯。

纯对乙酰氨基苯磺酰氯可通过重结晶获得，方法如下：把粗产品放入 250 mL 圆底烧瓶中，先加入少许氯仿，加热至沸，再逐渐加入氯仿直至固体全部溶解。将溶液迅速转移至 250 mL 锥形瓶中，并置于冰水浴中冷却，即有结晶析出。减压过滤，用少量氯仿洗涤晶体，抽干，称量。纯对乙酰氨基苯磺酰氯的熔点为 149℃。

注释

[1] 氯磺化反应常过分剧烈，所以将乙酰苯胺凝结后再反应。

[2] 氯磺酸有强腐蚀性，危害性极强，处理时需穿上实验服，戴上手套和防护眼镜，并且只能在通风橱

内进行。氯磺酸与水甚至水汽发生剧烈反应,在反应过程中产生 HCl 气体,吸入时会使人窒息。注意任何靠近氯磺酸的水源。反应中使用的仪器和试剂应非常干燥。含有氯磺酸的废液不可倒入水槽,应倒入废酸缸中。

[3] 本实验的关键是防止局部过热,否则会造成氯磺酸的分解。

思考题

(1) 本实验加入氯磺酸时为什么保持反应温度在 15℃ 以下?

(2) 为什么苯胺要乙酰化后再氯磺化?如果苯胺直接氯磺化可以吗?

参考文献

Coppock P, Park S H, Paredes J, et al. 2017. Enhancing research skills and attitudes in undergraduate organic chemistry with course-embedded undergraduate research experience(CURE) via green organic synthesis. Journal of Laboratory Chemical Education, 5(3): 41-47.

Hurdis E, Wang J. 1969. Preparation of sulfanilamide from aniline: an organic chemistry experiment. J Chem Edu, 46: 697-698.

实验 32 对氨基苯磺酰胺的制备

实验目的

(1) 掌握制备对氨基苯磺酰胺的原理和方法,同时掌握酰氯氨解和乙酰氨基衍生物水解的原理。

(2) 巩固回流、脱色、重结晶等基本操作。

实验原理

氨基苯磺酰胺又称磺胺,为白色颗粒或粉末状结晶,分子式为 $C_6H_8N_2O_2S$,分子量为 172.22,熔点为 164.5~166.5℃,无臭,味微苦,微溶于冷水、乙醇、甲醇、丙酮,易溶于沸水、甘油、盐酸、氢氧化钾及氢氧化钠溶液,不溶于苯、氯仿、乙醚和石油醚。最早的磺胺是一种红色染料,在某次偶然的机会,人们发现这种红色染料对细菌具有很强的抑制作用,从而将它用作药物。磺胺是磺胺类药物的最基本结构,也是其药性的基本结构,是一类用于预防和治疗细菌感染性疾病的药物,是现代医学中常用的一类抗菌消炎药。

反应式:

实验用品

仪器：锥形瓶、抽滤瓶、圆底烧瓶、布氏漏斗。

试剂：乙酰苯胺、氯磺酸、浓氨水、浓盐酸、碳酸钠。

实验步骤

1. 对乙酰氨基苯磺酰胺的制备

在通风橱中将实验 31 中制得的乙酰氨基苯磺酰氯加入 125 mL 锥形瓶中，在搅拌下慢慢加入 23 mL 浓氨水[1]，立即发生放热反应，并生成白色糊状物。加完后，继续搅拌 15 min，使反应完全。然后将反应瓶置于 70℃水浴中搅拌加热 10 min 以除去多余的氨。冷却、抽滤，用少量冷水洗涤晶体、抽干，得到粗产品。粗产品可直接用于下一步合成。

2. 对氨基苯磺酰胺的制备

将粗产品转移至 50 mL 圆底烧瓶中，加入 20 mL 10%盐酸，加热回流使固体充分溶解(约需 0.5 h)。反应溶液冷却至室温后，得到几乎澄清的溶液。如果有固体重新析出，测试溶液的酸碱性。若不呈酸性，则酌情补加适量盐酸，并继续将混合物煮沸 15 min，直到在室温冷却后没有固体析出。将滤液转移到烧杯中，在搅拌下缓缓地加入碳酸钠固体，直到 pH 为 7～9[2]。在中和过程中，会析出对氨基苯磺酰胺固体。用冰水浴冷却混合溶液，抽滤，用少量冰水洗涤。用水重结晶粗产品(每克产品需 12 mL 水)。所得对氨基苯磺酰胺为白色叶片状晶体，熔点为 165～166℃。

注释

[1] 由于氨有毒且具有非常强烈的刺鼻气味，因此整个过程必须在通风橱内进行。

[2] 用碳酸钠中和盐酸时有大量的二氧化碳气体产生，故需不断搅拌。

思考题

如何理解对氨基苯磺酰胺是两性物质？试用反应式表示磺胺与稀酸和稀碱的反应。

实验 33　N-溴代丁二酰亚胺的绿色合成

实验目的

(1) 掌握制备 N-溴代丁二酰亚胺的原理、工艺及操作。

(2) 了解绿色合成的概念和工艺标准。

实验原理

本实验利用溴酸钠、溴化钠在酸性条件下的原位氧化还原反应形成的溴，与丁二酰亚胺反应制备 N-溴代丁二酰亚胺(NBS)。该方法产率高，操作简单、安全，避免了直接使用溴而带来的污染和危险性。

反应式:

6 ⬡NH + 2NaBrO$_3$ + 4NaBr + 3H$_2$SO$_4$ ⟶ 6 ⬡N—Br + 6H$_2$O + 3Na$_2$SO$_4$

实验用品

仪器：三颈烧瓶、恒压滴液漏斗、漏斗、U 形玻璃管、冷凝管、烧杯。

试剂：丁二酰亚胺、溴化钠、溴酸钠、5% NaOH 溶液、50% H$_2$SO$_4$ 溶液。

实验步骤

如图 3.19 所示，在 100 mL 三颈烧瓶中加入 1.8 g(18 mmol)丁二酰亚胺、1.0 g(7 mmol)溴酸钠、1.4 g(13.8 mmol)溴化钠和 10 mL 水。在恒压滴液漏斗中装入 2 mL 50%硫酸溶液，在三颈烧瓶的两个颈上分别安装恒压滴液漏斗和水冷凝管，并塞住烧瓶的另一颈。在冷凝管顶部连接一个单孔塞子，将 U 形玻璃管的一端插入塞子中，并用倒置漏斗连接管子的另一端。将漏斗边缘悬挂在烧杯中 5% NaOH 溶液表面的顶部。在 25℃下搅拌溶解反应瓶中的固体，然后从滴液漏斗中慢慢逐滴加入 50%硫酸溶液到反应瓶内[1]，此过程溶液变为红棕色。加完后继续搅拌直至溶液的红棕色消失(约 0.5 h)。减压过滤，干燥，得到产品[2]。纯 N-溴代丁二酰亚胺的熔点为 178℃。

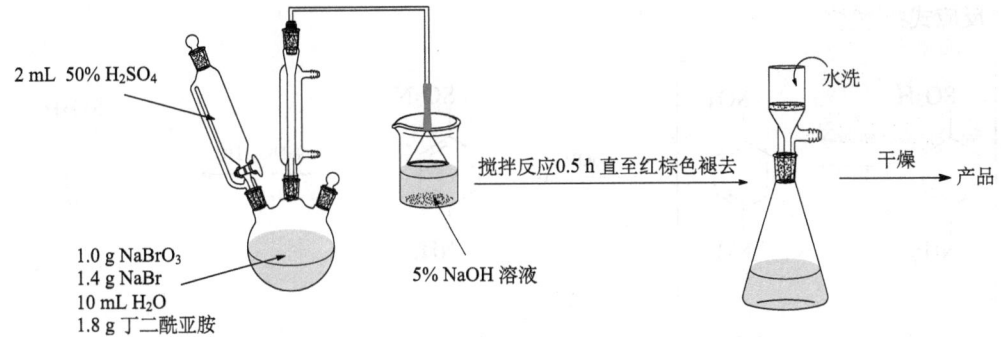

图 3.19 制备 NBS 的反应装置和流程

注释

[1] 硫酸的滴加速度需严格控制，应使反应溶液上端无较多溴蒸气产生。
[2] 反应废液不得直接倒入水槽，必须将其收集到废酸缸内统一处理。

思考题

为什么硫酸溶液需要慢慢滴加至反应瓶中，而不是一次性加入？

参考文献

王聪, 王利民, 王芳, 等. 2011. *N*-溴代丁二酰亚胺绿色合成工艺条件的探究. 精细化工中间体, 41(2): 63-65.

Fujisaki S, Hamura S, Eguchi H, et al. 1993. Organic synthesis using sodium bromate. Ⅱ. A facile synthesis of *N*-bromoimides and amides using sodium bromate and hydrobromic acid (or sodium bromide) in the presence of sulfuric acid. Bull Chem Soc Jan, 66: 2426-2428.

Kajigaeshi S, Nakagawa K, Fujisaki S, et al. 1985. A practical synthesis of *N*-bromo imides by use of sodium bromite. Bull Chem Soc Jan, 58: 769-770.

实验34　甲基橙的制备

实验目的

(1) 掌握通过重氮化反应和偶联反应制备偶氮染料的原理和方法。

(2) 制备甲基橙。

实验原理

甲基橙是一种偶氮染料，由对氨基苯磺酸重氮盐与 *N*, *N*-二甲基苯胺的乙酸盐在中性或弱酸性介质中发生偶联反应合成。甲基橙常用作酸碱指示剂。在 pH>4.4 的溶液中，甲基橙以橙黄色负离子存在。在 pH<3.2 的溶液中，甲基橙以红色的酸式甲基橙呈现。因此，甲基橙可用作滴定的指示剂，其滴定的 pH 范围为 3.2～4.4。指示剂通常配成 0.01%的水溶液。

反应式：

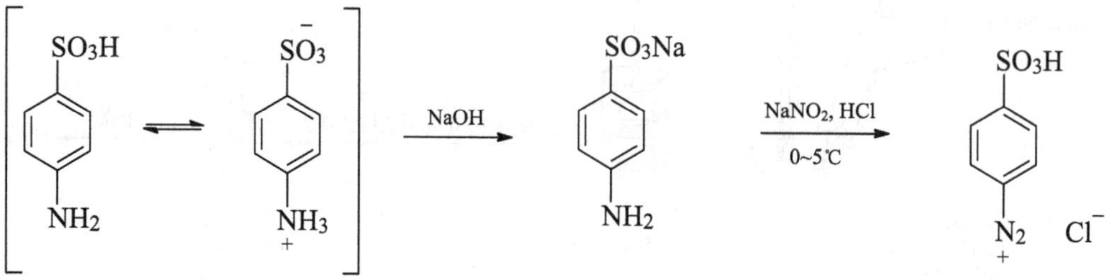

对氨基苯磺酸(两性离子)　　对氨基苯磺酸钠

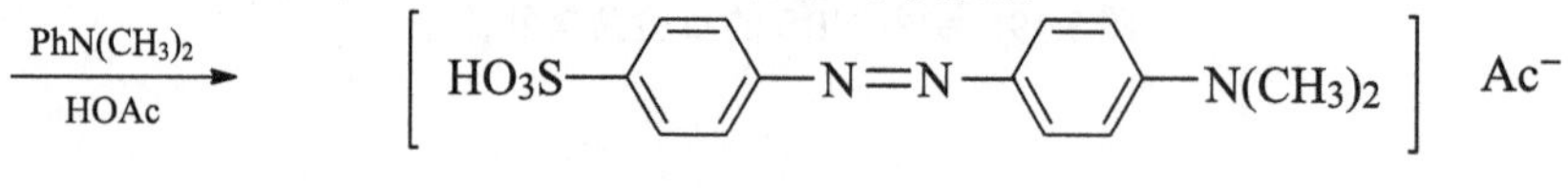

酸性甲基橙(红色)

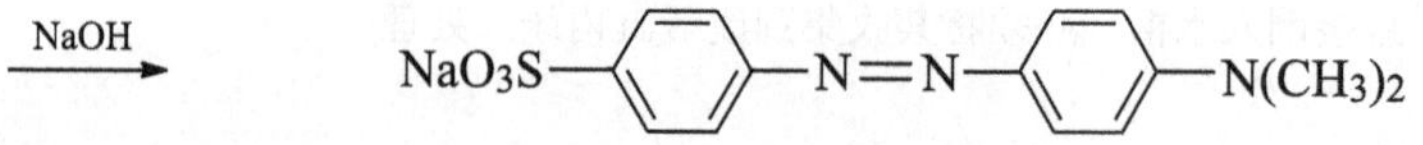

甲基橙(橙黄色)

实验用品

仪器：烧杯、试管、温度计、电热套、表面皿、玻璃棒、布氏漏斗、抽滤瓶、烘箱。

试剂：对氨基苯磺酸、亚硝酸钠、冰块、氢氧化钠溶液(1%、5%、10%)、冰醋酸、浓盐酸、N,N-二甲基苯胺、饱和氯化钠溶液、乙醇、乙醚、碘化钾-淀粉试纸。

实验步骤

1. 对氨基苯磺酸的重氮化反应

在 100 mL 烧杯中加入 2.0 g (0.012 mol) 对氨基苯磺酸和 10 mL 5%氢氧化钠溶液，温热至结晶溶解[1]。将溶液冷却至室温，在搅拌下添加亚硝酸钠溶液(0.8 g 亚硝酸钠溶于 3 mL 水中)，直到溶解完成。

缓慢地将溶液倒入含有 13 mL 冷水和 2.5 mL 浓盐酸的 250 mL 烧杯中。反应过程中，温度应控制在 5℃以下[2]。

用淀粉-碘化钾试纸监测水层。若不呈现蓝色，需补加亚硝酸钠溶液[3]。继续在冰水浴中搅拌 15 min，直至反应完全。白色悬浮液不经分离直接用于甲基橙的合成。

2. 偶联反应

将 1.3 mL N,N-二甲基苯胺和 1 mL 冰醋酸振荡混匀。在搅拌下将此混合液缓慢加入上述在冰水浴冷却的重氮盐溶液中，加完后继续在冰水浴中充分搅拌 10 min，形成红色染料的硬糊状物。

向混合物中加入 15 mL 10%氢氧化钠溶液，加热至沸腾，使固体完全溶解。当大多数甲基橙溶解时，用冰水浴冷却混合物使甲基橙晶体完全析出。依次用少量饱和氯化钠溶液、乙醇和乙醚洗涤。抽滤，收集粗产品。

3. 重结晶

粗产品用 1%氢氧化钠溶液重结晶[4]。抽滤，依次用饱和氯化钠溶液、乙醇和乙醚洗涤，压紧抽干，得片状结晶，产量约 2.5 g。将少许甲基橙溶于水中，加几滴稀盐酸，再用稀碱中和，观察颜色变化并记录。

注释

[1] 对氨基苯磺酸为两性化合物，其酸性强于碱性，它不溶于酸性溶液而溶于碱性溶液。然而，重氮化反应必须在酸(HNO_2)溶液中进行。为了解决这个问题，首先将对氨基苯磺酸溶解在碱性溶液中，而不是溶解在酸性溶液中。

[2] 重氮化反应过程中，必须严格控制温度，若反应温度高于 5℃，生成的重氮盐易水解为酚，降低目标产物的产率。

[3] 若碘化钾-淀粉试纸不变蓝色，则需补充少许亚硝酸钠溶液。

[4] 湿甲基橙在自然光照射下，颜色迅速变深。普通的粗甲基橙是一种红紫色(淡紫色)晶体。用乙醇和乙醚洗涤，可以使产品快速干燥。

思考题

(1) 本实验中，在重氮化反应前为什么还要加入氢氧化钠？如果直接将对氨基苯磺酸与盐酸混合后加入亚硝酸钠溶液进行重氮化反应可以吗？为什么？

(2) 对氨基苯磺酸的重氮化反应为什么要控制温度低于 5℃？

(3) 重氮盐偶联反应为什么不能在强酸或强碱介质中发生？

(4) N,N-二甲基苯胺与重氮盐的偶联反应为什么总是生成对位产物？

(5) 在重氮盐的制备中加入氯化亚铜作为催化剂会出现什么结果？

实验 35　二苯甲酮肟的制备及贝克曼重排

实验目的

(1) 掌握制备肟的方法。

(2) 了解贝克曼重排的原理。

实验原理

与醛和酮一样，二苯甲酮能与羟胺作用生成相应的肟。肟在酸的作用下发生分子重排生成酰胺。该反应 1886 年由贝克曼首先发现，称为贝克曼重排。

反应式：

$$R^1R^2C=NOH \xrightarrow{酸} R^1-NH-C(=O)-R^2$$

反应机理：

$$R^1R^2C=NOH \xrightarrow{H^+} R^1R^2C=N^+(OH_2) \xrightarrow{H^+} [R^1-N\equiv C-R^2]^+ \cdots$$

$$\longrightarrow R^2-C(=N-R^1)-O^+H_2 \xrightarrow{-H^+} R^2-C(=N-R^1)-OH \longrightarrow R^1-NH-C(=O)-R^2$$

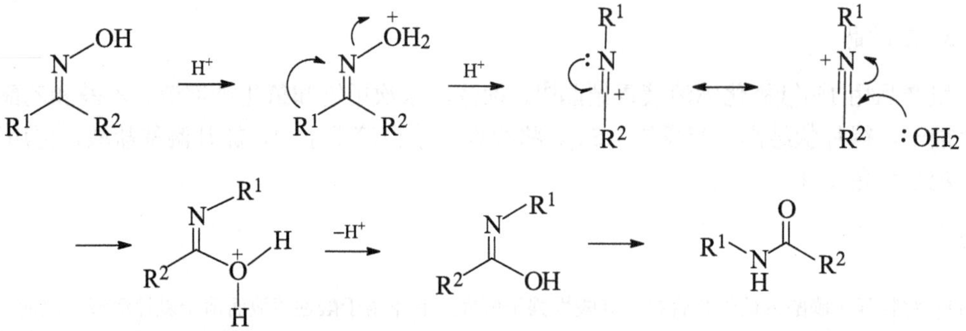

贝克曼重排通常用酸催化。在贝克曼重排中，肟氮原子插入醛和酮的 $C_肟—C_\alpha$ 键中。在大多数情况下，酮肟的贝克曼重排是立体专一的，N—O 键断裂的同时发生迁移，且迁移基团与氮上的离去基团处于反叠位。贝克曼重排的主要应用是通过肟将环己酮转化为己内酰胺。己内酰胺是生产尼龙-6 的原料。

本实验将羟胺和二苯甲酮作用生成二苯甲酮肟。将二苯甲酮肟在多聚磷酸(PPA)催化下进行分子重排，再倒入冰水中，即可得到苯甲酰基苯胺。多聚磷酸是一种浆状液体，与水快速反应生成磷酸溶液，是进行该反应的良好介质。当反应完成后，加水简单稀释就会析出产物，然后重结晶即可。

反应式：

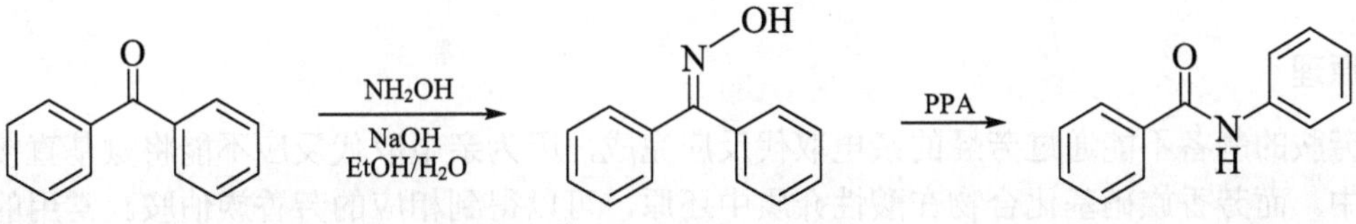

$$\underset{}{\text{Ph}_2\text{C=O}} \xrightarrow[\text{EtOH/H}_2\text{O}]{\text{NH}_2\text{OH, NaOH}} \underset{}{\text{Ph}_2\text{C=N-OH}} \xrightarrow{\text{PPA}} \underset{}{\text{PhC(O)NHPh}}$$

实验用品

仪器：锥形瓶、烧杯、布氏漏斗、抽滤瓶、熔点测定仪、电热套。

试剂：二苯甲酮、羟胺盐酸盐、氢氧化钠、多聚磷酸、乙醇、浓盐酸。

实验步骤

1. 二苯甲酮肟的制备

在 125 mL 锥形瓶中加入 2.5 g(0.0137 mol) 二苯甲酮、1.5 g(0.0216 mol) 羟胺盐酸盐、5 mL 乙醇和 1 mL 水，然后加入 18~20 粒固体氢氧化钠并充分振荡，使固体完全溶解。

将锥形瓶放在水蒸气浴上，温和煮沸约 5 min。在搅拌下将溶液转入装有 8 mL 浓盐酸和 50 mL 水的烧杯中，即有白色固体析出。

冷却后抽滤，并用少量冷水洗涤晶体，抽滤得到粗产品，粗产品放在滤纸上压干，用约 20 mL 乙醇重结晶提纯。纯二苯甲酮肟为无色针状晶体，熔点为 142~143℃。

2. 二苯甲酮肟的贝克曼重排

在 100 mL 烧杯中放入 25 mL 多聚磷酸和前面制得的二苯甲酮肟，烧杯内放一支 200℃ 温度计，不断用玻璃棒搅动反应液。

用电热套小火加热，慢慢升温到 100℃ 进行重排反应。保温 20 min 后，继续加热并搅动，直至温度为 125~130℃。停止加热，放置冷却 10 min 后，将黏稠液慢慢倒入盛有 350 mL 冷水的烧杯中[1]，不断搅拌。

抽滤固体后，用约 20 mL 乙醇重结晶。干燥，称量[2]，测定熔点。

纯苯甲酰基苯胺为银白色针状晶体，熔点为 163~164℃。

注释

[1] 将 PPA 倒入水中时会放出大量的热，因此这一步操作应注意安全。

[2] 产品最好先在滤纸上挤压，然后置于空气中晾干。

思考题

(1) 贝克曼重排的催化剂一般为何种类型？

(2) 试写出由环己酮制备己内酰胺的反应机理。

实验 36 苯胺的制备

实验目的

(1) 掌握硝基苯还原制备苯胺的原理和方法。

(2) 熟练掌握水蒸气蒸馏和简单蒸馏等基本操作。

实验原理

芳胺的制备不能通过芳烃的亲电取代反应完成，因为亲电取代反应不能将氨基直接导入芳环中。而芳香族硝基化合物在酸性介质中还原，可以得到相应的芳香族伯胺。常用的还原剂有铁-盐酸、铁-乙酸、锡-盐酸等。工业上用铁粉和盐酸还原硝基苯制备苯胺，但会产生大量含苯胺的铁泥，造成环境污染。现在逐渐改用硝基苯催化氢化的方法，常用的催化剂有镍、铂、钯等。实验室仍用铁粉还原法制备芳胺。

反应式：

$$4\,C_6H_5NO_2 + 9Fe + 4H_2O \xrightarrow{H^+} 4\,C_6H_5NH_2 + 3Fe_3O_4$$

实验用品

仪器：二颈烧瓶、冷凝管、水蒸气发生器、接液管、锥形瓶。

试剂：硝基苯、乙醚、碳酸钠、氢氧化钠、冰醋酸、铁粉、饱和氯化钠溶液。

实验步骤

将 9 g (0.16 mol) 铁粉、17 mL H_2O 和 1 mL 冰醋酸加入 150 mL 二颈烧瓶中，振荡混匀，装上回流冷凝管和装有 7 mL 硝基苯的滴液漏斗。用滴液漏斗逐滴加入硝基苯，每次加完后用力振摇[1]。硝基苯加完后，将反应物在热水浴中加热回流 0.5 h，并时常摇动烧瓶[2]。将反应液冷却至室温，加入碳酸钠至反应物呈碱性[3]。改为水蒸气蒸馏装置[4]，继续蒸馏至馏出液变澄清。向馏出液中加入氯化钠至溶液饱和[5]。

将溶液转移至分液漏斗中，分出苯胺层，水层用乙醚萃取三次(每次 7 mL)。合并有机层，用粒状氢氧化钠干燥[6]。将干燥后的醚溶液移入 50 mL 圆底烧瓶中，在 70℃水浴上蒸去乙醚。换用空气冷凝管[7]，烧瓶中插入一支 250℃温度计，用电热套加热，收集 182~185℃馏分[8]。

纯苯胺的沸点为 184.4℃，n_D^{20} 为 1.8563。

苯胺的红外光谱和核磁共振氢谱分别见图 3.20 和图 3.21。

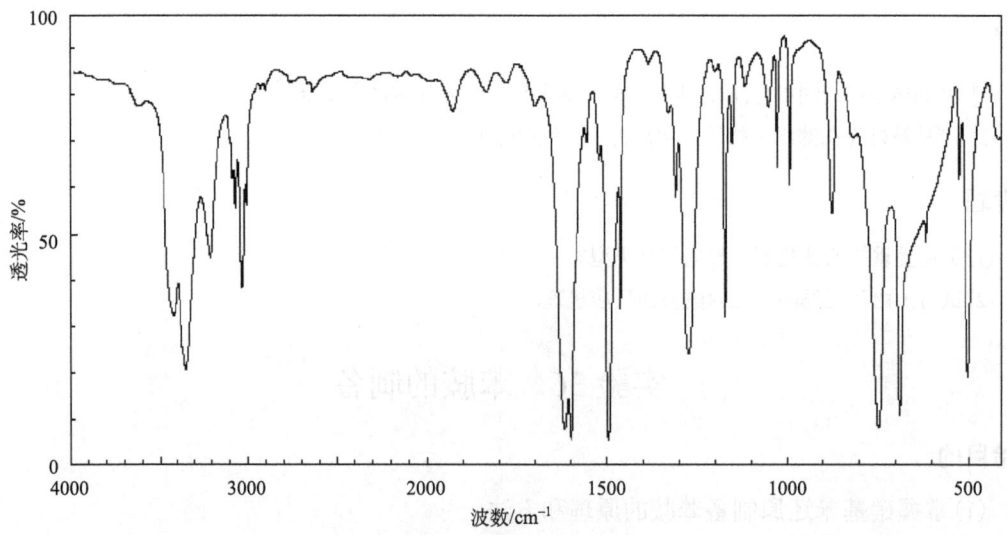

图 3.20 苯胺的红外光谱

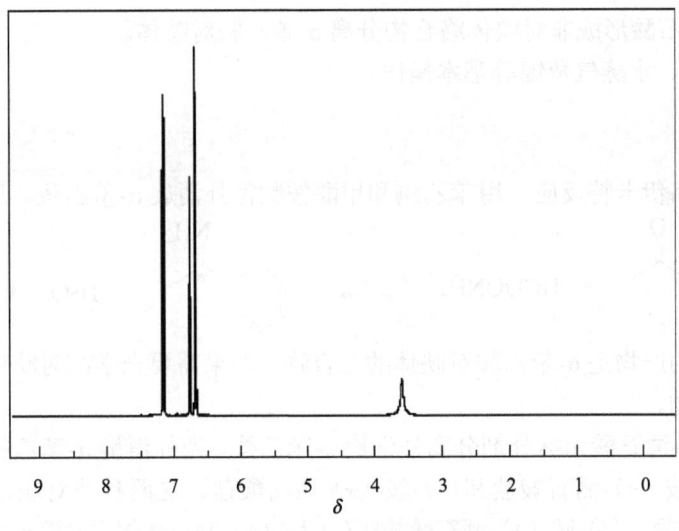

图 3.21 苯胺的核磁共振氢谱

注释

[1] 本实验是放热反应,也是非均相反应(油/水/固三相)。每次加入硝基苯时均有剧烈的反应发生,所以需注意控制反应温度,防止温度过高或聚热而引起暴沸;充分振荡,保持瓶内反应物之间充分接触。

[2] 硝基苯为黄色油状物,如果回流液中黄色油状物消失,转变成乳白色油珠,表示反应已完全。

[3] 当反应完成后,部分苯胺以盐酸盐的形式存在。将溶液调节至强碱性,使苯胺盐酸盐完全以苯胺形式游离出来。

[4] 进行水蒸气蒸馏时,苯酚盐及联苯胺等都留在残液中,而馏出液中含苯胺和硝基苯。

[5] 在 20℃时 100 g 水中可溶解 3.4 g 苯胺。根据盐析原理,加氯化钠使溶液饱和,可以减少苯胺的损失。

[6] 因为氯化钙与苯胺能形成复合物,所以本实验用氢氧化钠干燥。

[7] 除蒸馏外,从硝基苯中分离出苯胺也可以用下列方法:将水蒸气蒸馏得到的混合物用盐酸酸化,形成苯胺盐酸盐而溶于水中,用乙醚将硝基苯萃取出来。然后将水层碱化,使苯胺游离出来,再用乙醚萃取苯胺。注意:苯胺有毒,操作时避免与皮肤接触或吸入蒸气。

[8] 反应完毕后,圆底烧瓶上黏附的黑褐色物质可用 1∶1 盐酸水溶液稍加热除去。

思考题

(1) 什么有机物可以采用水蒸气蒸馏分离?本实验为什么选择水蒸气蒸馏把苯胺从反应混合物中分离出来?

(2) 精制苯胺时,为什么用粒状氢氧化钠作干燥剂而不用硫酸镁或氯化钙?

实验 37　外消旋 α-苯乙胺的制备和拆分

实验目的

(1) 掌握洛伊卡特反应制备 (±)-α-苯乙胺的原理和方法。

(2) 与(+)-酒石酸形成非对映体络合物分离 α-苯乙胺对映体。
(3) 巩固萃取、水蒸气蒸馏等基本操作。

实验原理

本实验利用洛伊卡特反应,用苯乙酮和甲酸铵制得外消旋 α-苯乙胺。反应式为

$$\underset{}{C_6H_5COCH_3} + HCOONH_4 \xrightarrow{\triangle} \underset{}{C_6H_5C^*H(NH_2)CH_3} + H_2O + CO_2$$

该反应得到的产物是 α-苯乙胺对映体的混合物。如果需要光学纯对映体,则需要进一步分离对映体混合物。

本实验以(+)-酒石酸为拆分剂分离外消旋 α-苯乙胺。当外消旋 α-苯乙胺与(+)-酒石酸反应时,形成(+)-胺-(+)-酒石酸盐和(−)-胺-(+)-酒石酸盐。这两种非对映体络合物在甲醇中的溶解度有显著差异。(+)-胺-(+)-酒石酸盐比(−)-胺-(+)-酒石酸盐更易溶于甲醇。(−)-胺-(+)-酒石酸盐可以通过选择性结晶从甲醇溶液中分离出来。分离的(−)-胺-(+)-酒石酸络合物与过量的氢氧化钠水溶液反应,将盐转化为游离胺(溶于有机溶剂)和酒石酸二钠(溶于水)。分离出的盐经氢氧化钠处理后,通过萃取分离转化为光学活性的游离碱。利用(+)-酒石酸对 α-苯乙胺进行化学拆分的流程如图 3.22 所示。

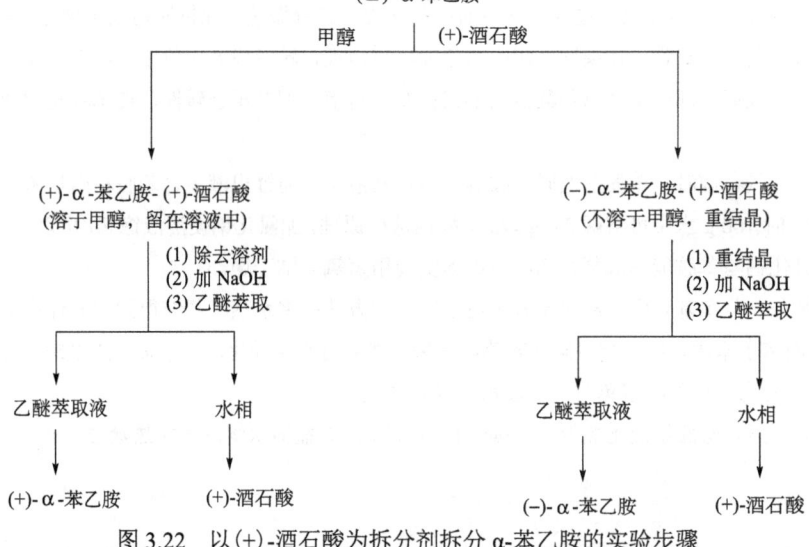

图 3.22 以(+)-酒石酸为拆分剂拆分 α-苯乙胺的实验步骤

用来测定产品旋光度的仪器称为旋光仪。从旋光度可以测定对映体的过量并评价其拆分效率。

实验用品

仪器:水泵、旋光仪、折光仪、接液管、分液漏斗、三颈烧瓶、蒸馏头、圆底烧瓶、容量瓶、球形冷凝管、抽滤瓶、锥形瓶、克氏蒸馏头。

试剂:苯乙酮、甲酸铵、氯化钠、甲醇、盐酸、苯、氢氧化钠、(+)-酒石酸、乙醚、无水硫酸钠。

实验步骤

1. (±)-α-苯乙胺的制备

在 100 mL 圆底烧瓶中加入 12.0 g (0.1 mol) 苯乙酮、20.0 g (0.32 mol) 甲酸铵及 2~3 粒沸石，温度计插入反应混合物中，装配简单蒸馏装置。

用电热套缓慢加热，当温度达到 150℃时，反应混合物分为两相，两相间有一条明显的界线。继续加热，反应物成一相，当温度升至 185℃时，停止加热。在此过程中，水和苯乙酮被蒸出。

将馏出物转入分液漏斗，分层，有机层倒回反应瓶。继续加热并保持 185~190℃反应 1.5 h。冷却至室温，将反应混合物转入分液漏斗，分层，用饱和氯化钠溶液洗涤有机层两次，每次 10 mL。将有机层倒回原反应瓶，加入 10 mL 浓盐酸和 2 粒沸石，加热至沸，回流 50 min。冷却至室温，将混合物转入分液漏斗，用甲苯萃取两次，每次 10 mL。保留有机层，水层转入烧杯中。

将水层在冰水浴中冷却，在搅拌下缓缓加入 20 mL 50%氢氧化钠溶液。转入分液漏斗，分出有机层，水层用乙醚萃取三次（每次 10 mL）。合并有机层和乙醚层，用粒状氢氧化钠干燥。将清液转入干燥的蒸馏烧瓶。

蒸馏收集 189~190℃的馏分，称量，计算产率，(±)-α-苯乙胺的折光率为 1.5238。

2. (±)-α-苯乙胺的拆分

在 100 mL 圆底烧瓶中加入 50 mL 甲醇和 3.8 g (0.119 mol) (+)-酒石酸，在温水浴上加热使其溶解。

将溶液加热至近沸，加入 3.0 g (0.025 mol) (±)-α-苯乙胺[1]。加完后，振荡并自然冷却至室温。塞住烧瓶，将溶液室温放置 24 h，应析出白色菱形晶体（若析出的晶体不是菱形而是针形或无定形，应重新加热再次结晶直至析出白色菱形晶体）[2]。抽滤，用少量冷甲醇洗涤晶体 3 次，抽干，即可得到(−)-α-苯乙胺-(+)酒石酸盐。称量，计算产率。

将所得晶体置于锥形瓶中，加入 10 mL 水和 2 mL 50% NaOH 溶液。搅拌混合物至固体完全溶解。将溶液转入分液漏斗，用乙醚萃取三次（每次 10 mL），萃取液用无水硫酸钠干燥。

将清液转入圆底烧瓶，常压蒸馏除去乙醚，再减压蒸馏，收集 81~81.5℃/2.4 kPa 的馏分，即得(−)-α-苯乙胺。称量，计算产率，并测量折光率。如欲得(+)-α-苯乙胺，可提纯(−)-α-苯乙胺-(+)酒石酸盐晶体的滤液。将滤液中的溶剂除去，可得到(+)-α-苯乙胺-(+)酒石酸盐晶体。再用相同方法处理，即可得(+)-α-苯乙胺。

3. 比旋光度的测定

取洁净、干燥的 50 mL 容量瓶，用电子天平准确称量后，将所制得的(−)-α-苯乙胺移入，再次准确称量后，计算(−)-α-苯乙胺的质量。加入甲醇稀释至刻度。

取一个干净的 1 dm 旋光管，用准备好的样品溶液润洗旋光管后倒出。然后加入样品溶液满至管口，确保没有气泡。如果仍有气泡，可将旋光管带凸颈的一端向上倾斜，将气泡逐入凸颈部位，而不是旋光管的顶部或底部。然后，打开旋光仪，将旋光管放入样品室内，盖好盖子[3]。开始读数，测量样品的旋光角。读数和零点之间的差异即为旋光角的度数。计算比旋光度。

注释

[1] 需小心将(±)-α-苯乙胺加入热溶液中，以免液体暴沸。
[2] 必须得到菱形晶体，这是实验成功的关键。
[3] 小心使用旋光仪。

思考题

(1) 本实验用苯乙酮制备 α-苯乙胺的方法存在哪些优点？
(2) 本实验中关键步骤是什么？
(3) 什么是外消旋体？有多少种方法可以用来分离光学对映体？

实验 38　反-1,2-二氨基环己烷的手性拆分

实验目的

(1) 了解异构体手性拆分的原理。
(2) 学习利用 L-酒石酸与反-1,2-二氨基环己烷形成非对映异构体进行手性拆分的方法。
(3) 学习比旋光度的计算方法。

实验原理

邻二胺(或 1,2-二胺)是催化剂和药物制备中使用最广泛的结构单元之一。其中，易得的 1,2-二氨基环己烷(DACH)是广泛使用的手性二胺。

手性拆分是从外消旋体中分离对映体的方法。作为一种简单、实用的方法，手性拆分被广泛使用。最常见的手性拆分方法之一是通过将外消旋化合物和光学拆分剂反应得到非对映异构体，利用非对映异构体之间物理性质的差异将其分离。

本实验中，L-酒石酸用作分离反式-1,2-二氨基环己烷的拆分剂。L-酒石酸与外消旋 1,2-二氨基环己烷反应，形成 DACH-L-酒石酸络合物的非对映异构体混合物。(R,R)-DACH 与 L-酒石酸生成的络合物在水中的溶解度比(S,S)-DACH 与 L-酒石酸生成的络合物低得多。因此，只有(R,R)-DACH-L-酒石酸络合物白色晶体从水溶液中析出，在溶液中剩下另一个络合物。分离得到(R,R)-DACH-L-酒石酸盐络合物。最后，可以通过加碱化从盐中得到(R,R)-DACH。实验流程参见图 3.23。

实验用品

仪器：电热套、圆底烧瓶、温度计、量筒、磁力搅拌器、分液漏斗、布氏漏斗、旋光仪。
试剂：反-1,2-二氨基环己烷、冰醋酸、L-酒石酸、二氯甲烷、氢氧化钾、无水硫酸钠、乙醇。

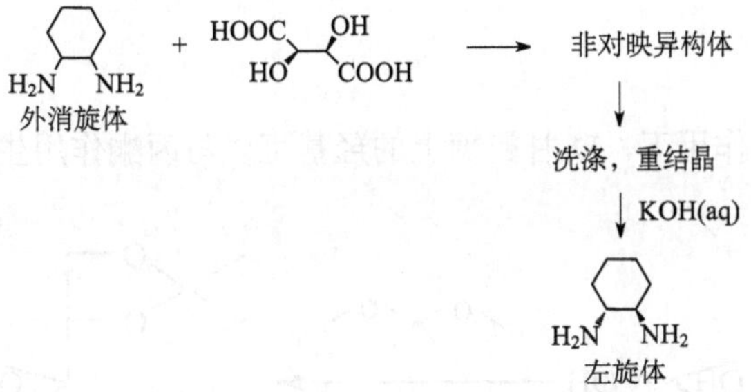

图 3.23 利用 L-酒石酸手性拆分反-1,2-二氨基环己烷

实验步骤

1. 反-1,2-二氨基环己烷的拆分

取 50 mL 圆底烧瓶,将 5.0 g(0.0438 mol)反-1,2-二氨基环己烷溶于 8.5 mL 水中,在搅拌下慢慢加入 3.3 g(0.0220 mol)L-酒石酸,并使温度不超过 70℃。

在搅拌下慢慢滴加 2.2 mL 冰醋酸直至酒石酸完全溶解,并控制反应温度在 90℃左右。不断搅拌使反应液冷却,直至有沉淀析出[1]。

停止搅拌,用冰水浴冷却 30 min,过滤得到晶体,依次用 2 mL 冰水和 4 mL 乙醇洗涤晶体,抽干后用水洗涤,得到白色固体。

将白色固体转入烧杯中,缓慢加入 3 mL 4 mol/L 氢氧化钾溶液,待晶体完全溶解后转入分液漏斗,用乙酸乙酯(2×15 mL)萃取。

合并有机相,在干燥的锥形瓶中用无水硫酸钠干燥。用旋转蒸发仪除去溶剂,收集产物(−)-反-1,2-二氨基环己烷。称量,计算产率。

2. 比旋光度的测定

精确称取 500 mg 产物,配成 5 mL 乙醇溶液。用旋光仪测定旋光度。记录旋光度并计算比旋光度。

注释

[1] 不断搅拌使反应冷却,直至有沉淀析出。

思考题

L-酒石酸与反-1,2-二氨基环己烷反应时,为什么温度不能超过 70℃,而且冰醋酸要逐滴加入?

实验 39 2,3;5,6-二-O-异丙叉基-α-D-呋喃甘露糖的制备

实验目的

(1) 了解保护糖上羟基的原理和方法。
(2) 学习利用异丙叉基保护 D-甘露糖的实验操作。

实验原理

在对甲苯磺酸催化作用下，D-甘露糖上的羟基可以与丙酮作用生成缩酮，从而实现羟基保护。反应式如下：

实验用品

仪器：圆底烧瓶、磁力搅拌器、冷凝管、电热套、分液漏斗、锥形瓶、温度计、量筒、布氏漏斗。

试剂：D-甘露糖、DMF、2,2-二甲氧基丙烷、对甲苯磺酸、氢氧化钠、乙酸乙酯、氯化钠、无水硫酸钠、乙醚、正己烷。

实验步骤

在 50 mL 圆底烧瓶中加入 1.0 g (0.0056 mol) D-甘露糖和 10 mL DMF，在磁力搅拌器上搅拌使其完全溶解[1]。加入 5 mL 2,2-二甲氧基丙烷及 10 mg 对甲苯磺酸。安装回流装置，冷凝管上端安装干燥管。

将反应液加热至 80℃，搅拌反应约 1 h 后，稍微冷却，加入适量 10% NaOH 溶液中和[2]。

将中性溶液转入分液漏斗，用 20 mL 水稀释后用乙酸乙酯(2×15 mL)萃取。合并有机相，用饱和氯化钠溶液洗涤三次。加入无水硫酸钠干燥，过滤，用旋转蒸发仪除去溶剂。

纯 2,3;5,6-二-O-异丙叉基-α-D-呋喃甘露糖为无色晶体，熔点为 122～123℃。实验流程参见图 3.24。

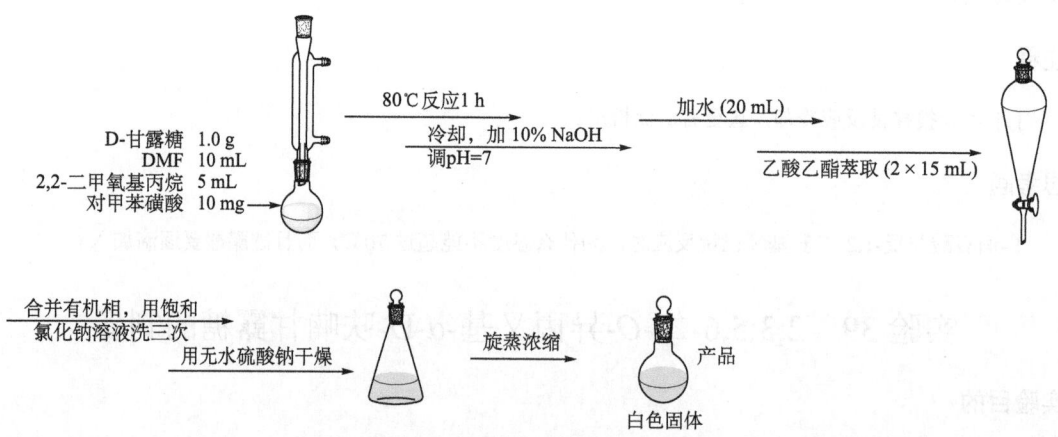

图 3.24 制备 2,3;5,6-二-O-异丙叉基-α-D-呋喃甘露糖的实验流程

注释

[1] 用磁力搅拌器进行搅拌时应控制转速,防止液体暴溅。
[2] 加入 10% NaOH 溶液时需控制量,切忌过量。

思考题

(1) 带干燥管的冷凝管有什么作用?
(2) 旋转蒸发后若无晶体析出,可采取什么措施?

实验 40　从茶叶中提取咖啡因

实验目的

(1) 掌握用索氏提取器提取咖啡因的原理和方法。
(2) 熟悉萃取、蒸馏、升华等基本操作。

实验原理

咖啡因是一种甲基黄嘌呤,IUPAC 命名为 1,3,7-三甲基-3,7-二氢-1H-吲哚-2,6-二酮。

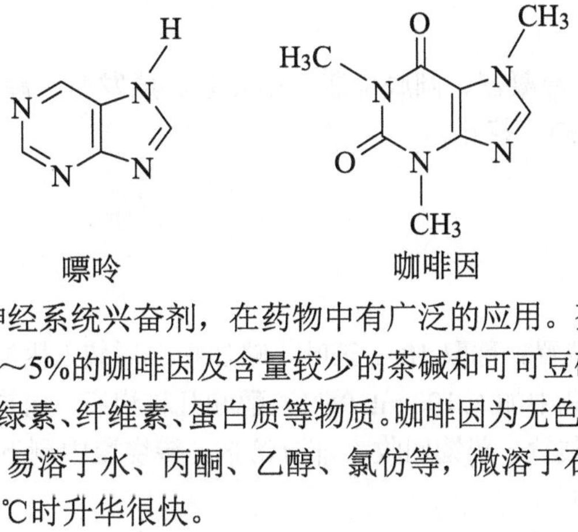

嘌呤　　咖啡因

咖啡因可作为中枢神经系统兴奋剂,在药物中有广泛的应用。茶叶中含有多种生物碱,其主要成分是含量为 1%～5% 的咖啡因及含量较少的茶碱和可可豆碱。此外,茶叶中还含有 11%～12% 的丹宁酸及叶绿素、纤维素、蛋白质等物质。咖啡因为无色针状结晶,熔点为 234℃。含结晶水的咖啡因味苦,易溶于水、丙酮、乙醇、氯仿等,微溶于石油醚、苯和乙醚。120℃时升华相当显著,至 178℃ 时升华很快。

本实验采用索氏提取器从茶叶中提取咖啡因。粗咖啡因溶液(使用有机溶剂,如乙醇或氯仿)首先通过索氏提取器的连续萃取分离出来,进一步升华得到纯咖啡因。

索氏提取器是 Soxhlet 于 1879 年发明。它最初设计用于从固体材料中提取脂质。如图 3.25 所示,索氏提取装置由冷凝管、索氏提取器和圆底烧瓶组成。

当用索氏提取器提取时,将固体样品放在用厚滤纸制成的筒中,然后整个筒放在索氏提取器的套管中。当溶剂加热回流时,溶剂蒸气上升至冷凝管冷凝后流下来进入套管,浸润样品。套管逐渐被热的溶剂充满。一些待提纯物质逐渐溶解在热溶剂中。当套管几乎被充满时,溶剂顺着虹吸管自动流出,重新进入烧瓶进行蒸馏。该循环可视为一次索氏提取。这个循环可进行多次,提取时间甚至可能达到数小时或数天。

每次提取循环中都有一部分非挥发化合物溶解在溶剂中,经过多次提取后这些化合物在蒸馏烧瓶中浓缩。该装置的优点是仅使用一份溶剂循环,即可达到多次提取。

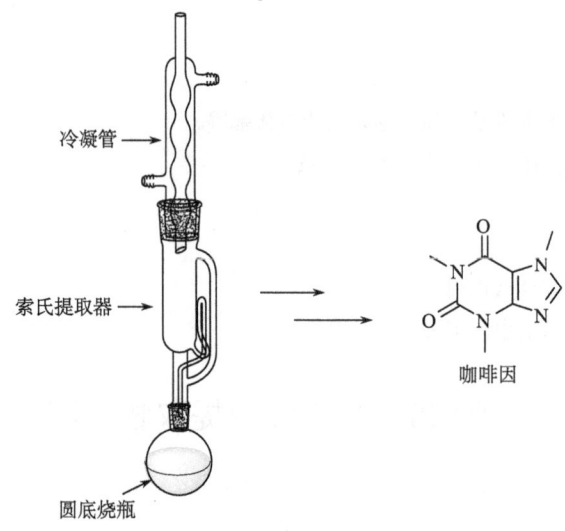

图 3.25 索氏提取装置

提取结束后蒸出溶剂(通常用旋转蒸发仪),即可得到被提取的化合物。不溶的固体物质则留在提取器中,弃去。

实验用品

仪器:索氏提取器、冷凝器、圆底烧瓶、电热套、蒸发皿、漏斗。

试剂:95%乙醇、CaO、茶叶。

实验步骤

1. 咖啡因的初提取

按图 3.25 安装实验装置。称取 10 g 茶叶,略加粉碎后装入用滤纸做成的套袋中[1],装入索氏提取器[2]。在圆底烧瓶中加入 100 mL 95%乙醇和几粒沸石,加热回流,连续索氏提取两三次[3]。稍冷后,改成蒸馏装置,蒸馏回收大部分溶剂,待烧瓶中剩下约 5 mL 时[4],停止加热,趁热将残液转入蒸发皿中。

2. 升华提纯咖啡因

在蒸发皿中加入 4 g CaO[5]。将混合物用研杵磨成灰泥,搅拌成淡黄色粉末。在淡黄色粉末上方放一张准备好的有几十个孔的滤纸(孔由滤纸底部扎到上部,直径约为 2 mm)。如图 3.26 所示,在滤纸(滤纸直径略大于漏斗直径)上放一个倒置玻璃漏斗,漏斗颈部塞一小团疏松的棉花。把蒸发皿放在电热套上。

小火加热,适当控温,当发现有大量白色蒸气出现时,停止加热,将蒸发皿从电热套上移开,并让升华继续进行 5 min[6]。当蒸发皿自然冷却至室温时,用刮刀将滤纸上和器皿周围的咖啡因刮到一张已称量的称量纸上。称量并测定升华过程的回收率。记录咖啡因粗产品和升华后产品的颜色和形状。

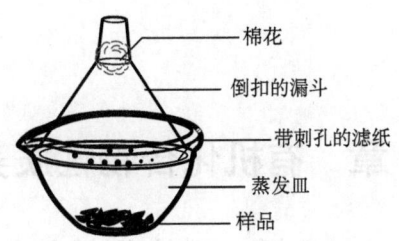

图 3.26 升华法提纯咖啡因的装置

注释

[1] 滤纸套大小要紧贴器壁,其高度不得超过虹吸管。滤纸包茶叶末时要仔细严密,防止漏出堵塞虹吸管。

[2] 索氏提取器的虹吸管极易折断,装置仪器和取拿时须特别小心。

[3] 若提取液颜色很淡时,即可停止提取。

[4] 瓶中的乙醇不可蒸得太干,否则残液很黏,转移时损失较大。

[5] CaO 用来吸水和除去部分酸性杂质。

[6] 升华过程中要控制好温度。若温度太低,升华速度较慢;若温度太高,产物将分解。

思考题

(1) 升华咖啡因时用到 CaO,它的作用是什么?

(2) 为什么采用升华法可以提纯咖啡因?

第4章 有机化合物性质实验

实验41 乙烯、乙炔的制备和性质

实验目的

(1) 掌握乙醇脱水制备乙烯的原理和方法。
(2) 掌握用碳化钙制备乙炔的原理和方法。
(3) 学习乙烯、乙炔的性质及鉴别方法。

实验原理

实验室中常用乙醇和酸(硫酸、磷酸等)脱水制备乙烯。浓硫酸起催化剂和脱水剂的作用。还存在生成乙醚的副反应。反应式如下：

$$CH_3CH_2OH \xrightarrow[170℃]{H_2SO_4} H_2C=CH_2 + H_2O$$

$$2CH_3CH_2OH \xrightarrow[140℃]{H_2SO_4} CH_3CH_2OCH_2CH_3 + H_2O$$

实验室常采用碳化钙与水反应制备乙炔。除乙炔外还生成一分子氢氧化钙。反应式如下：

$$CaC_2 + 2H_2O \longrightarrow HC \equiv CH + Ca(OH)_2$$

实验用品

仪器：圆底烧瓶、洗气瓶、蒸馏头、试管、温度计、恒压滴液漏斗、带支管的连接头、酒精灯。

试剂：浓硫酸、95%乙醇、10%硫酸、10%氢氧化钠溶液、2%溴的四氯化碳溶液、1%高锰酸钾溶液、2%氨水、饱和氯化钠溶液、10%硫酸铜溶液、氯化亚铜氨溶液、2%硝酸银溶液、碳化钙。

实验步骤

1. 乙烯的制备和性质

制备乙烯的实验装置如图 4.1 所示。在 250 mL 圆底烧瓶中加入 4 mL 95%乙醇，小心加入 12 mL 浓硫酸，再加入几粒沸石。在洗气瓶中加入 6 mL 10%氢氧化钠溶液，用来洗涤乙烯气体。加热，使温度控制在 160～170℃[1]，即可产生乙烯气体。实验结束时，先将导管从水中取出，再移走酒精灯，防止倒吸现象。

(1) 加成反应：取一支试管，加入 2 mL 2%溴的四氯化碳溶液，通入生成的乙烯气体，观察试管中溶液的颜色变化。

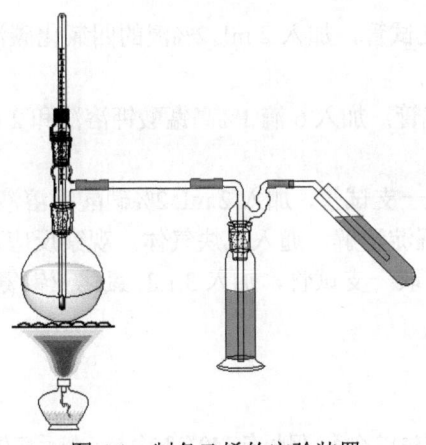

图 4.1 制备乙烯的实验装置

(2) 氧化反应：取一支试管，加入 6 滴 1%高锰酸钾溶液和 2 mL 10%硫酸，将乙烯通入溶液中，观察现象。

(3) 点燃实验：在尖嘴导管口点燃乙烯气体，观察燃烧的火焰。实验结束后，用盛有冷水的试管套在导管口熄灭火焰。

2. 乙炔的制备和性质

制备乙炔的实验装置参见图 4.2。在 250 mL 圆底烧瓶中加入 5 g 碳化钙，装上恒压滴液漏斗，在恒压滴液漏斗中加入 40 mL 饱和氯化钠溶液[2]。圆底烧瓶的支管连接盛有 10%硫酸铜溶液的洗气瓶。旋开恒压滴液漏斗的活塞，使饱和氯化钠溶液缓慢滴入圆底烧瓶中。产生的乙炔气体可用排水集气法收集。

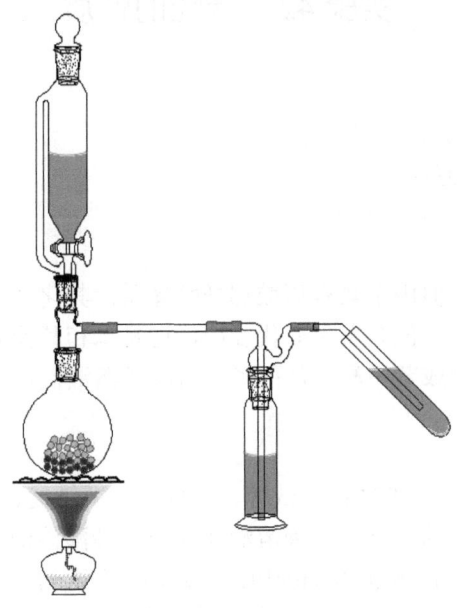

图 4.2 制备乙炔的实验装置

(1) 与卤素反应：取一支试管，加入 2 mL 2%溴的四氯化碳溶液，通入生成的乙炔气体，观察试管中溶液的颜色变化。

(2) 氧化反应：取一支试管，加入 6 滴 1%高锰酸钾溶液和 2 mL 10%硫酸，将乙炔通入溶液中，观察现象。

(3) 乙炔银的生成[3]：取一支试管，加入 2 mL 2%硝酸银溶液和 1 滴 10% NaOH 溶液，再滴加几滴 2%氨水，振荡使沉淀溶解，通入乙炔气体，观察反应现象。

(4) 乙炔亚铜的生成[3]：取一支试管，加入 3 mL 氯化亚铜氨溶液，通入乙炔气体，观察反应现象。

注释

[1] 硫酸氢乙酯在 170℃分解生成乙烯，但是在 140℃时会与乙醇反应生成乙醚，因此实验中要求迅速升温至 160℃以上，减少乙醚的生成。

[2] 水与碳化钙反应很剧烈，用饱和氯化钠溶液可以使反应平稳而均匀地产生乙炔气流。

[3] 干燥的乙炔银和乙炔亚铜受到震动或受热极易发生爆炸。为了防止爆炸发生，实验完毕应立即用稀硝酸或稀盐酸处理乙炔银和乙炔亚铜。

思考题

(1) 浓硫酸催化乙醇脱水制备乙烯时，产物中可能有哪些杂质？对实验有什么影响？
(2) 乙烯制备过程中如果不迅速升高温度会有什么影响？
(3) 用碳化钙制备的乙炔常带有臭味，如何除去？
(4) 分析并解释乙烯、乙炔加成和氧化反应的速率不同。

实验 42　芳烃的性质

实验目的

(1) 掌握芳烃的化学性质。
(2) 应用实验技术鉴别芳烃。

实验原理

苯虽然含有不饱和键，但由于具有离域(共轭)体系，其化学性质比烯烃、炔烃稳定，不易发生加成反应和氧化反应，但易发生取代反应，这是苯和其他芳香烃的特征反应。但甲苯等烷基苯能被高锰酸钾氧化成苯甲酸，从而使高锰酸钾溶液褪色。

实验用品

仪器：60 W 以上的灯泡、玻璃棒、试管、酒精灯、试管夹、烧杯。

试剂：苯、甲苯、环己烯、高锰酸钾溶液(0.5%)、硫酸溶液(10%)、浓硫酸、浓硝酸、萘、乙酰苯胺的乙酸-水(9∶1)溶液(0.2 mol/L)、氯苯的乙酸-水(9∶1)溶液(0.2 mol/L)、对硝基苯酚的乙酸-水(9∶1)溶液(0.2 mol/L)、苯酚的乙酸-水(9∶1)溶液(0.2 mol/L)、苯的乙酸-水(9∶1)溶液(0.2 mol/L)、溴的乙酸-水(9∶1)溶液(0.05 mol/L)、溴的四氯化碳溶液(3%)、苄氯、氯仿、无水三氯化铝、硝酸银溶液(50 g/L)。

实验内容

1. 与高锰酸钾溶液反应

在三支试管中分别加入 0.5 mL 苯、甲苯、环己烯,再分别加入 0.2 mL 高锰酸钾溶液 (0.5%)和 0.5 mL 硫酸溶液(10%),剧烈振荡(必要时在 60~70℃水浴上加热几分钟),观察并比较三种化合物与氧化剂作用的现象,并解释原因。

2. 磺化反应

1) 单环芳烃的磺化反应

在两支分别盛有 1 mL 苯和 1 mL 甲苯的干燥试管中各加 3 mL 浓硫酸,用单孔塞塞住,孔中插入一根玻璃管。将试管放在水浴(不超过 30℃)中加热,时常摇动试管[1]。仔细观察哪支试管内的分层先消失。当分层都消失后,将试管内的反应混合物分别倒入两个装有 10 mL 水的小烧杯中,稍轻搅拌后,观察现象并解释。

2) 萘的磺化反应

取一支干燥试管,加入 1 g 萘,水浴加热熔化。稍冷后加 1 mL 浓硫酸,小心加热 1~2 min,加热时不断振摇,直至混合物变成均匀液体。冷却后,在制得的深色黏稠液中加 2 mL 水,微热溶解,再冷至 15~20℃,观察现象并解释。

比较上述两个实验中三个磺化反应的结果,确定哪一个最容易,并解释原因。

3. 硝化反应

在两支干燥的试管中分别加入 1.5 mL 浓硝酸和 2 mL 浓硫酸,充分混合后,将热的混酸用冷水浴冷却至室温。慢慢分别滴加 1 mL 苯、甲苯,不断振荡使溶液混匀。如果放热过多,应用冷水冷却,控制温度不超过 50℃。3~5 min 后把混合液倾入盛有 20 mL 冷水的烧杯中,静置,观察现象。

4. 卤代反应

1) 芳环卤代反应

在五支试管中分别加入乙酰苯胺、苯、氯苯、对硝基苯酚、苯酚的乙酸-水(9∶1)溶液 (0.2 mol/L)各 2 mL,将这些试管放在 400 mL 装有温水[(35±2)℃]的烧杯中。另将一支装有 10 mL 溴的乙酸-水(9∶1)溶液(0.05 mol/L)的试管放在同一温水中。将这些试管放置 5 min 以达到水温。分别加入 2 mL 溴的乙酸-水(9∶1)溶液(0.05 mol/L)于五支试管中,用玻璃棒迅速混合,并注意褪色的时间(以秒计)。记录所用时间(溶液变为无色或淡黄色即到达终点)。比较反应速度。

2) 侧链卤代反应

取两支干燥试管,各加入 1 mL 甲苯、10 滴溴的四氯化碳溶液(3%),摇匀后将一支放在阳光下(如果没有阳光,可用镁光或 60 W 以上的灯光照射),另一支用黑纸包住放在黑暗处(如柜子中)。稍等片刻,取出,向试管口吹气,观察现象。

取两条滤纸条分别插入上述两支试管中,将滤纸一端用溶液浸湿,然后取出在空气中稍晾干，小心嗅其气味，描述有何不同（苄溴有刺激性气味，有催泪性）。

3）苯环卤代反应和侧链卤代反应比较

在两支干净试管中分别加入 2 mL 蒸馏水，其中一支加入 0.5 mL 氯苯，另一支加入 0.5 mL 苄氯，加热至沸，再分别滴加 0.3 mL AgNO$_3$ 溶液（50 g/L），观察并比较现象。

5. 傅-克反应

在一支干燥洁净的试管中加入 2 mL 氯仿和 3 滴无水苯，摇匀，斜执试管，使管壁润湿。再沿管壁加少许无水三氯化铝，使一部分粉末粘在管壁上。观察粘在管壁上的粉末和试管中溶液的颜色[2]。

注释

[1] 做好这个实验的关键在于要充分摇动，因为芳烃和混酸很难互溶。

[2] 芳香族化合物（包括芳香族卤化物）在三氯化铝催化剂存在下能与氯仿反应生成三芳基甲烷，呈各种颜色。例如，苯及其同系物、芳香族卤化物为橙至红色，萘为蓝色，蒽为绿色，而联苯和菲为紫红色等。要及时观察颜色，否则易发生变化。

思考题

(1) 为什么烷基苯和环己烯能被高锰酸钾氧化，而苯却不行？
(2) 用什么方法鉴别苯环上卤素和侧链卤素，用什么方法鉴别苯和甲苯？
(3) 根据实验结果，按使取代苯溴化的活性降低的次序排列 NHCOCH$_3$、Cl、OH 等基团。

实验 43　卤代烃的性质

实验目的

(1) 理解卤代烃的亲核取代反应机理，熟悉消去反应性质及反应条件。
(2) 了解利用亲核取代反应的活性区别伯、仲、叔卤代烃的方法。

实验原理

亲核取代反应是卤代烃的主要化学性质。卤代烃中 C—X 键是极性共价键，比烃类的 C—H 键活泼得多，其活性因卤原子和烃基结构的不同而异。对于相同烃基的卤代烷，其反应活性次序为：RI>RBr>RCl；不同烃基结构的卤代烷，其活性次序又因反应机理的不同而不同。在单分子亲核取代（S$_N$1）反应中，如卤代烷与硝酸银的乙醇溶液反应，卤代烷的活性次序是：叔卤代烷(3°)>仲卤代烷(2°)>伯卤代烷(1°)；而在双分子亲核取代（S$_N$2）反应中，如溴（或氯）代烷与碘化钠的丙酮溶液作用生成碘代烷的反应，卤代烷的活性次序是：伯卤代烷(1°)>仲卤代烷(2°)>叔卤代烷(3°)。卤代烯烃、卤代芳烃的化学性质与卤代烷烃有较大的区别，它们的反应活性不活泼，相比卤代烷烃的活性次序为：烯丙型或苄基卤代烃>伯、仲卤代烷>乙烯型卤代烃或卤代苯。卤代烃在强碱作用下，发生 β-消去反应，生成烯烃。多卤代烃（如 CHCl$_3$）在强碱作用下，发生 α-消去反应，生成卡宾。卡宾是重要的有机合成中间体。

实验用品

仪器：试管、酒精灯、试管夹、烧杯。

试剂：饱和硝酸银乙醇溶液、1-溴丁烷、2-溴丁烷、2-甲基-2-溴丙烷、溴苯、3 mmol/L 硝酸溶液、1-氯丁烷、1-碘丁烷、碘化钠丙酮溶液、苄氯、氯苯、2,4-二硝基氯苯、氢氧化钾、20%氢氧化钠溶液、氯仿、2%高锰酸钾溶液。

实验步骤

1. 与硝酸银乙醇溶液反应

(1)将 4 支干燥的试管依次标上记号并各加入 1 mL 饱和硝酸银乙醇溶液[1]，然后分别加入 2 滴 1-溴丁烷、2-溴丁烷、2-甲基-2-溴丙烷、溴苯，振摇后注意观察沉淀出现的先后次序。若未见沉淀析出，可在 70℃水浴中加热 3 min。在每支有沉淀的试管中各滴入一滴 3 mmol/L 硝酸溶液，振摇，若沉淀溶解则不是卤化银。根据实验现象，排列出四种溴代物的反应活性次序，说明烃基结构不同对 S_N1 反应速率的影响。

(2)在 3 支干燥的试管中各加入 1 mL 饱和硝酸银乙醇溶液，然后分别加入 2 滴 1-氯丁烷、1-溴丁烷、1-碘丁烷，充分振摇后，观察沉淀出现的先后次序。若未见沉淀析出，可在 70℃水浴中加热。根据实验现象，排列出三种卤代丁烷的反应活性次序，说明不同卤原子对 S_N1 反应速率的影响。

2. 与碘化钠丙酮溶液反应

在试管中加入 1 mL 碘化钠丙酮溶液，分别加入 3 滴 1-溴丁烷、苄氯、氯苯、2,4-二硝基氯苯，振摇后静置。观察是否出现沉淀或浑浊。若不见浑浊，可将试管于温热水浴中加热数分钟，再观察现象。浑浊出现说明有溴化钠或氯化钠生成[2]。

3. 消去反应

1) 生成烯烃：β-消去反应

在试管中加入 1 g 固体氢氧化钾和 4~5 mL 乙醇，稍微加热，当 KOH 全部溶解后，再加入 1 mL 1-溴丁烷。振摇混匀，塞上带有导管的塞子，导管另一端插入盛有溴水或酸性高锰酸钾溶液的试管中。试管中有气泡产生，溶液褪色，说明有乙烯生成。

2) 生成卡宾：α-消去反应

在试管中加入 3 mL 20%氢氧化钠溶液，再滴入 8 滴氯仿，在振摇下小心加热 1~2 min 至溶液沸腾，然后把试管浸在冷水中冷却至室温，加入 2~3 滴 2%高锰酸钾溶液，观察现象，写出反应式[3]。

注释

[1] 在 18~20℃时，硝酸银在无水乙醇中的溶解度为 2.1 g/100g。由于卤代烃能溶于乙醇而不溶于水，所以用乙醇作溶剂，能使反应处于均相，有利于反应顺利进行。

[2] 碘化钠溶于丙酮，而溴化钠和氯化钠不溶于丙酮。

[3] 氯仿在强碱条件下生成二氯卡宾,它在碱性溶液中易水解生成甲酸盐,后者把高锰酸钾还原成锰酸盐,使溶液变成绿色。

思考题

伯、仲、叔卤代烷与硝酸银乙醇溶液和碘化钠丙酮溶液作用的活性次序有何不同?试解释原因。

实验44 醇和酚的性质

实验目的

(1)巩固醇和酚的化学性质。
(2)掌握区别醇和酚的实验方法。

实验用品

仪器:试管。

试剂:正丁醇、2-丁醇、叔丁醇、苯甲酰氯、苄醇、环己醇、乙醇、甘油、苯酚、间苯二酚、对苯二酚、硝酸铈铵试剂、冰醋酸、盐酸、氯化锌、5%重铬酸钾溶液、10%氢氧化钠溶液、溴水、三氯化铁。

实验步骤

1. 醇的性质

1) 与苯甲酰氯反应

取三支配有塞子的试管,分别加入 0.5 mL 正丁醇、2-丁醇、叔丁醇,再加入 1 mL 水和 10 滴苯甲酰氯,然后加入 2 mL 10%氢氧化钠溶液。每次加完后,把瓶塞塞紧,剧烈摇动,使试管中溶液呈碱性。观察试管溶液是否分层,闻一下每支试管的气味。

2) 与硝酸铈铵试剂反应

取乙醇和甘油各 5 滴于两支试管中,用 1 mL 水稀释,分别加入 0.5 mL 硝酸铈铵试剂并摇动,观察反应现象[1]。

3) 与卢卡斯试剂反应

在三支干燥的试管中分别加入 5 滴正丁醇、2-丁醇、叔丁醇,再各加入 1 mL 卢卡斯试剂。塞好试管,振荡,冷却至室温(最好在 26~27℃)。静置 5 min,观察现象[2]。

4) 与铬酸试剂反应

在三支干燥的试管中分别加入 5 滴正丁醇、2-丁醇、叔丁醇,先分别加入 1 mL 丙酮溶解样品,再分别加入 5 滴铬酸试剂,振荡,观察并记录现象。

2. 酚的性质

1) 酚的酸性

在一支干净试管中加入 6 mL 饱和苯酚水溶液,用玻璃棒蘸取 1 滴在广泛 pH 试纸上试验其酸性。然后把上述溶液分成两份,一份做空白对照,另一份滴入 5%氢氧化钠溶液,边加边振荡至溶液澄清,再滴加 1 mol/L 盐酸至酸性,观察现象并与空白试剂对比。

2) 与溴水反应

取一支干净试管,加入 2 滴饱和苯酚水溶液,用水稀释至 2 mL,滴加溴水。滴加过程中观察是否有沉淀析出和溴水褪色情况。

3) 与三氯化铁反应

在一支试管中加入几滴苯酚及 2 mL 水,再加入 1~2 滴 1%三氯化铁溶液。另取一支试管,用蒸馏水及三氯化铁试剂做空白实验。比较两个溶液的颜色。

用间苯二酚和对苯二酚代替苯酚进行上述实验,观察现象。

注释

[1] 硝酸铈铵检验醇类化合物时,通常只适用于 10 个碳原子以下的醇类。反应溶液颜色变红。

[2] 卢卡斯试剂只适用于鉴别 3~6 个碳原子的伯、仲、叔醇。C_6 以上的醇不溶于卢卡斯试剂,甲醇、乙醇所得产物易挥发,现象不明显。

思考题

(1) 有 6 瓶无标签试剂,已知分别为叔丁基氯、环己醇、乙醇、叔丁醇、2-氯丁烷和正氯丁烷。试选择合适的试剂进行鉴别。

(2) 苯酚溶液中加入过量的溴水,会产生什么现象?解释原因。

实验 45 醛和酮的性质

实验目的

(1) 理解醛和酮的特征性质。
(2) 掌握区别醛和酮的实验方法。

实验用品

仪器:试管。

试剂:95%乙醇、丙酮、丁醛、苯甲醛、苯乙酮、托伦试剂、费林试剂、I_2-KI 溶液、NaOH 溶液。

实验步骤

1. 与 2,4-二硝基苯肼反应

取三支试管,各加入 1 mL 2,4-二硝基苯肼,然后分别加入 1~2 滴丙酮、丁醛和苯甲醛。观察并记录实验现象[1]。

2. 与托伦试剂反应

在三支试管中各加入 1 mL 托伦试剂,分别滴加 2 滴丙酮、丁醛和苯甲醛。在室温下放置几分钟,如果试管上没有银镜生成,在 50~60℃水浴中稍热几分钟[2],观察是否有银镜生成。

3. 与费林试剂反应

取三支试管，分别加入 1 mL 费林试剂 A 和 1 mL 费林试剂 B 制成混合溶液，再分别加入 2 滴丙酮、丁醛和苯甲醛，摇动后放入沸水浴中 3~5 min，观察反应现象。

4. 与碘仿反应

取三支试管，分别加入 4 滴丙酮、乙醛和苯乙酮，各加入 1 mL 碘-碘化钾溶液，慢慢滴加 3 mol/L 氢氧化钠溶液。注意观察反应液的颜色变化，并观察是否有黄色结晶析出[3]。

注释

[1] 析出结晶的颜色一般与醛、酮分子中的共轭体系有关，非共轭酮生成黄色沉淀，共轭酮生成橙至红色沉淀。

[2] 加热时间不可过长，否则将生成易爆炸的雷酸银。

[3] 除乙醛和 α-甲基酮外，α-甲基醇也能发生碘仿反应，因为它能被次碘酸钠氧化成 α-甲基酮。

思考题

根据上述实验鉴别下列各组化合物：①苄醇、苯甲醛与环己酮；②2-丁酮与 3-戊酮。

实验 46　羧酸及其衍生物的性质

实验目的

熟悉羧酸及其衍生物的化学性质。

实验用品

仪器：试管。

试剂：甲酸、乙酸、草酸、0.5%高锰酸钾溶液、乙酰氯、2%硝酸银溶液、乙酸乙酯、硫酸、30%氢氧化钠溶液、乙酸酐、乙酰胺、羟胺盐酸盐乙醇溶液、95%乙醇。

实验步骤

1. 羧酸的性质

1）酸性

取三支试管，分别加入 5 滴甲酸、乙酸及 0.5 g 草酸，再加入 2 mL 蒸馏水。摇动试管，用洁净的玻璃棒分别蘸取相应的酸溶液，在同一刚果红试纸上画线[1]，比较线条的颜色和深浅程度并给予解释。

2）氧化反应

在三支试管中分别加入 0.5 mL 甲酸、乙酸及 0.2 g 草酸，再加入 1 mL 水，分别加入 1 mL 稀硫酸(1:5)及 3 mL 0.5%高锰酸钾溶液，加热至沸，观察现象。

2. 羧酸衍生物的性质——水解

1）酰氯的水解

在盛有 1 mL 蒸馏水的试管中加入 3 滴乙酰氯，略微摇动，观察现象。反应结束后，再加 3～4 滴 2%硝酸银溶液，观察有何变化。

2）酯的水解

取三支洁净的试管，各加入 1 mL 乙酸乙酯和 1 mL 水。第一支试管中加入 2 滴 15%硫酸，第二支试管中加入 2 滴 30%氢氧化钠溶液。将三支试管同时放入 70～80℃水浴中，摇动试管，比较三支试管中酯层的消失速度。

3）酸酐的水解

在盛有 1 mL 水的试管中加入 3 滴乙酸酐。乙酸酐不溶于水，呈珠粒状沉于管底。稍加热，观察现象并辨别产生的气味。

4）酰胺的水解

酸性水解：在试管中加入 0.5 g 乙酰胺和 3 mL 6 mol/L 硫酸，加热至沸。辨别是否有酸的气味。

碱性水解：在试管中加入 0.5 g 乙酰胺和 3 mL 6 mol/L 氢氧化钠溶液，加热至沸。用湿润红色石蕊试纸鉴别是否有氨。

3. 异羟肟酸反应

取一支试管，加入 1 mL 羟胺盐酸盐的乙醇溶液，再加入 1 滴液体样品或 50 mg 固体样品。摇动，加入 0.2 mL 6 mol/L 氢氧化钠溶液，加热至沸，稍冷后加入 2 mL 1 mol/L 盐酸。若溶液变浑浊，加入 2 mL 95%乙醇，再加入 1 滴 5%三氯化铁溶液，继续逐滴加入至溶液显紫红色且颜色不再消失[2]。

注释

[1] 刚果红是一种指示剂，变色范围 pH 3.0（蓝色）～5.2（红色），与弱酸作用显蓝黑色，与强酸作用显稳定的蓝色。

[2] 这个反应不能用来鉴别含有三氯化铁显色的官能团的样品。可先用三氯化铁溶液进行显色试验，若不显色，溶液应呈黄色。

思考题

比较羧酸及其衍生物水解和醇解反应的活性。

实验 47　胺 的 性 质

实验目的

(1) 熟悉胺的碱性。

(2) 掌握用简单的化学方法区别伯、仲、叔胺。

实验用品

仪器：试管。

试剂：苯胺、二苯胺、N-甲基苯胺、N,N-二甲基苯胺、无水乙醇、β-萘酚、苯磺酰氯（或对甲苯磺酰氯）、浓盐酸、氢氧化钠、亚硝酸钠。

实验步骤

1. 碱性实验

取一支试管，加入 1～2 滴苯胺和 0.5 mL 水，振荡试管，观察现象。然后滴加 1～2 滴浓盐酸，观察结果。再用水稀释，观察稀释后的现象。

取一支试管，加入数粒二苯胺晶体和 0.5～1 mL 无水乙醇，振荡试管使二苯胺完全溶解。然后加入 0.5～1 mL 水，振荡，观察反应现象。再滴加数滴浓盐酸，振荡，观察溶液是否转为透明。最后用水稀释，观察结果。

2. 胺的亚硝化

在一支试管中加入 10 滴苯胺及 4 mL 20%盐酸，用玻璃棒搅拌使其溶解。将试管置于冰浴中冷却至 0℃，逐滴加入 25%亚硝酸钠溶液，搅拌直至混合液遇碘化钾-淀粉试纸立即呈蓝色为止[1]，得澄清溶液。取此溶液数滴加到 β-萘酚溶液中[2]，观察有无橙红色物质生成。

在一支试管中加入 5 滴 N-甲基苯胺及 2 mL 20%盐酸，搅拌使其溶解。将试管置于冰浴中冷却至 0℃，逐滴加入约 10 滴 25%亚硝酸钠溶液，边加边搅拌，观察有无黄色油状物生成。

在一支试管中加入 2 滴 N,N-二甲基苯胺及 0.8 mL 20%盐酸，搅拌。将试管置于冰浴中冷却至 0℃，逐滴加入 4～5 滴 25%亚硝酸钠溶液，搅拌，然后滴加 10%氢氧化钠溶液至呈碱性，观察是否有绿色固体生成[3]。

3. 兴斯堡实验

取三支试管，分别加入 3 滴苯胺、N-甲基苯胺、N,N-二甲基苯胺。再加入 5 mL 10%氢氧化钠溶液，充分混合。然后各加入 6 滴苯磺酰氯（或 0.2 g 对甲苯磺酰氯）[4]，用塞子塞住管口，间歇振荡 3～5 min。打开塞子，将试管置于水蒸气浴上 1 min[5]，检查溶液是否呈碱性。如果不显碱性，则逐滴加入 10%氢氧化钠溶液。观察是否有固体或油状物析出，若有析出，则将它们分离出来（过滤或使用滴管等方法），并将它们分别置于 5 mL 水和 5 mL 5%盐酸中，试验其溶解性。如果无沉淀析出，则用 20%盐酸酸化至 pH 为 6，用玻璃棒摩擦管壁并使试管冷却，再观察有无沉淀析出。根据实验结果得出结论。

注释

[1] 用玻璃棒蘸一点反应液，与碘化钾-淀粉试纸接触，观察试纸的颜色变化，以检查重氮化反应的终点。

[2] β-萘酚溶液的配制：将 0.1 g β-萘酚溶于 1 mL 5%氢氧化钠溶液中。

[3] N,N-二甲基苯胺的亚硝化产物在盐酸作用下生成红棕色化合物，必须滴加碱以后才能生成绿色物质。

[4] 芳磺酰氯有毒并具有腐蚀性，应避免与皮肤接触，也不能吸入其蒸气。此反应应在通风橱内进行。

[5] 某些 N,N-二烷基苯胺与苯磺酰氯共热时会生成紫红色染料。一旦发生这种情况，则重新在 15~20℃ 水浴中进行反应。

思考题

比较苯胺和二苯胺的碱性强弱。

参 考 文 献

蔡会武, 曲建林. 2007. 有机化学实验. 西安: 西北工业大学出版社.
陈东红. 2009. 有机化学实验. 上海: 华东理工大学出版社.
蒋华江, 朱仙弟. 2018. 基础实验Ⅱ(有机化学实验). 2版. 杭州: 浙江大学出版社.
刘红英. 2008. 有机化学实验. 北京: 中国农业出版社.
刘峥, 丁国华, 杨世军. 2010. 有机化学实验绿色化教程. 北京: 冶金工业出版社.
庞金兴, 袁泉, 刘军, 等. 2014. 有机化学实验. 武汉: 武汉理工大学出版社.
王莉贤. 2009. 有机化学实验. 上海: 上海交通大学出版社.
郗英欣, 白艳红. 2014. 有机化学实验. 西安: 西安交通大学出版社.
曾和平. 2014. 有机化学实验. 4版. 北京: 高等教育出版社.
章鹏飞, 强根荣, 赵华绒, 等. 2013. 有机化学实验. 杭州: 浙江大学出版社.
赵斌. 2013. 有机化学实验. 修订版. 青岛: 中国海洋大学出版社.
朱文庆, 李红. 2011. 有机化学实验. 西安: 西北工业大学出版社.
J.A.米勒, E.F.诺齐尔. 1987. 现代有机化学实验. 上海: 上海翻译出版公司.
Mohrig J R, Alberg D G. Hofmeister G E, et al. 2014. Laboratory Techniques in Organic Chemistry. 4th ed. New York: W. H. Freeman and Company.